nothingness

•

nothingness

·

THE SCIENCE
OF EMPTY SPACE

·

HENNING GENZ

Translated by Karin Heusch

PERSEUS PUBLISHING

Cambridge, Massachusetts

English translation copyright© 1999 by Perseus Books Publishing, L.L.C.
Copyright 1998 © by Perseus Books Publishing

Originally published in German as *Die Entdeckung des Nichts*© 1994 Carl Hanser Verlag Müchen Wien

Cataloging-in-Publication Data available from the Library of Congress

ISBN 0-7382-0610-5

Perseus Publishing is a member of the Perseus Books Group

Perseus Publishing books are available at special discounts for bulk purchases in the U.S. by corporations, institutions, and other organizations. For more information, please contact the Special Markets Department at the Perseus Books Group, 11 Cambridge Center, Cambridge, MA 02142, or call (800) 255-1514 or (617) 252-5298 or email j.mccrary@perseusbooks.com.

Text design by Jean Hammond
Set in 10.5 Minion by Modern Graphics

First paperback printing, December 2001
2 3 4 5 6 7 8 9 10—05 04 03

CONTENTS

•

PREFACE
Things, Just Things   vii

Acknowledgments   xi

1   PROLOGUE   1
Physics and Metaphysics

2   NOTHING, NOBODY, NOWHERE, NEVER   33
Philosophical, Linguistic, and Religious Ideas on Nothingness

3   PROBLEMS WITH NOTHINGNESS   97
How to Make It a Physical Reality

4   MATTER IN THE VOID   145
Ether, Space, Fields

5   CROWDED SPACE   179
Movement All Around—the Quantum Vacuum

6   SPONTANEOUS CREATION   209
Particles and Fields

7   LET NATURE BE AS SHE MAY   227
Special Systems

8   NOTHING IS REAL   257
The Universe as a Whole

9   EPILOGUE   305
Physics and Metaphysics

Notes   317
Figure Sources   325
Bibliography   327
Index   337

PREFACE

•

THINGS, JUST THINGS

•

THIS BOOK IS DEDICATED TO THE QUESTION "CAN THERE BE SPACE independent of things?" Space that is immutable, like a stage that remains the same no matter what is being played on it? Space that may be empty and for all time remain empty? Such a space would be a "void" proper, or what we call "nothingness"—concepts that in antiquity were found by natural philosophers to be so controversial as to be unthinkable. Over the millennia, this void has evolved into what physicists call *the vacuum*—their term for empty space. Physics fills this vacuum with the progeny of quantum mechanics, of the general theory of relativity, of whatever the Big Bang left for us. This question then poses itself: How empty can space be and still remain in consonance with the laws of nature?

In the seventeenth century, Galileo's disciple Evangelista Torricelli was the first to succeed in the removal of everything material from an otherwise empty container. This experimental success made it harder for skeptics to insist that there couldn't be such a thing as empty space—a skepticism maintained by the followers of Aristotle, who clung to the tradition of the *horror vacui,* nature's supposed abhorrence of a vacuum. It was Isaac Newton who considered the meaning of space in the laws of nature; he also discussed the concept of *the ether*—not the substance used to anesthetize but some mysterious matter that was assumed to fill the universe. Newton found that these concepts helped him describe (if not understand) the motion of a planetary system through the long-distance action of gravity across empty space.

From the seventeenth century on, science has made tremendous progress based on the notion that there are just two fundamentals: matter and empty space. In the back of their minds, the natural philosophers—who slowly evolved into natural scientists in the modern sense—had long nurtured the concept of the mysterious ether, a substance that could not be chopped up into atoms but instead was strictly continuous.

It took Albert Einstein to do away with this idea. The details are complicated, and we will discuss them. Today, we know that the laws of nature do not permit space that is absolutely empty. At high temperatures, space at its emptiest will be filled with thermal radiation. At low temperatures, structures will form in the void. According to quantum mechanics—more specifically, to *Heisenberg's uncertainty relation*—we can never precisely fix the amount of energy that fills a certain region of space in a certain amount of time. The amount of energy will fluctuate. Consequently, we will never be able to define a *zero-scale* for energy. One might say that the vacuum of physics emits energy—more of it the shorter the time span we define, less of it for longer times.

According to Albert Einstein's famous formula $E = mc^2$, energy and mass are the same thing. Mass therefore also fluctuates, and empty space will see a constant emergence and disappearance of particles that carry this mass. These particles don't last, and physicists call them *virtual particles*.

The physical vacuum is by no means empty and devoid of characteristics. Rather, anything that can exist at all will oscillate and spin in it in a random, disordered fashion. In this vacuum, quantities will emerge that, in an abstract space of particle properties, will define directions; these quantities, which in their abstract space act somewhat like magnets in real space, are called *fields*. Although these fields influence the way in which we perceive the physical world on all levels, the discipline of physics needs to examine minuscule regions in order to confront them directly. It might appear paradoxical—but the huge accelerating machines of elementary particle physics not only examine the particles they accelerate but also explore the emptiest of spaces we can imagine, and thereby some of the questions that the Greek natural philosophers bequeathed to us as problems still to be solved. But it is not only with huge accelerators that we investigate the void—we can perform less spectacular but enormously precise and complex experiments based on nothing but light.

Most of this book is devoted to what we know about the void. There is, however, another question intimately related to this void, a question that has fascinated humankind across the millennia, that has spawned myths and legends of creation: How did it happen that at some point in time, something appeared out of nothing? To this day, physics cannot give a definite answer to this question. The standard models advanced by cosmologists and elementary particle physicists permit them to reconstruct the history of the universe back to fractions of a second after the Big Bang. So much is for certain. But the closer we get to the Big Bang, the less certain our knowledge. We can only speculate about the Big Bang itself, and what happened immediately after it. This is the subject of the last two chapters of this book. Which ideas about creation do the laws of nature admit? These laws always and everywhere have been the same. This is a central tenet of physics: It is just the state of our world that has changed over time.

The main purpose of this book is the transmittal of scientific insight. Only such insight can foster the realization that nature is, in fact, understandable to humankind. I stand convinced that this is the most noble aim of basic scientific research. By this I do not mean to assert that the world in all its complexity will one day be completely understood. That, to be sure, will not happen. Rather, the implication is that natural phenomena are not the work of spooks and demons but are due to rationally explorable causes. This is the attitude that launched Western cultural and scientific thinking six centuries before Christ. We owe it to the so-called pre-Socratic philosophers in Ionic Greece, who, as Erwin Schrödinger, recipient of the 1933 Nobel Prize in physics, has put it, "saw the world as a rather complicated mechanism, acting according to eternal innate laws, which they were curious to find out. This is, of course, the fundamental attitude of science up to this day."

My physics colleagues who browse through this book may be amazed at the long passages I devote to the ancient naturalists. I do this for two reasons: first, there is my curiosity, which goes back to my own school days; second, there is my conviction that those ideas from antiquity do not differ much from what determines many of our contemporaries' notions of the natural sciences. It is therefore appropriate that we take them as a starting point for the communication of today's scientific insight. It should be possible to start from the views of the ancient Greek naturalists and move to those of modern science. This is what I have tried to accomplish here.

In the process, I have had to ask myself how far I can pursue this path without losing contact with the notions actually developed by those ancient naturalists. Take the ideas of Anaxagoras as an example: I started with his *divisibility of ur-matter* and interpreted it in terms of modern ideas on the formation of structures of matter. In doing so, I did not dare to go as far as taking what he calls the *seeds* of, say, hair, which are hairs themselves, to be an early form of self-replicating fractals.

In my original manuscript, I limited myself to the interpretation of those ideas from antiquity that were directly related to the topics I am covering. Other ancient notions I merely renamed in modern terms. This was not enough for Eginhard Hora, my meticulous editor at Hanser Publishing: He insisted on the insertion of minireviews on many physics subtopics, from A (as in antimatter) to Z (as in zero-point energy).

The philosophical and historical passages are based on an unsystematic perusal of the available literature. The term *nothingness* invariably evokes mythical psychological connotations. The reader will not find these connotations here, nor will terms such as *nirvana, black-out, immersion, the nothing that acts out of nothingness* be found here. I hope I have managed to write this book in such a way that these connotations cannot even be read into my text.

But I cannot be sure. The great English astrophysicist Sir Arthur S. Eddington wrote in the preface to his popular science book *New Pathways in Science*, published in 1935: "A book of this nature has to evoke precise thinking by means of imprecise turns of language." Therefore, the author may not always succeed in "evoking in the reader's mind the very ideas he is trying to convey. He certainly cannot do so unless the reader joins in an active effort." Eddington demands that the author manage to "relegate secondary complications to the background." Those complications will become obvious to anybody trying to describe the facts in a straightforward fashion. But what do we mean by "secondary"? That may well be a matter of opinion. Certainly, some scientists who have spent years chasing down one of those complications will be unhappy to see it classed as "secondary."

Given such a book, how will it be read, and by whom? All authors naturally hope to see their readership riveted on their books, breathlessly engaged from start to finish, but such readers are rare. Most pick and choose, looking for what elicits their interest, their happy concurrence, or their violent protest. I tried to write the prologue and epilogue so that those readers who read nothing else will still gain an overview of the substance of the entire book. The prologue and epilogue should stand on their own as an intelligible and, I hope, captivating synopsis of our subject.

ACKNOWLEDGMENTS

•

Part of this book is the result of a six-month stay at the TRIUMF Laboratories in Vancouver, British Columbia, from 1991 to 1992. The author wishes to thank his Canadian colleagues for their hospitality. He acknowledges the support of the Volkswagen Foundation, which made that visit possible. He has greatly profited from discussions with, and from the reactions of, friends and colleagues who may have wondered about some of the topics that came up in the process. This book has benefited considerably from the helpful comments of Eginhard Hora, editor at Carl Hanser Editions, and Jeffrey Robbins of Helix Books. To all of them, and to many whom I cannot name but who are well aware of their contributions to this project, I feel deeply beholden.

CHAPTER 1

PROLOGUE

•

PHYSICS AND METAPHYSICS

•

LET'S ASSUME WE CAN REMOVE ALL MATTER FROM SOME REGION OF space. What will we be left with? A region of empty space? Not necessarily.

In the universe, between galaxies, each atom is at a distance of about one meter from its next neighbor. Still, the space between those atoms is not empty; it is bright with light and other radiation from very different sources. It is only in the absolutely empty space of our imagination that no light, no radiation penetrates—that space is as dark as the legendary rooms of Schilda, in the German fairy tale. A region of space is not really empty simply by virtue of not containing matter.

BLACKBODY RADIATION

If we wanted to produce a region of really empty space, we would have to remove from it not only all matter but also all radiation. To keep it from exchanging matter and radiation with the space around it, we would have to shield it effectively—say, by surrounding it with walls. We might then take an ideal pump to evacuate this enclosed space, hoping that the radiation it contained would gradually be absorbed by the walls and that the final result would in fact be a truly empty space.

Unfortunately, that is not how it works. First of all, walls not only absorb radiation but also emit it. Every enclosed space is filled with the radiation absorbed and emitted by its walls. That is why a space free of matter is not necessarily empty space.

The radiation we are discussing here might be thermal radiation; at higher temperature, it might be red light, like that emitted by an overheated electric stove; and at still much higher temperature, it might be the light of the Sun. This radiation weakens as the temperature of the emitting body decreases, but we would have to go to what the physicists call *absolute zero*—that is, -273

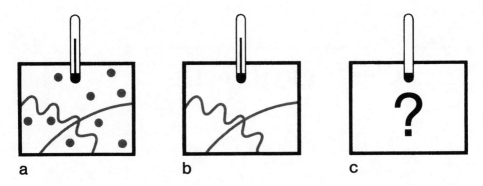

a   b   c

Figure 1 A container filled with air at temperature above absolute zero (−273 degrees Celsius) contains molecules and thermal radiation (1a). When we remove the molecules with an ideal pump, the thermal radiation remains (1b). The space that remains when, in a gedankenexperiment ("thought experiment"), we cool it to the (inaccessible) temperature of −273 degrees Celsius is what we call the physical vacuum (1c): Everything that the laws of nature permit to be removed has been taken out.

degrees Celsius (C)—to have it die off altogether. It follows that above the unreachable temperature of absolute zero, the radiation emitted by the walls will never permit an enclosed space that is truly empty (see fig. 1).

THE TORRICELLI REVOLUTION

There is no way to produce a space emptier than the one that we approximate by lowering the temperature in, say, a box we have previously evacuated by pumping out its air content. It is not a matter of course that we use the words *evacuation* and *pumping* in this context. We owe it to what may well be the most significant event in the historical development of our subject. In the seventeenth century, the Italian naturalist Evangelista Torricelli, a student of Galileo's, was the first to formulate a correct answer to the question of why it is possible to suck the liquid content out of a vessel through a straw. He discovered that the weight of all the air above weighs down on the surface of a liquid. If we remove this weight inside the straw by sucking, that same weight, which continues to exist outside the straw, will push the liquid upward inside the straw.

The Torricelli experiment seems simple enough to today's reader: Fill an 80-cm-long tube with mercury, invert it, and set its open end in a bowl of mercury. (See fig. 2.) Mercury from the tube will now flow into the bowl through the open end of the tube—but not all of it. The mercury level inside the tube will drop to 76 cm, and will stay there. The 4 cm above that level are empty of air and appear as a vacuum from which the mercury column is suspended.

In reality, of course, the mercury does not dangle from the vacuum above;

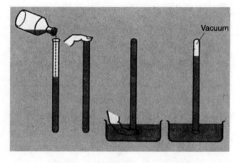

Figure 2 Torricelli's experiment. It showed that air can be removed from a given volume. No air can be contained in the space denoted "vacuum," since air can penetrate neither glass nor mercury. The space is not, however, entirely empty, because it contains such prosaic substances as the vapor of mercury. (At the present stage of this book we are not taking into account the complications arising from present-day physics.) Right after Torricelli's experiment it was argued by Scholastic philosophers that the so-called vacuum was a plenum—a space full of a hypothetical, ever-present fluid that can penetrate even glass and mercury. This fluid was invented in order to save the hypothesis of *horror vacui,* that emptiness cannot be created.

rather, it is supported by the external air pressure, just like the liquid inside a straw. The Torricelli experiment showed, above all, that nature opposes the formation of a space devoid of air with a finite force that we can overcome. In other words, the experiment shows that space without air is possible after all.

When Torricelli's experiment became known to the scientific world, it moved to center stage immediately (see fig. 3); many investigations were triggered by it. Up to that time, there had been no way to link the characteristics of empty space with the consequences of our special position in the universe—our existence on the surface of Earth, but at the bottom of an ocean of air.

About one century before Torricelli, the Copernican revolution began to gain acceptance; it stipulated that we live on a planet that orbits the Sun. This made it possible to separate the influence of our specific position in the universe on our astronomical observations from the consequences of the natural laws that govern our solar system—and opened up the question of how physics explains the motion of the Moon, the planets, and comets around a Sun at rest. The progress in our thinking owed to this "revolution" is clearly demonstrated in today's planetaria, which often simulate a space voyage from Earth out into interstellar space: What we perceive as the irregular motion of the planets gradually settles into simple elliptical orbits about the Sun.

Where our topic, empty space, is concerned, the Torricelli revolution is analogous to the upheaval caused by Copernicus. The Torricelli revolution would not have been possible without the knowledge extant in Greek antiquity: that air is "something" rather than "nothing." A room full of air had long been known not to be an empty room. When we now discuss a space as empty as possible inside an evacuated box at ever-decreasing temperatures, we are in fact taking Torricelli's empty space one step further. It soon became clear that even after a given volume had been evacuated by pumping action, material from the walls

Figure 3 This 1698 engraving illustrates the activities of the French Academy of Science founded by Louis XIV. The choice of subjects shows that the main interest of the academy was in the applications of science. The final influence of Torricelli's experiment on the development of the sciences and their applications was more important than any of the practical devices shown here. Although in the seventeenth century there was no foreseeable application of that experiment with the exception of the barometer, it finds its rightful place near center stage in this scene. I was unable to discover any depiction concerning electricity or magnetism in this picture. If there is really none, this is a serious omission.

evaporated into it—in Torricelli's case, this was mainly mercury vapor. If we now insist on creating a space we can call empty, we have to pump out these vapors too. The realization that the universe beyond Earth's atmosphere should be a matter-free space in that sense began to take hold soon after Torricelli.

MATTER AND SHAPES

How do we characterize space in a state we approximate gradually by pumping out a container and lowering its temperature as best we can? Physics says that this container is as empty as it can get, but adds that it is not empty in the sense of a true, absolute vacuum. So what will necessarily remain in a space after we have taken out of it all that can be taken out?

The stories, myths, and legends of creation ask the opposite question—probably the oldest question of natural philosophy: How did the world get started? From what did it originate? That would probably be the same as whatever remains after all has been removed that can be removed from a region of space. This question has been answered in the most diverse fashion through the ages, star with the earliest Greek philosophers before Socrates. Thales of Miletus, the earliest Greek natural philosopher we know of, thought that the universe originated from water. In his opinion, water filled the universe in its different forms—solid, liquid, gaseous. There is no such thing as empty space, since water pervades everything. If we interpret his ideas in terms of modern physics, Thales' space is as empty as it can get when it contains nothing but water in its primal form—presumably in its liquid form, as in the oceans. To remove all matter from the world would then be tantamount to reducing everything to this primal form, melting down the rocks, condensing the air. If all that exists is some form of water, the opposite must then also be true: Water contains every potential form of matter.

SOMETHING, NOTHING, AND THE VOID

In reality, Thales and his successors did not ask about empty space. Rather, they inquired about the concept of nothing, the opposite of something. The physics question about the existence and properties of empty space can, in principle, be answered by experiment. The philosophical concept of *nothing*, of course, is accessible only to logical analysis: Is it possible to imagine this *nothing*, the absence of anything, without violating the laws of logic? Will not this nothing become something simply by virtue of its being imagined? Is our problem in speaking of this nothing perhaps simply a problem of language, or is it one of logic?

No successor of Thales has ever given us an answer to what exactly defines *nothing*, other than characterizing it simply by negatives. The most important was the argument that *nothing* cannot exist—clearly a tautology. In the creation myths and in the ideas of the earliest philosophers, some "ur-matter" existed. Whence it came and what exactly it was—these questions were of less concern than the question of how it had acquired the form it possesses. This question remains today. By posing it, the originators of creation myths and the early philosophers developed modes of thought, and these are more important than the specific concepts they frame. Physical models of what is something and what is nothing can have only coincidental value if developed without the knowledge available to modern physics. Still, the modes of thought that hail back to the creation myths and to the early philosophers retain their significance.

WORLDS IN A VOID

Just as the metaphysical nothing presents us with a conceptual conundrum, so does empty space. A whole sequence of naturalists and natural philosophers, from Aristotle through the medieval Scholastics to the great French mathematician

né Descartes and on to Albert Einstein, has tried to come to
e process, we have come to see that truly empty space cannot
concept has to be further clarified before we can even enter
se discussion of its meaning.

Let _ _ _ _ th a dilemma: Let a box be so big, and let its walls be so far
removed from us, the observers, that our experiments on the properties of empty
space cannot be influenced by the walls. Taking this approach to the extreme,
we might remove the walls altogether; then the box in its infinite expanse has
become a universe that contains nothing but us and our experimental gear. In
that case, what can we find out about this space all around us, and about our
motions inside it?

First, we want to find out whether the result of an experiment can be
influenced by the location where it is performed. Suppose we have a toy planetary
system at our disposal. In the first experiment, we situate the Sun in space
somewhere in front of us, add the planets, give them a set of well-defined initial
velocities, and record their subsequent motions with a movie camera. After a
little while, we stop the experiment and redo it somewhere else. The great German
philosopher and mathematician Gottfried Wilhelm Leibniz, a contemporary of
Newton's, believed that the scenario I described does not permit the definition
of "somewhere else." If everything that exists is embedded in an infinite empty
space, any imaginable universe is identical with itself—so argues Leibniz in what
he calls his principle of the "identity of the indiscernibles." It would be senseless
to differentiate between a world "here" and another world "there."

BODIES AS PROBES

If we consider only logical differences, this is certainly correct. Physics, however,
differentiates between objects that influence others and objects that simply mark
a space without influencing others; let us call these latter objects "probes." In
the language of physics, such probes are mostly idealizations that approach, but
never quite reach, reality. Let two examples serve to explain this. Take Earth,
with its magnetic field. If you measure that field with a compass carried from
place to place, the influence of the needle itself on Earth's interior, which generates
the magnetic field, can be ignored. The influence is there, but it is so small that
we cannot really measure it. The compass needle is a probe that is exposed to
the influence of Earth's magnetic field, but its own magnetic field is so small
that its action on Earth's field is negligible. The same would hold if we distributed
a whole series of compass needles to map out the entire magnetic field of Earth.

If we now tried to map out the magnetic field of a compass needle instead
of Earth's field, it is clear that we would need much smaller probes; otherwise,
the field to be measured would collapse under the influence of the new probes.
In high school physics, we tend to use small iron filings for this purpose.

Next, let us try to measure the gravitational field of Earth by approximate probes; for this measurement, we utilize the probes' mass. From the motion of the tides, we know that the Moon deforms the oceans and thus influences Earth's gravitational field. This means that the Moon, due to its own large mass, is not a very precise probe for measurement. On the other hand, the masses of communications satellites are so small that their presence exerts no noticeable influence on the shape, and therefore on the gravitational field, of Earth. The implication is that the closer we get to an ideal probe for the measurement of Earth's gravitational field, the smaller we must make that probe.

Leibniz does not differentiate between markings that do or do not influence physical behavior. He considers logical differences only, so that, for him, marked spaces are always different from unmarked spaces. But as physicists, we may imagine that a small probe has no influence on a faraway experiment. Thus we leave a mark at the position (and with the velocity) of our toy planetary system before moving it to another location—satisfied that the probe has no other effect than to indicate where the system has been. At the new location, we perform the same experiment on our toy planetary system and compare the results with a movie of the previous experiment. There should be no difference—this is what the reader expects. For Leibniz, however, once a mark has been left at the location of the first experiment, the second experiment will be entirely different. If no mark has been left, the first and second experiments, as Leibniz would see them, are entirely the same experiment; the ensuing courses of events must then be identical, on logical grounds. But if a mark has been deposited, Leibniz will not admit any relation between the two experiments. However small and far away the mark may be, the mark has altered the logical situation of the second experiment as compared with the first. Consequently, the two experiments have no relation whatsoever to each other; for Leibniz the first implies nothing about the second.

THE VELOCITY OF EMPTY SPACE

Based on these experiments, we would imagine that there is absolute space, immutable everywhere. We can cover it with a network of probes that we call a *coordinate system*. Let these probes be at rest with respect to each other, and let their characteristics be indistinguishable from those of the space they cover, which they define. So far, we see no dilemma. That will arise only when we start talking about the *velocity* of space.

Let us now imagine we set in motion our toy planetary system for the second experiment relative to the markings of the first one. If we now ask again whether our observations in the second experiment concur with those of the first, we run into trouble: If they do, it would not be possible to tell a space in motion from a space at rest. To be able to talk about space, we would have to choose arbitrarily one space with a certain velocity relative to all others from infinitely many physically equivalent spaces, and define the one we chose as *the* space. Physics does not distinguish any one of our choices of spaces from any other one. However, if the second experiment does *not* look like the movie we took of the first, then space must be able to tell one form of motion from the other; we would therefore have to say that space *acts* in this fashion, and this action makes it observable, if only indirectly. If we apply this train of thought to the space in an evacuated box—the one we want to define as "empty"—at ever-decreasing temperature, then we have to concede that this space behaves like a medium.

Newton's mechanics unites these two possibilities, in a curious way. As long as the second experiment moves with a velocity that is constant in both magnitude and direction when compared with the first one, there will be no difference between the actual event and the movie. Therefore Newton's mechanics can serve to define empty space in a given *state of motion*: All "spaces" that move at constant speed relative to each other are physically equivalent. In each one of them, the same laws of Newton's mechanics will apply. If, however, our motion in the second experiment is accelerated with respect to the first, then what is recorded on the film is not the same as what we observe in actuality.

This is so because we need forces to effect accelerations. All rotations belong to the class of accelerated motions, since the direction of the motion is constantly changing. The force active here is the centrifugal force. Let us assume we are located on a rotating platform when we position our toy planetary system. Initially, it will share with us not only location and velocity but also acceleration. The initial acceleration will, however, not influence its further behavior. Once released in empty space, each system moves at a constant—that is, unaccelerated—speed. Had we released another observer together with the system, that observer would see what our film shows. We, on the other hand, being further accelerated, will notice a different behavior. It is not the velocity but the acceleration of an observer that figures in the laws of nature that describe what this observer will see. Therefore, neither the magnitude nor the direction of the constant velocity of some object has absolute significance; what does count is the simple statement that it moves in an unaccelerated state of motion.

Historically, the most important example of an object's moving at constant speed if not influenced by external forces was provided by Galileo. A stone released by a sailor from the top of a moving ship will fall parallel to the mast, just as though the ship were at rest (see fig. 4). The velocity in question is parallel to the surface of the ocean. The stone is being accelerated by Earth's gravity perpendicular to it.

Later, we will consider systems and their observers that are jointly subject to free fall. For instance, the toothbrush of an astronaut in a space capsule is suspended in front of him. A movie might show that he himself can turn freely,

Figure 4 The rock dropped by a sailor off the mast of a ship moving at constant velocity will drop vertically along the mast irrespective of the ship's motion.

as though he were weightlessly suspended in space. If he then moves in accelerated fashion, his toothbrush will not move with him if he leaves it to its own devices. This example is not unlike what we demonstrated with the toy planetary system earlier. The laws of nature that apply to an observer subject to rotation are more complicated than those that apply to an observer at an unchanging velocity.

The forces appearing in accelerated systems are well known. They cause cars to move out of lane on road curves, they lock seat belts, they distort the faces of astronauts at launch, and they lift the seats of rotating carnival rides. Newton's mechanics distinguishes not a particular *state of motion* but rather a particular *state of acceleration*. This physical reality, which figures in Newton's law of motion, is described in a remarkable article of Einstein's as the *ether of mechanics*. Instead of ether—according to Einstein—one could just as well speak of the "physical properties of Space."

Leibniz, like Descartes, thought that motion could be defined only by observing the motion of material objects with respect to other such objects. This is correct for motions at constant velocity. Because of the forces that oppose acceleration, it is possible to differentiate accelerated from unaccelerated motion using the characteristics of the system at hand; there is no need to look elsewhere. Also, "empty space" acts like a medium. If there is a system accelerated with respect to this empty space, there will be centrifugal force. That it is otherwise impossible to physically define "space" creates a paradox, which Immanuel Kant addressed in the following way:

> *The circular motion of two bodies about a common center (including the axial rotation of the Earth) can be recognized even in empty space. In other words, there is no need for comparisons to external space based on experience; still, a movement comprising a change in the properties of external space can be expressed empirically, even though said space is itself neither expressed empirically nor deserves to be an object of experience: here is a paradox that needs to be resolved.*

Let's admit that this tortuous statement of Kant's bears reading two or three times.

SPACE AND THE LAWS OF NATURE

Does it make sense to define empty space as a space from which "everything" has been removed? The answer to this question depends on the laws of nature. For example, we might never be able to remove substance A together with substance B; our choice might be to remove either A or B. Alternatively, those laws might prescribe that the remaining amounts of A and B form a product whose value cannot be less than some given quantity. They could also say that a remnant of A must remain in all enclosures. The first choice is, in fact, approximately the way it turns out in reality. Furthermore, the supply of a substance in any one vessel might be inexhaustible; for that reason alone, it would be impossible to empty it altogether: It is, so to speak, bottomless from the start.

The definition of the vacuum given by the great nineteenth-century Scottish physicist James Clerk Maxwell is the following: "The vacuum is that which is left in a vessel after we have removed everything which we can remove from it." This definition encompasses our question but still leaves an opening: What is it we are unable to remove, and how do we know we have removed "everything we can possibly remove"?

QUANTIFYING "SOMETHING"

Physics quantifies "something" through its energy. A bowl is empty in the physics sense when it has given off all the energy it can. Since air molecules with their masses m, according to Einstein's formula $E = mc^2$, stand for an amount of energy, we remove energy from a vessel when we pump air out of it. In this case, Torricelli's definition of empty space coincides with that of physics. But the latter surpasses Torricelli's. Energy is the "special stuff" of physics. Any system left to its own devices will give off as much energy to its surroundings as those surroundings can absorb. Let the pendulum serve as an example. In its state of lowest energy—which we might call the *ground state* or the *vacuum state*—the pendulum hangs motionless. Whatever its initial motion, it will eventually pass into this state. This applies to any pendulum that can give off its energy through any mechanism—for instance, through friction.

And so it goes with all physical systems: Left to their own devices in an environment to which they can pass their energy, they will eventually assume a state of lowest energy—hence the special position that energy holds in the characterization of "something" in physics. Wherever Torricelli's definition of "empty" is applicable, it coincides with its definition in physics. In some other cases, the physical definition of emptiness will lead to a surprising result. Take, for example, a glass filled with water at 0 degrees Celsius. Let this be our physical system; never mind Thales for now. Let's look into this system's state of lowest energy, its "vacuum." At 0 degrees Celsius, when water passes from the liquid state to the frozen state, it surrenders energy in the form of heat. When it melts, it absorbs energy—the so-called heat of melting—so that water in the state of lowest energy at 0 degrees Celsius is solid, not liquid.

Obviously, we can take the ice out of the vessel. That means we lower its energy again, according to Einstein's formula $E = mc^2$. Could it not be that there is something—some substance A (which surely could be neither ice nor water)—that we cannot take away from a given system without *raising* that system's energy? Or that we cannot remove from a given system at all? That could indeed be the case, and that is in fact what happens. This substance A is, to be precise, the subject of our book. At this point, we are ready to give it a name (or, rather, two names, since there are two variants to our substance). In the first variant, we speak of the *Higgs field*. Once this field appears in a vessel that has been pumped empty and whose temperature has been lowered as much as possible, its energy will be further lowered. This is similar to what we observe in water: When we lower the temperature of water from above 0 degrees Celsius to below 0, it will turn into ice at precisely 0 degrees. But in the process of turning from liquid to solid, it will give off "melting heat"; that is, we have to remove more energy so that the temperature can drop *below* 0 degrees Celsius.

PHASE TRANSITIONS

The freezing of water is called a *phase transition*. The temperature at which this happens is called a *critical temperature*. A different phase transition is the evaporation of water, at another critical temperature of 100 degrees C. Similarly, iron undergoes a phase transition at its critical temperature of 763 degrees Celsius: At this temperature it loses its magnetization; when cooled below this temperature, it will regain its magnetization. The critical temperature of the phase transition at which the Higgs field appears is so high that only at a very early phase of the existence of the universe, fractions of a second after its creation, did higher temperatures exist. The Higgs field has pervaded the entire universe ever since, including our container—our so-called black box—the temperature of which we have been attempting to lower.

THE HIGGS FIELD AND VIRTUAL PARTICLES

There is no "something," no "water" whose "ice" could be tied to the Higgs field we mentioned above. This field pops up in our empty space simply because in its presence this space is in a state of lower energy than in its absence. How is it possible for a "something" to contain less energy than a "nothing"? I will explain shortly. But first I must return to the second variant of what we called substance A above, and give it a name: We will call it *virtual particles.* Such virtual particles exist always and everywhere. To remove them is impossible. The reader should think of them as droplets in a steam bath—droplets that materialize and dissolve in saturated steam. I will explain their physical meaning as we go on. I mention them here simply to complete the concepts that contribute to our picture of empty space—of space that is as empty as possible.

NEGATIVE ENERGIES

Now let's talk about a "something" that is characterized by an overall negative energy. Let's start with a rock so small that it fits into a single point in space, and let's place it into the gravitational field of an Earth that we also imagine to be pointlike. In our gedankenexperiment, we can pretend the rock is suspended by a thread from a cog on the axis of a generator; when we let it drop in the gravitational field, it will set the generator in motion. The closer it gets to our pointlike Earth, the larger the force with which it is attracted by the gravitational field; this means that the system consisting of rock and Earth is able to give off more energy the smaller the relative distance. We can extract an almost arbitrary amount of energy, and might take it away in electrical form by the use of an accumulator. Obviously, the more energy we take out of the system of rock and Earth, the more the energy of the system itself has to decrease—such that eventually it must become negative. On the other hand, when rock and Earth are also taken into account, Einstein's formula $E = mc^2$ sees to it that they add a very large positive amount of energy into the balance. And yet we can imagine things to be such that rock and Earth are so close together that the energy balance of positive and negative contributions adds up to zero or even to a negative value.

Negative energies are the focus of a unique fascination. Physicists disagree once the concept of negative energies is extended to the universe as a whole—a disagreement that emerged pointedly in a dialogue that two recently deceased Nobel laureates, Eugene P. Wigner and P. A. M. Dirac, engaged in: If energy can become arbitrarily negative, then negative energies may appear, regardless of which energy level has been defined as having value zero. Einstein's general theory of relativity, in fact, permits us to define that zero level as the energy that does not cause a gravitational force.

The thought process that led us to negative energies was touched on in previous paragraphs; it will reappear as we proceed in this book. Let us recapitulate: A space that contains a pointlike rock in the gravitational field of a pointlike Earth at close distances may well be at a lower energy level than if it were truly "empty." If only energy counted, rock and Earth could then spontaneously emerge in empty space. In reality, this won't work for rock and Earth—but it will work with the Higgs field: Below a certain very high temperature the Higgs field can,

Figure 5 Every water molecule consists of two hydrogen atoms (small empty circles) and one oxygen atom (large gray circles). In the fluid state, the water molecules move at random. In ice crystals, they oscillate around their rest positions.

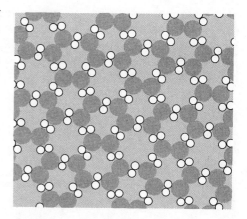

and indeed must, appear out of the void, simply because the resulting contribution to the total energy is negative; there is no characteristic tied to the Higgs field that would impede its appearance.

MOTION AND ORDER

Why should that be true only below a given temperature? The reason is that this mysterious field connotes a highly *ordered state*—just as water molecules become ordered when they undergo a phase transition into ice. As long as they move in the liquid water phase, they wander about in random motion; in the crystalline ice phase, they oscillate about a *point of rest* (see fig. 5): Their state of lowest energy is always the ordered state in the ice crystal. But they will reach this state only below the critical temperature, which is 0 degrees Celsius in this case. Their own thermal motion will not permit an ordered state at higher temperatures: The hotter a substance, the faster and the less ordered its molecular motion, and the larger the momentum with which molecules collide. Above 0 degrees Celsius, the water molecules collide so frequently and so violently that no structure can appear or be sustained. Should there be somewhere in the water a small piece of ice, its molecules will be rent asunder by the momentum of the surrounding molecules. It is only at 0 degrees Celsius that the number of these hits and their impact no longer suffice to destroy incipient structures: The water can now freeze into ice.

The water molecules in the ice phase form a regular pattern; we might think of it as a pattern on a wallpaper, except that it is three-dimensional. Therefore, ice (as distinct from water) does not share the high degree of symmetry, with respect to arbitrary translations and rotations, that empty space possesses: Ice has only very limited translational and rotational symmetries, leaving its crystal

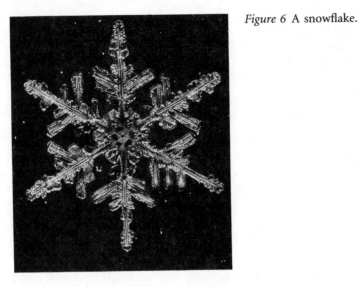

Figure 6 A snowflake.

structure unchanged (see fig. 5). This means that certain distances and certain directions in geometrical space are singled out as special. The same is true for the Higgs field; this field also distinguishes certain directions in its own abstract space—the space spanned by the properties of elementary particles.

SYMMETRIES OF OBJECTS

We will return to the Higgs field farther down the road. But we will keep talking about symmetries of individual objects, as well as of the laws of nature, as we go. Take the definition of the mathematician Hermann Weyl, as formulated by the physicist Richard P. Feynman: "[A] thing is symmetrical if there is something we can do to it so that after we have done it, it looks the same as it did before." In this fashion, a snowflake is symmetrical with respect to the operation "rotation by 60 degrees," because a 60-degree rotation does not change its appearance (see fig. 6). Moreover, snowflakes are *mirror symmetric:* If we reflect a snowflake on any one of six different straight lines across the center, its appearance will not change.

Figure 7 shows part of a pattern that extends to infinity toward both right and left. We call it *translation symmetric,* because the pattern can be shifted right or left by discrete amounts (which we can read off the figure) without changing it. Going a step further, we can apply the same thinking to figure 8: If we consider it part of a wallpaper pattern that continues to infinity in all directions in its plane, then the pattern we see enjoys multiple translational symmetry. Moreover,

Figure 7 This figure shows an excerpt of a pattern that extends infinitely to both sides symmetrically. It is translationally, rotationally, and mirror symmetric. The symmetry with respect to reflection at the horizontal median line of the pattern is obvious. It is also easy to note the two sets of turning points and vertical mirror lines. (The figure comes from a computer program written by the author.)

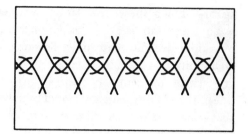

Figure 8 Excerpt of a large symmetrical pattern that is invariant under a number of different translations, rotations, and mirror imagings. (This image was also created by the author's computer program.)

the patterns of both figure 7 and figure 8 are invariant when reflected on certain lines, or rotated at given angles about certain points.

ORDER, CHAOS, AND SYMMETRY

Although the symmetries we discern in figures 6, 7, and 8 are the most obvious features in these illustrations, they show evidence of the *breaking* of higher symmetries: those of the circle, the straight line, the plane. The circle remains unchanged if we turn it around its origin, or center, by any angle; the snowflake, by contrast, has to be turned by 60 degrees or a multiple thereof to remain the same. A horizontal straight line can be shifted any distance right or left without change. Figure 7 can be shifted only by *certain given amounts*. The same goes for the planar pattern in figure 8: While an empty plane can be rotated without change by any angle about any point, the pattern will remain the same only if we rotate it by certain angles about certain points. That should be enough: Patterns with symmetries that are based on the ordering of finite elements will, by that very definition, break higher symmetries based on continuous parameters.

Since the water molecules in liquid water move around without any order, water possesses, on the average, all the symmetries of empty three-dimensional space—among these, translation symmetry by arbitrary amounts. As soon as water freezes into ice, this translation symmetry is reduced to the lesser symmetry, by certain given amounts only. This obviously implies that the crystal can occupy an infinity of positions that can be distinguished by comparing the displaced crystal with the same crystal in its original position. The position of figure 5 is one of them. Others are generated by subjecting the crystal to any transformation that is *not* one of its symmetry transformations.

As ice forms in the water, the emerging configuration has to decide which of these possible positions it will occupy. True enough, there may be prior asymmetries or external influences that influence this decision—the formation of the ice will make these influences visible by magnifying their effects. But for all practical purposes, the formation of ice breaks the continuous symmetries of water nondeterministically. This nondeterministic symmetry breaking is called *spontaneous symmetry breaking.*

SYMMETRY AND EMPTY SPACE

Given that empty space, as we understand it, contains nothing that could be changed by translation, rotation, or mirror imaging, we may call it symmetric under translation, rotation, and mirror imaging. "The void as such," according to Aristotle, "is unable to differentiate." In it, a stone cannot know what is above or below. Consequently, the stone cannot begin to fall. Newton called that the stone's *state of rest*—and this is certainly what Aristotle meant. The symmetries of physical empty space are not a matter of course. This space, the way we see it, remains the same when seen from a rotating merry-go-round. But here is the hitch: The merry-go-round is accompanied by centrifugal force; there is no such thing as a symmetry under a change of the velocity of rotation of our merry-go-round.

CLOSED SYSTEMS

The symmetries of a physical empty space depend on those of the laws of nature. We will be discussing their multiple relations as we go along. The symmetries of the laws of nature can be read from the characteristics of self-contained physical systems. The toy planetary system we experimented with was one of those, since we set it in an otherwise empty space. We can interpret it just as we do any closed system, as a probe for the properties of "empty" space. An arrow sent off by an observer on Earth's surface is not a closed system: It will not continue in a straight line, simply because Earth attracts it (and we haven't talked about the effects of air friction on its trajectory). The arrow's path provides enough information to observers, at least in principle, to tell them about their own elevation above sea level. At higher elevations, the force of gravity is smaller, so that the path of the arrow will be less strongly

Figure 9 The path taken by an arrow depends on how and where the archer shoots it. Suppose he practices on the banks of Lake Erie (9a). In a subsequent competition at the top of Independence Pass, he will miss his target, because of diminished effects of gravity, which doesn't bend the path as strongly as at the elevation of Lake Erie (9b). In empty space, in the absence of gravity, the arrow will fly in a straight line (9c). The archer can therefore infer certain facts about his location simply by the success of his marksmanship, without bothering to look at his surroundings.

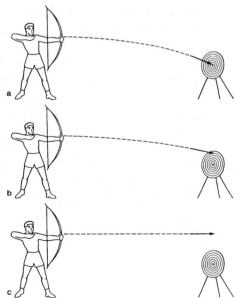

deflected. That applies equally to an arrow sent off inside a closed hall. To find out whether he or she is on the plains of Nebraska or in Independence Pass in Colorado, the archer need not look out at the surrounding landscape. The *inner characteristics* of the arrow's trajectory—that is, its curvature—will provide that information (see fig. 9). To hit a bull's-eye, the archer has to aim differently on the plains from the way he or she does in Independence Pass.

I do not claim that the effect is so large that the archer needs to consider it here on Earth, but it does exist. The same point can be made more strongly: It would make little sense to practice archery on Earth for a subsequent contest on the Moon—or, say, on the tiny planet of Saint-Exupéry's Little Prince. Where there is no gravity to speak of, the arrow will take off in a straight line.

That means the observer with bow and arrow in a gym on Earth does not form a closed system. Earth influences archery. The arrow behaves differently in Earth's vicinity from how it would behave in a hall somewhere in otherwise empty space. The laws of nature that are valid for the flight of the arrow close to Earth's surface are not fundamental: They depend on the presence of Earth, which we did not include in our system. Therefore, they are not translation symmetric. If we move the system from, say, the plains of Nebraska to Independence Pass in Colorado, we will have changed the locally valid laws of nature such that the trajectory valid on the plains is no longer possible in the Rockies, and vice versa. This fact makes the elevation of the imagined hall in which we practice archery an observable quantity. We don't have to look out the window to ascertain the elevation of the hall and the relative change between the two locations: The laws that govern the trajectory of the arrow tell us about the altitude of the hall.

Let's return to the closed systems: A closed system behaves exactly as though there were nothing else in the universe. The universe as a whole clearly is such a system. Can systems that do not include the whole universe but do not show the influence of the remainder of existing matter be considered closed?

We cannot shield any system from the fundamental laws of nature. They act on every system, everywhere, at all times. Suppose those laws of nature that we consider fundamental based on our limited experience actually depend on the existence of distant galaxies somewhere out there in space: This would mean that these laws are in fact not fundamental and that only the universe in its entirety is a closed system. It does not preclude the potential existence of other laws beyond our knowledge, more fundamental than those we observe; they might be independent of the state of our universe, and therefore truly fundamental.

SYMMETRIES OF THE LAWS OF NATURE

The hall in otherwise empty space we have been imagining for our archery experiments is, according to Newton's mechanics, a closed system. In it, the arrow, once sent off, moves with a constant speed in a straight line—unless, of course, it hits some obstacle. This is independent of the hall's position in otherwise empty space. When we say a law of nature has translational symmetry, we mean that translations transform every process allowed by the law into a process that is also allowed. Rotational symmetry, on the other hand, means symmetry with respect to rotations. It applies to the laws of nature, such as Newton's law for the motion of a planet around the Sun, but almost never to the individual processes the law allows, such as the motion of the planet on an ellipse, which is clearly not rotation symmetric. The type of motion that the arrow adopts in empty space is independent of the direction in which the archer shoots it. Its trajectory will always be a straight-line path, at constant velocity—conforming to Newton's first law. This law is therefore symmetric with respect to both translation and rotation.

Now let's place the hall on the surface of Earth. The laws of nature that now apply are no longer fundamental. They describe the trajectory of our arrow in a way that lacks both translational and rotational symmetry. Moving the scene of the shooting from the plains of Nebraska to Independence Pass would yield equivalent paths of the arrow only if the laws of nature allowed the arrows to travel the same path in different topographies—but we know they don't. Even more obvious is the laws' lack of rotational symmetry close to Earth's surface: The trajectory of an arrow shot off parallel to Earth's surface clearly differs from one that starts its motion vertically, either up or down.

The *fundamental* laws of nature—to which, as Newton was the first to realize, we can reduce the effective laws that are valid close to Earth's surface—*are* both translationally and rotationally symmetric. We can visualize this by including Earth in our observed system. In the joint system of archery hall and Earth, the laws of nature no longer depend on where Earth happens to be in its orbit around the Sun, and how it is oriented in space on its own axis. It should be emphasized that we are not concerned with the rotation of Earth but rather its orientation in space at any moment due to that rotation. Strictly speaking, we should try to imagine that the motion of Earth about the Sun, as well as Earth's rotation, stop at certain points, at which we then inquire about the trajectory of the arrow. The laws of nature that are valid for the system containing Earth, arrow, and bow are subject to Newton's laws of gravity and of motion under the influence of forces. In contrast to the archery hall on its own, the combined system is a closed one, if we ignore a few corrections. One such would be the gravitational pull of the Sun's mass on the arrow—and to eliminate it, we would have to include the Sun in our system also. What we are really after is an idealized system consisting of arrow, bow, and all celestial bodies in an otherwise empty space. In this system, which is truly a closed one, Newton's rotationally and translationally symmetric fundamental laws will be valid. They specify how different components of the system attract each other. They also imply how gravity influences the curvature of the path of the arrow. The precise shape of Earth with its mountains and valleys makes no difference as far as the laws of nature are concerned—it is simply part of the definition of the system.

ONCE AGAIN: SYMMETRY AND EMPTY SPACE

Although it may appear that we have moved away from our principal topic— empty space—this is not really so. Between the symmetries of empty space and the symmetries of the natural laws, there are close connections. Both empty space and natural laws are symmetrical with respect to translation, rotation, and certain changes of velocity. These changes in velocity exclude, in Newton's mechanics, acceleration or deceleration. When we turn, we experience centrifugal force; this means the applicable laws of mechanics are not symmetrical with respect to changes in the velocity of rotation (which is, technically, called *angular velocity*). The reason, according to Newton, is that we rotate with respect to absolute space—that space is therefore equally asymmetrical under the transformation of our example: When an observer who rotates in a space is subject to centrifugal force, the observer's relation to that very space implies the possibility of defining *the fact that he or she is turning.*

Position, direction, and constant velocity of space in Newton's physics can be freely chosen, because of the symmetries the laws of nature possess with respect to changes of position, direction, and constant velocity. Newton's laws do not depend on these three properties of his imagined space. To what extent can we then say that a space with these very properties is, in fact, *existent*? No doubt, it can be defined by a set of coordinates, but those coordinates are not implied by the space. The only property of space that can be defined by itself objectively and free of arbitrary choices is the physical concept of its acceleration; this is true *because* Newton's laws are not symmetrical under changes in acceleration. Let us stress that the acceleration we are discussing is not a change of velocity with respect to some predefined space; rather, it describes the change of velocity of systems under the influence of external forces when compared with others that move in the absence of such forces. This comparison illustrates why we use the term *acceleration;* we could equally well describe the difference of the two systems in terms of the forces that pull on each of them.

We have no doubt that there is space in which we can embed things, objects, matter. But its status, ontologically and physically speaking, is far from clear. Our senses do not tell us whether the space is "curved," even less whether centrifugal force would act on a merry-go-round in a space devoid of stars or galaxies. These are physics questions that we will address—but we will not be able to answer in which sense space as such "exists."

In the physical void or vacuum—that is, in the lowest-energy state permitted by the laws of nature—the two properties with which we endow this very vacuum in our imagination do not necessarily coexist. These properties are that there is no matter in the vacuum's volume and that it shares the symmetries of the laws of nature prevailing in it. The postulate that our space be empty in a conventional sense, in fact, uniquely defines the state of the void. Symmetry transformations of the laws of nature cannot change the void—after all, there is nothing whatever in it that could be changed. Therefore any such transformation will leave our system unchanged: It shares the symmetries of the laws of nature. For the state of lowest energy, the empty space as defined by physics, this is not necessarily true. This state may, in fact, have some structure that is not invariant under symmetry transformations of the fundamental laws of nature. In other words, we can infer from its property of *not* sharing the symmetries of the fundamental laws of nature that there must be several possible states with the same lowest energy. Symmetry transformations of the laws of nature cannot change the value of the lowest possible energy.

We'll return to this subject in more detail later on. The basic idea is Aristotle's: In empty space, no stone can begin to fall, because it does not know which way is up, which way is down. In this sense, the vacuum of physics—the space that contains the minimal permissible amount of energy—cannot be empty in a general sense. If a space of lowest energy is in a state that is not invariant under symmetry transformations of the basic laws of nature, then it contains "something"; it is not a void. The transformations act on this "something" and change it. The very fact that this space has distinguishable states, equivalent to each other through symmetry transformations of the laws, permits the definition of a direction, a distance, a concept that defines a particular rotational motion, and these very transformations can take the space from one such state to another. Take a rotation by 180 degrees of a space that includes our Earth, and let this rotation happen about a horizontal axis that goes through, say, a church spire. This rotation transforms our system into a different state—on in which our Earth stands, so to speak, on its head. We can now tell "up" from "down": It is because our space contains Earth that the symmetries of the fundamental laws of nature no longer apply—that we can tell that the space is not empty. Our rotation of space by 180 degrees permits us to introduce two states of that space, connected by the rotation of 180 degrees. They fix the up and down directions for a stone we may drop. Any space that distinguishes directions and accelerations cannot be "empty," in Aristotle's meaning.

Aristotle, in fact, is the outstanding figure in the history of the topic we are discussing. He was the first to deny flatly the existence of empty space on Earth's surface, elevating this denial to the concept of the *horror vacui* ("fear of the void") and determining the direction of occidental thought on this topic for more than two thousand years. In addition, he established the first image of our present definitions of a physically empty space with his concept of *ur-matter*, which his translators into Latin called *materia prima*. From this empty space of physics, as empty as the laws of nature will permit, anything may arise as an excitation, just as from Aristotle's *materia prima*.

Aristotle thought that the universe must have a center toward which all massive bodies gravitate. Were that so, empty space could not exist in the sense that there is no preferred direction in it. Newton's physics knew no such concept as a center of the universe. A space that included Earth's gravitational pull could not be empty. Aristotle, in fact, had somewhat contradictory ideas: The center of Earth, for one thing, defined the center of the universe; for another, massive Earth itself must be moving in this very direction. But in no way did he think that Earth with its surroundings was a special system that hid the existence of the fundamental laws of nature—rather, Earth and its surroundings made up the universe. He thought there was nothing whatsoever, not even empty space, outside the shell of visible fixed stars. By contrast, our modern concept takes Earth and its surroundings—what we now call our solar system—as just one unit, which we integrate into "empty space." It serves as a kind of probe for the fundamental laws of nature, which must be valid always and everywhere.

FORCES IN A VOID

Aristotle would have agreed with Einstein, who did not want to call Newton's space that contained no "bodies," and therefore no gravitational force, an empty space, but rather the *ether of mechanics*. Let's look at another example, in order to pin down the difference between real physical objects that act on their surround-

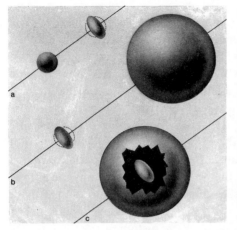

Figure 10 Of two identical spheres in an otherwise empty space, one rotates about the common axis, the other one does not. How is it possible that the rotating one (the upper one along line a) knows that it is turning and that its neighbor is not, and that it should deform in response to centrifugal forces, while the other should not? Newton's mechanics alleges that this is precisely what happens. Mach's principle, as formulated by Einstein, denies it. It says, instead, that all the masses of the universe cooperate in determining what rotational motion, or the lack thereof, is. If, in an otherwise empty space, a very massive sphere and a very small one rotate with respect to each other, the small one will deform, but the large one will not (10b). Furthermore, according to this view, a single sphere in an otherwise empty space will never deform. In 10c, we have placed the small sphere inside a hollow space surrounding the center of the large one: The larger one surrounds the smaller one just as the masses of the universe surround a centrifuge. The general theory of relativity, incidentally, is not in complete agreement with Mach's principle. Rather, it allows for many different realizations of empty space.

ings and idealized probes that merely register the forces acting on them. Take two identical spheres made of elastic material in an otherwise empty universe. Let them rotate about an axis that coincides with the straight line connecting their centers of gravity, but at different rates (see fig. 10a). Could it be that one of them will be deformed in some way but the other will not? According to Newton, that can be so. Let's leave the true answer open for a minute. Newton's rules of mechanics tell us which (if any) of the spheres remains free of centrifugal force—it is the sphere that does not rotate with respect to absolute space. If there is no centrifugal force, there will be no deformation of the sphere—no flattening at the poles, no bulging at the equator. Newton would interpret figure 10a as showing that the sphere on the upper right-hand side rotates with respect to absolute space but not the one on the lower left.

What exactly would happen to the spheres in our gedankenexperiment was beyond Newton's knowledge. He simply assumed that whatever his mechanics predicted would in fact happen. In reality, the planets, the moons, and the rocks whose behavior his mechanics describes are not situated in an otherwise empty space: The space surrounding them also contains distant stars and galaxies, and who can tell whether they alternately define the meaning of rotation? If that were the case, Newton's mechanics would constitute nothing beyond an effective description of our world; it would become invalid once we removed those distant stars. One fact speaks in favor of the latter interpretation (unless we want to

Figure 11 A long photographic exposure of the sky with a camera fixed in location.

treat it as a mere coincidence): The system in which we observe no centrifugal forces is identical to that in which the firmament—that is, the observed set of fixed stars—does not rotate.

The reader can verify this easily. Figure 11 shows that our Earth rotates relative to the firmament. There is centrifugal force. We learn about its action in high school: Since Earth makes only one full rotation per day, we usually do not notice this force. If a reader in a swivel chair rotates five times per second, the stars in his view will perform the circular motion indicated in figure 11, five times per second. What does he notice? First, he himself rotates relative to the firmament. Second, he notices that his arms are pulled up and out by centrifugal force. Could it be a mere coincidence that this force acts on him only while he rotates with respect to the stars?

Now back to our spheres: Up to this point, we have treated one of them merely as a marker for the other one—a system of reference that permits us to speak of relative rotation. Let's stick with the notion that space contains only these two spheres and that, contrary to Newton's mechanics, one of them does not realize that it is the one that rotates, that will deform. We now build up the universe, with its totality of physical masses, in a very special fashion: We add more and more mass to one of the spheres (say, to the one in the upper right of figure 10a); we will not change the other sphere—we leave it small and at constant low mass. If in fact it is the masses of the universe that determine the definition of rest versus rotation, it will be the upper-right sphere in figure 10a that does so in our gedankenexperiment. Anything that rotates exactly like this reference mass is defined as not rotating. Consequently, the small sphere will be deformed if there is relative rotation—the greater the deformation, the higher the rate of this rotation and the larger the mass of the upper-right sphere (see fig. 10b). We can treat the sphere with small mass as though it did not act on its surroundings, as though it were only a probe. The same is true in figure 10c:

Just as the firmament surrounds a centrifuge, the massive sphere surrounds, in this figure, the (nearly) massless one.

A related experiment to that in figure 10b can be performed with a top to be spun inside a satellite in orbit around Earth. After all, Earth is part of the overall mass distribution in the universe, all of which together determines the meaning of what rotates and what doesn't. The top on its path around Earth will therefore have to find a compromise that takes into account the centrifugal force of the firmament and the forces that the rotating Earth exerts on its motion.

UR-MATTER AS A PHYSICAL CONCEPT

It is not clear what physics will come up with as a substitute for the concept the creation myths call *ur-matter*. That is because there has not been a final synthesis of the two great theoretical developments of our century: Quantum mechanics and the general theory of relativity each make statements about the vacuum, but each requires allowances to be made for the other. We think that the radical statements of quantum mechanics and of the special theory of relativity don't say enough about the vacuum of physics. We have already talked about them: First, if we remove from some space as much energy as possible, there may be a "spontaneous" materialization of structures that we have named Higgs fields without really understanding their nature. Second, virtual particles incessantly emerge and vanish in this space. Because of the indeterminism of quantum mechanics, we cannot say exactly how many of these particles exist in a given interval of time and space, nor what their energies and momenta are. But nothing prevents us from marking points in space and time to our liking. A theory that unifies quantum mechanics and the general theory of relativity will presumably differ in this respect: Very small distances in time and space will no longer be individually identifiable. Maybe the theory will even keep us from telling spatial from temporal distances. Maybe, as we go to very small distances, we will no longer be able to say how many dimensions there are to time and space, just as the number of particles becomes indeterminate in ordinary quantum mechanics.

If this sounds less than clear, don't attribute it to my inability to formulate. Rather, we will have to concede that, so far, there has been no successful way of expressing these ideas in a compelling theory. It is out of the question that the modifications to which the concept of empty space has been subjected by quantum mechanics and the special theory of relativity might be reversed by future developments. The vacuum of physics contains in its faculties everything that the laws of nature will permit. It fluctuates—the virtual particles come and go. The only thing they may be missing is the energy it would take to make them appear as real particles. All that can appear in reality must be present as a possibility—as a state of virtual particles—in a vacuum. Add energy to the vacuum and those virtual states may appear as particles.

Figure 12 The two spirals are the traces of an electron and a positron. A photon enters from the left; it is electrically neutral; therefore, it does not leave a track. As it collides with an atom, the latter absorbs it and emits an electron-positron pair. An external magnetic field makes the resulting tracks spiral in opposite directions, where we see a reduction of the radius of curvature as the particles slow down. The electron and positron lose motion energy and, thereby, velocity, so that the curvature increases. It is the charge of a particle that determines which way the spiral turns, in opposite directions for electron and positron. Another track, appearing at the same location as the electron-positron pair, is that of an atomic electron pushed out of its atomic orbit by the incident photon. During this collision, the particle-antiparticle pair appears as a by-product. The photon vanishes in the process; its energy is converted into the mass energy of the particle pair plus the motion energy of the particles in the final state of the entire process.

In coming chapters, the puzzling phenomenon of *fluctuation* will be described further. Let's give an example here to illustrate what we mean by the a priori mysterious concept of "vacuum fluctuations" (see fig. 12).

In the figure, we see two particles of opposite but equal electric charges emerge from the vacuum on the left-hand side. One of these is an electron, the other a positron. Since they are charged, these particles leave tracks. They were kicked out of the vacuum by a photon—a particle of light. This light quantum is electrically neutral and therefore leaves no track. What it did to make the particle pair materialize is this: It collided with an atomic nucleus. But the nucleus acts only as a catalyst in this process; it absorbs the recoil of the emerging particles. Because of that, the photon is able to transfer its energy to the electron-positron pair in the vacuum. Thus a virtual excitation of the vacuum becomes real—one of the possibilities that are inherent in the vacuum is thus being realized.

Real systems are, in this sense, "excitations of the vacuum"—much as surface waves in a pond are excitations of the pond's water. Just as the properties of water determine what amplitude and speed of the waves are permitted, so the properties of the physical vacuum define the possible excitations—the possible systems that can emerge from the physical vacuum. Water, just like the vacuum, contains the final reality as an initial possibility. The vacuum in itself is shapeless, but it may assume specific shapes: In so doing, it becomes a physical reality, a "real world."

UR-MATTER AS A MODEL OF THOUGHT

We could use almost identical words to characterize both Plato's *space* and Aristotle's *materia prima*. Plato attributes actual existence not to matter as such, but to the structural plan of matter, to its *form*, its *idea*—today we would say to "the laws of nature." To recognize this plan, Plato thinks, we don't need to know the specifics of matter. The human spirit is part of the higher reality of forms and basic laws: That is why it is able to recognize these through a mere thought process. Matter is part of actual existence only as far as it can be understood. According to Plato, geometry finds the properties of space by means of human thought, human insight; that is why space is more "real" than the material world that we observe. He interprets the latter by means of a fairly convoluted construction as empty space within four of the five regularly shaped bodies that we associate with his name (see fig. 23a in chapter 2). He divides empty space into such bodies in order to make us see the world as we observe it in reality, in his terms. Plato's notion is that the world is part of our reality by virtue of the fact that his special construction subjects it to mathematical forms.

The fundamental aspects of the question of whether or not empty space exists were of no great interest to Plato. His philosophy knew empty space, and that was it. What we said about the vacuum of physics does, in fact, apply to his views: Space has no shape by itself, but it can assume form or shape—and that is how it becomes the observable universe. That also goes for the *materia prima* of Aristotle, the other great Greek philosopher, whose philosophies dominated Western thinking for two thousand years. The starting point of Aristotle's construct is always a specific concrete object with its own properties: This object he calls a *substance*. A substance is matter in a specific shape—for instance, in the shape of a statue of Socrates. The *materia prima* is defined as the remainder of a step-by-step process of analytic thought that starts from that substance, like a stage setting. All specific properties it possesses are disregarded one after the other—its specific shape, its material strength, whether it is made out of marble or out of some other material, and so on. In reality, matter and form always appear simultaneously. One form of matter may be replaced by another form. But there can be neither form without matter nor matter without form. The *materia prima* is therefore an abstract object: Real objects may approximate it, but they cannot turn into it.

By reversing the steps that we took to advance from the concept of substance to that of *materia prima*, we can imagine building up our universe: The homogeneous basic substance, *materia prima*, contains all the possibilities that the laws of nature will permit. There is no such thing as empty space into which our material world is introduced—that is what Aristotle preached, and modern physics agrees with him. The emptiest space known to it—the physical vacuum—has characteristics that Aristotle gave his *materia prima*.

Werner Heisenberg characterized Aristotle's matter in the following way: "Aristotle's matter is certainly not a given substance such as, say, water or air; also it is not just simply space. Rather, it is an indeterminate physical substrate which contains the faculty to assume some given form, and thus to enter into physical reality."

EXPERIMENTS ON THE VOID

After chance observations in antiquity and a few selective model attempts, the experimental investigation of the void started with Evangelista Torricelli's production of an airless space above a mercury column. Blaise Pascal continued Torricelli's investigations; he tried to prove by straightforward experimentation that it is the external air pressure that is responsible for the phenomena ascribed to nature's "abhorrence of the void," the *horror vacui* of the ancients. Otto von Guericke in Germany and Robert Boyle in England were the first to develop pumps that permitted the removal of air from larger containers and, thereby, experimentation in a vacuum. These experiments contributed to important insights for any discussion that attempts to define empty space: first, that empty space is translucent; and second, that magnetic fields are not dampened by it. Beyond that, evacuated volumes were used to produce some spectacular effects. Guericke's so-called Magdeburg hemisphere was the first of those. Pascal redid Torricelli's experiment in front of an audience of five hundred, with water and wine instead of mercury. In place of Torricelli's meter-long tube, Pascal used hoses of 10 meters' length hoisted on a ship's mast for his experiment. He built a water barometer in Rouen (see fig. 13), and soon it was discovered that empty space could be used to do actual work (see fig. 14).

Torricelli's vacuum lost its philosophical interest soon after its discovery; it was seen as space devoid of air, nothing else. Gradually, the philosophical questions about the void became questions of physics. It was recognized that the space between the planets is empty like Torricelli's vacuum—but then how is it possible that the gravitational force of the Sun acts over such large distances?

Figure 13 Seventeenth-century street scene of the city of Rouen. It is here that Blaise Pascal measured the air pressure with a water barometer around the year 1650.

How does its light travel through empty space all the way to Earth? How does this empty space transmit magnetic and electrical effects? Newton's experiments on heat conduction demonstrated that space also appears to transmit thermal energy; how can that be?

To answer these questions, the theory of the ether emerged, and it took Einstein's 1905 special theory of relativity to finally rid us of it. Today, there is still a good deal of experimental and theoretical investigation dealing with space that is as empty as possible. One of these is an experiment on the so-called Casimir effect: It shows that supposedly empty space actually exerts pressure (see fig. 15). As an experiment, it may not be spectacular, but its importance for our understanding of empty space is obvious. The tremendous machines of experimental particle physics in today's research centers also investigate the vacuum, in their own way. They include the LEP accelerator at the CERN (European Council for Nuclear Research) laboratory for particle physics, in Geneva, which we will discuss in chapter 6. While they cause electrons and their antiparticles, which we call positrons, to collide at energies a hundred thousand times that which corresponds to their rest mass, it is still the vacuum that they are really helping us to understand.

When electrons and positrons collide, they annihilate in a flash of pure

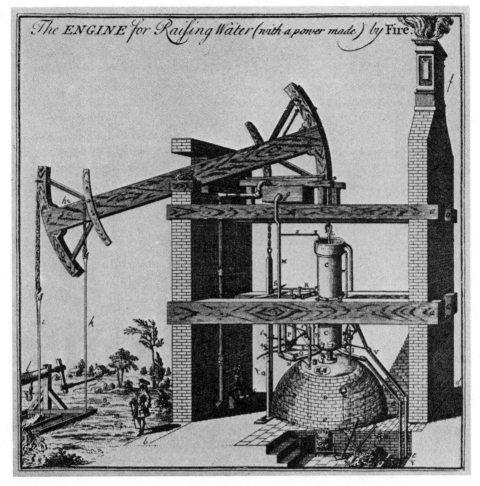

Figure 14 This contraption is the steam engine of the English inventor Thomas Newcomen, from the year 1712. It pumps water from a well. The pumping rods on the left are actuated by external air pressure. Water vapor generated in the atmospherical volume B fills the cylinder C, where it is cooled down by the outside application of water. The external air pressure then forces down the piston. The machine was able to pump 1000 liters of water per minute, in twenty strokes.

energy. In the process, they concentrate their energy in a small region of space, the properties of which do not differ from the vacuum in any respect. This flash furnishes the energy that makes particles appear as "real," whereas in the normal, not energetically excited vacuum they are so-called virtual particles. As a consequence, these virtual particles emerge from the vacuum and interact, causing an annihilation that produces a flash, which can be seen by a detector. That is how they tell us about the structure of the vacuum.

Figure 15 Two electrically conducting parallel plates, at a temperature of absolute zero (− 273 degrees Celsius), attract each other. The space around them, said to be "empty as possible," therefore cannot be empty: Whatever remains in it exerts pressure on the plates. They attract each other, so the pressure outside them is greater than the pressure inside. Hence the space between the plates is "emptier" than the vacuum reaching to infinity on either side.

The excited state of the vacuum "knows" which particles it can generate with what probabilities. Among its many possibilities is the creation of an electron-positron pair just like the one whose annihilation created it. But this is only one of many possibilities; the annihilation process makes the electron and positron lose their identity and generates a whole spectrum of possibilities from the vacuum.

THE VACUUM AND THE BIG BANG

On a very small scale, the accelerator experiment emulates the state of the universe as it was fractions of a second after its beginning. There is general consensus among physicists about how the world developed after this point in time, but not before it. One model for the evolution of the universe from the very start is the so-called hot Big Bang. According to this model, the universe began its existence as an infinitely hot and dense flash of pure energy. Because of this infinity, physicists speak of the *singularity* that marked the beginning of the universe. From then on, according to the model, the universe became colder and colder. I'm not saying that the model is uniquely compelling: There may never have been a hot Big Bang; maybe it took the universe a few small fractions of a second to reach the very high temperature—very high but not infinite—that we know it once possessed. But there is agreement both on that hot temperature's being much higher than any temperatures we will ever be able to duplicate in our laboratories and also on the steady cooling the universe has undergone ever since. The higher the energy density we can create in an experiment, the closer we approximate the state of the universe at that initial high temperature. Its state at that point, when there was nothing except that energy, is best described as "the physical vacuum plus energy."

One of the most important questions we expect future accelerator experiments to answer concerns the structure of the early universe. The hotter the universe is, the less structure it can have. This is just like what happens when ice melts into water, and water evaporates into steam. In a very early phase of the universe, it was so hot that no structure could evolve. But the cooling process began immediately. Theory tells us that only fractions of a fraction of a second after the Big Bang, the Higgs field emerged as an early structure—just as ice develops in water as it cools down. Inversely, it should be possible to melt the structures that are frozen into the vacuum. If the temperature is high enough, the physical vacuum must undergo a *phase transition*. It passes from a state with frozen structures to a state devoid of them. It is likely that the energy densities that can be produced at the LEP machine in Geneva will not suffice—although we don't know just how much energy we need to effect this transition. Neither theory nor experiment has given us a convincing indication for the structure of the vacuum. The inquiry about the state of the universe within fractions of a second after its very beginning is as exciting a question as I can imagine.

TIME AND/OR SPACE

If we knew the state of lowest energy ascribed to the world by a theory that unites both quantum mechanics and the general theory of relativity, we would be able to say much more about the origin of the universe than we presently can. This is so because our world has most probably had its origin in this state of lowest energy, or *ground state,* which may be identified with the absolute vacuum. We tend to associate with this origin a particular moment in time—the time at which our world started to exist. The origin of time is an almost ineffable idea; nevertheless, it might well be that time came into being together with our universe, instead of our universe starting at some point in a preexisting time. This is an important topic of this book. Via the quantum mechanical uncertainty relations, time and space are related to the fluctuating energy and momentum of matter so that all three—time, space, and matter—may share a common origin. However, as of yet, nothing definite is known on this topic. A theory of the physicists James Hartle and Stephen Hawking implies that time and space are not separable at the origin of the world. If we move backward in time, time and space blur into each other at some distance from today. Instead of speaking of "former times," we should speak of another place in space-time. We never reach some previous earliest point in time, because we can move in space-time just as in space alone. This means that we may have already started a forward motion in time while we still think we are moving back in this coordinate.

In spite of its finiteness, the surface of our Earth knows no limits: We can move on it without ever running into such a limit. The same goes for the combination of time and space according to Hartle and Hawking. If we substitute "north" for an advance in time and "south" for backward motion in time, moving across the South Pole on Earth's surface is equivalent to changing the direction of time from backward to forward. We can imagine that time and one dimension of space together form a surface like the surface of Earth. In this case, nothing distinguishes one point on Earth's surface from any other. According to Hartle and Hawking, the same goes for time and space. We can draw arbitrary curves into diagrams that depict space-time, and we can follow these curves in arbitrary directions—we will never hit a limit. It is a condition of our world that it has no confining margin, be it in space or in time. In Hartle and Hawking's model, the question of the earliest time is as pointless as it would have been to ask Roald Amundsen why, when he reached the South Pole on December 14, 1911, he did not go farther south.

NAIVE REALISM

The physical vacuum, which in the course of history has made its appearance in human thought under as varying a set of names as nothing, the void, space, *materia prima* (Aristotle), matter (Plotinus), and the ether, carries within itself the possibilities of everything that can exist in the physical world. Once we attain true knowledge of this vacuum, we will have a comprehensive knowledge of everything, including the laws of nature. It is as in the thinking of the ancient philosophers: The knowledge of the void, the "nothing," is intimately connected with the knowledge of the "something." Let the world image of the atomists serve as an example: There is nothing except atoms and the empty space between them. The void is part of our reality.

The worldview of the atomists has been a good starting point for a description of reality and continues to be so today. We take it as a matter of course that empty space is not really empty; also, that atoms are not indivisible, continuous, compact physical structures. Quantum mechanics rules the universe. If we start from quantum mechanics alone, we have no way of understanding the origin of the quasi-classical laws that govern our world as we know it. Why, and in what manner, did an early cosmos develop in such a way that our Earth orbits our Sun on a well-defined trajectory?

CHAPTER 2

NOTHING, NOBODY, NOWHERE, NEVER:

•

PHILOSOPHICAL, LINGUISTIC, AND RELIGIOUS IDEAS ON NOTHINGNESS

•

GREEK PHILOSOPHY STARTS WITH A DOUBLE NEGATION: IT DENIES THE concept of "nothing." Thales of Miletus states flatly that a "something" cannot issue from a "nothing"; neither can it vanish into nothingness. He thereby denies the *creatio ex nihilo*—the world's creation out of a void. This concept played no part in the Greek mythology of creation. True, the ancient Greek Orphic poets sang songs of "chaos, from which all else issued"; but soon "black-winged night is delivered of a still birth": It then has "Eros couple with the winged nocturnal chaos." Thus, Greek mythology prefers images to philosophical concepts—images often violent and grotesque, even baudily fantastic. Take the example of Gaia, the Earth Mother, who has her son Kronos castrate her husband, the god Uranus. For fear they might depose him, Kronos himself devours four of the five offspring he has with his wife, Rheia. Only his last son, Zeus, is spared: In his stead, Kronos wolfs down a rock wrapped in diapers, falling for another ruse of Rheia's—this is as raucous a tale as you might invent and, fortunately, not the subject of our discussion. But it forms the backdrop for Thales' idea that the world arose from the waters. This ur-ocean, however, to him is not one more god among others, as is Hesiod's and Homer's "deeply vortexed Okeanos"; for Thales this ocean is the one and only source from which the world springs, by condensation and evaporation. And all the while the total amount of this ur-matter remains the same; it cannot be created or annihilated. It carries the force that moves both the cosmos and all organisms—so pervasive and so multi-faceted that it makes up the stuff of the world and fills it to the brim. In this worldview, there is no room for nothingness.

MYTHS OF CREATION

Most stories of creation deal with the development of ur-matter into the world we see, leaving aside the question of where the ur-matter came from. Thales, atypically, makes a point of denying the creation of his ur-matter out of the void. His successors simply assume the presence of ur-matter *ab initio*. Heraclitus, who does discuss development and contradictions in this edifice, still leaves out the question of where his ur-matter, fire, comes from; he simply lets our world come from fire and turn back into fire cyclically. But fire itself is eternal—it is identical with the deity, with the basic cause of our world.

More highly developed creation myths—those that can make do with gods who do not share in the more animalistic functions of humankind—can be divided into two categories: those that have a god create the world out of nothing and/or out of some ur-matter; and those where creation happened spontaneously. "Nothing" stands for ur-matter in a completely amorphous state. A spontaneous creation—we might call it a self-creation—appears in the writing of the Taoist Chuang-tzu as follows:

> If there is a beginning, there must be a time before the beginning; there furthermore must have been a time that preceded this very time before that beginning. If there is being, nonbeing must precede it; and before this nonbeing, there must have been a time where not even that nonbeing had started. Furthermore, another time before that which had not even seen the not-beginning of the nonbeing.

This is where, all of a sudden, nonbeing turns itself into being—"one cannot even say whether this being of the nonbeing is part of an overall being or nonbeing." Be that as it may, we may take it for granted that the perplexing paradox of being/nonbeing in the Taoist's tale will finally turn out to be the world.

The idea of something appearing out of nothing without the mediation of some creator is also present in the *Kojiki*, the oldest document in Japanese literature, dating from the year 712: "There was chaos—but who can say what shape it had? There was no shape; nothing moved, there was not even a name for it. But in all this emptiness, Earth and Heaven parted and something emerged between the two. What did emerge is a god—but a god who had no part in that creation, not even in his own."

Then there are the creation myths that contain a creator god. This god exists independent of the world, and the myths have nothing to say about his origin. Saint Augustine, in his interpretation of the biblical Genesis—the creation story of the Old Testament—puts it this way: "Heaven and Earth emerge. They call out in a loud voice that they have been created. . . . They also sing out that they did not create themselves. Thou, oh Lord, didst create them." I will not quote Genesis in this context, since it is one of the most diversely interpreted and commonly quoted books of the Bible. The creation story according to the Gospel of Saint John begins as follows: "In the beginning was the Word, and the Word was with God, and the Word was God. . . . And the Word became flesh, and dwelt among us." Whether or not we replace, as Goethe did, the term *Word* with *Sense, Force, Deed,* what the evangelist means to stress above all is the origin not so much of substance, of matter, but rather of form and structure.

The creation myth of the Maori aborigines of New Zealand starts with a god who, in the beginning, finds water in existence, and who then also creates structure—in time rather than in space: "In the beginning, there was darkness, and there was water everywhere. There was no light, and Io lived alone in this immeasurable space. And from the deepest darkness came Io's voice that said: Darkness, light up! And there was light. Then the voice said: Light, turn into darkness! And it was dark again." The change of day and night was thus fixed. The creation myth of the Popul Vuh—the holy book of the Maya—starts with the universe at rest. "Not a breath—not a sound. The Earth was motionless and silent. And the skies were empty. There were only the gentle oceans and the vast spaces of the skies." From all of this, six gods create the world "in water, flooded with light." Among the six gods, the first was Tzakol, who created the world in the beginning; the second was Bitol, who subsequently formed it.

Considering the care with which they detail the concepts of the creation of shape, structure, and ultimately of life, creation philosophies and myths deal only lightly with the initiation of the substance that forms or takes shape. Saint Augustine, our authority for Genesis, spends many pages of his *Confessions* wrestling with the fact that the first book of the Bible does not even mention the creation of matter as such. Did God, when he "created Heaven and Earth in the beginning," just give shape to matter that was already present? And is there such a thing as matter without shape? Let's listen to Saint Augustine again: "I would rather judge that something that has no shape at all does not exist in the first place; I cannot imagine a Something that provides a link between shape and nothing—a Something that is neither formed into some shape nor really nonexistent, Something I might call a shapeless almost-nothing." He adds to the question about substance and shape a theological one: Did any time elapse between the creation of substance and the creation of shape? His answer reads like what we would call today a carefully crafted press release. True enough, there is on one side the matter that makes up our Earth and fills our skies; there is also the very shape of Earth and the skies. "But Thou, o Lord, didst create both. Thou didst create matter out of the void, shape out of shapeless matter; both were created in one act: Matter took shape with no time loss in between."

Let's not haggle over the question of whether or not the various histories of creation have much, or anything, in common with today's scientific knowledge. That is not our point. Fundamentalists who claim today on biblical grounds that only a span of a few thousand years have elapsed since the creation of the world clearly expose themselves to ridicule. We might say that they put themselves in the tradition of James Ussher, archbishop of Armagh and primate of Ireland, who fixed the time and date of creation as the evening of October 23, 4004 B.C. Dating of this kind was done by counting the generations that, according to the Bible, creation followed. Actually, the universe is older by a factor of a million. But whoever forgoes science would have no trouble finding the world fashioned by the creator 6000 years ago so that today it looks precisely like a world that is some fifteen billion years old. The Grand Canyon, in this scenario, would have appeared simultaneously with the first totem pole—about 4000 B.C.—as would Crater Lake. In the same vein, the light that appears to originate from the Andromeda Galaxy, two million light-years from our solar system, would have been created in midflight. The same goes for the light from most stars in our galaxy, the Milky Way. We will also have to put aside our knowledge of fossils, which date as far back as three billion years (unless God went to the trouble of changing the laws of nature in the meantime).

Clearly, the various stories about creation don't elicit our interest because of the empirical truths they tell us about our world. Rather, they deserve our attention because they introduced the conceptual categories that still serve us today when we try to make sense of the world. Chief among them is the differentiation of "something" and "nothing"—about which the creation myths tell us very little. Another pair of concepts we owe them is that of the juxtaposition of chaos (absence of shape) and shape. Saint Augustine's musings, quoted earlier, are an instance of the usefulness of these categorizations.

WAS THERE TIME BEFORE THE BIG BANG?

Saint Augustine starts his discussion of the time when the world was created with a question: "What did God do before He created Heaven and Earth? Had He been idle . . . and not active, why did He not remain so for all times just as He had been idle and inactive before the creation?" Augustine's answer to his own question has theological as well as logical arguments, but it also opens up a new train of thought: Time started when the universe began its existence. If we track the development of the universe backward in time, we reach a point when both world and time made their appearance jointly. To seek a time before that makes no sense. In today's language, we would say: Events do not follow one another in a time that would also flow if they were absent; rather, they define the very concept of time. If there is nothing that can change, there is no way to make time an observable quantity. Take, as an example, the 100-meter dash at a track meet: A runner's time over the distance is fixed by comparison with other processes, like the ticking of a stopwatch. In so arguing, we beg the question of whether, as Newton postulated, there is such a thing as an absolute, true mathematical time that, of its own accord and by its own nature, takes its course uniformly, unrelated to any external object. Time, in this reading, would not be defined by clocks; clocks would only read it, and make it observable through their reading. The highly speculative attempts at a modern quantum theory of gravitation don't know time in Newton's sense. Instead, that part is taken over by a parameter called "the content of the universe." Were this book dealing with time, I would now have to investigate whether time thus defined prefers something like an "arrow of time" over its inverse. But time is clearly also observable in the absence of such arrows—say, if 50 percent of all clocks run with time moving forward and the other 50 percent with time moving backward.

The hypothesis of the Big Bang accords with a model in which time and the universe originate jointly. This is the standard model that physics has of the beginning of the universe. The galaxies are not distributed in the universe in a static fashion, like the cities on the surface of Earth; rather, they move apart as though they had been propelled from one and the same point in space by a catastrophic explosion some fifteen billion years ago.

The Big Bang hypothesis concludes from this distribution of matter that all of it was packed with infinite density at the instant of the explosion. The laws of physics say nothing about matter in that state. We can approximate its description, however, through a sequence of states in all of which the laws of physics are valid. To illustrate this sequence, think of all the universe's matter being concentrated, first, into a space the size of our galaxy, the Milky Way; second, into our solar system; then on to our Earth, to Mount Rushmore, to a nutshell, to a grain of salt. . . . In this fashion, we build up a sequence of states where the laws of physics apply; if we go on with such a sequence, we will approximate the state of matter at Big Bang time to arbitrary precision.

I am not claiming that we know the laws of physics for matter in that state; this isn't so. But we do think that there must be laws, albeit unknown to us, that do apply. Certainly, as the volume is reduced more and more, the matter must have moved faster and faster. We know that the galaxies, as they are moving apart, are slowed down by the force of gravity. This implies that the less space all this matter had at its disposal, the faster it must have moved. We also know that this motion must have been in the form of random relative motion. After all, the galaxies have different velocities; consequently, their building blocks must have moved at different rates in the beginning.

We conclude that immediately after the Big Bang, when all matter was infinitely closely packed, its components moved randomly with huge velocities. We do not know the makeup of these components, nor whether it makes sense even to speak of individual components. The important feature that characterized matter in the very early universe was its state of rapid random motion, whether we describe this motion in terms of discrete parts or as that of a liquid, with all the known features of fluid motion.

For the sake of the present argument, let's stick with the image of discrete particles of matter. The faster their random motion, the hotter this matter gets. Temperature is, in fact, defined by the random motion of particles. Hot water dissolves sugar better than cold water because the water molecules move faster at a higher temperature and are thus better able to knock the sugar molecules from their configurations. The closer we get to the Big Bang, going backward in time, the hotter all matter must have been. At the instant of the Big Bang, the universe was infinitely hot and infinitely dense. This is the state of maximal disorder or, as we say, of complete chaos. If there was such a thing as a "before," that "before" cannot have influenced the "after." The infinitely high temperature and density at Big Bang time erased all information about whatever state the universe might have been in "before." Whether or not it makes sense to speak of time "before" the Big Bang is immaterial for the development of the "after." Anything that we speculate about the "before" cannot be checked and is therefore not in the realm of science.

If the universe was, at one time, infinitely hot, that point in time can be reached from our vantage point only by following our own timescale in a backward direction. Obviously, it makes no sense to speak of an actual, physical motion backward in time. What we can construct is a table of states that describes the sequence through which the universe must have advanced in the course of time, and then read it backward. This backward reading will have to stop at the state marked by the description "infinite temperature." A table that does not stop at that point cannot be based on observation. It would have to rely on revelation, a concept well beyond the topic of our discussion.

If we take different substances with equal atomic composition and heat them until their atoms move individually, we lose, in the process of thermal breakup, all information on what fraction of these atoms originally belonged to any given substance. Take graphite and diamonds: Both are made up of carbon. When we heat a mixture of graphite and diamonds for a few hours to a temperature above 1500 degrees Celsius, there will be only graphite: All information about the original amounts of the two components has been lost, as have the diamonds we started with.

Back to the Big Bang: There is no harm, no contradiction to empirical fact, if we assume that there was a time "before" or that there was no such thing. The same goes for space. Space, rather than time, is closest to the topic of this book; we will therefore return to the question of whether there can be space all by itself—space without matter, without radiation, without energy. Is it legitimate to imagine the universe before the Big Bang as empty space in the context of considering the Big Bang as *creatio ex nihilo*—creation out of nothing? If this is not legitimate, maybe empty space before the Big Bang was more than an absolute nothing. Time and again, artists have tried to portray the "mighty Fiat of the

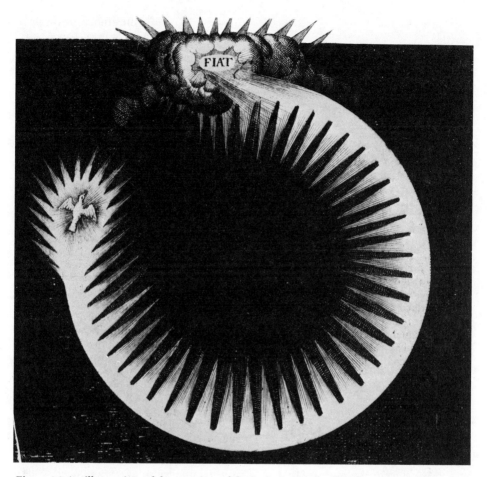

Figure 16 An illustration of the creation of the universe by the "mighty Fiat of the Creator Spirit."

Creator Spirit" of which Pope Pius XII spoke in a remarkable speech on November 23, 1951. See figure 16 as an example.

PROPERTIES OF SPACE

We have to establish the rules of the game before we try to answer these questions. The scientific method demands that as we pose a question, we introduce a procedure that can be used, at least in principle, to answer it. If that proves impossible, the question is not a scientific one. For our topic, the questions that can be asked in this scientific way are cloaked in intuition. When we think about

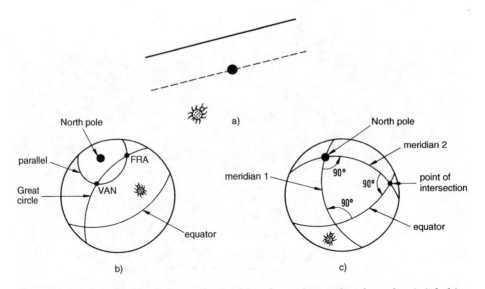

Figure 17 (17a) A flat bug believes the world to be a plane—the plane that it inhabits. This bug, coincidentally called Euclid, has just figured out that of all the straight lines that can be drawn through one point it is contemplating, there is one, and only one, that will not cross the line drawn in the figure. (17b) Another bug, called Non-Euclid, is two-dimensional, just like its cousin Euclid. But it is not flat: It has curvature, so that it hugs the spherical surface on which it lives. Having wandered across the markings we drew into the picture, it is pondering whether there is any straight line that does not cross the one connecting the points marked FRA and VAN. The solution it finds, illustrated in figure 17c, is described in the text.

space, we start from notions that originate in our everyday experiences. These notions contain a variety of ideas about the nature of space and were reduced to a few axioms as early as 300 B.C. by Euclid, in his treatise on the "elements." Among these axioms, a few deal in an obviously correct way with points, straight lines, and planes in the space that we inhabit. Here are two of them.

One axiom says that there are at least four points in space that do not lie in the same plane. Another one, equally obvious, states that out of three different points on a straight line, one and only one will be located between two others. But there is also Euclid's (in)famous axiom of parallel lines: Consider any straight line and a point not on that line. Obviously (fig. 17a) there will always be one and only one plane that contains the line *and* the point—the plane of the page, in the case of figure 17a. The axiom of parallel lines then says that there is precisely one line within the plane that goes through the point and that does not intersect the first line. This new line is called the parallel of the first one.

The points, straight lines, and planes of our imagination make up the ingredients of these axioms. Our experiences, however, are restricted to the limited spaces of our observation, and we cannot examine the general truth of these axioms insofar as they deal with the infinite. After all, over what distance would we have to observe the two straight lines in figure 17a to make sure they would not intersect? To infinity, for sure—but that we cannot do.

This axiom of parallel lines is notorious, since it makes a statement that reaches out into infinity. Over the centuries, there have been attempts to reduce it to obviously correct statements involving only finite regions, but to no avail. Around 1800, it was postulated that there might be spatial structures where all of Euclid's axioms apply with the exception of the last one, the axiom of parallel lines.

Such a structure would have to involve a curved space of three dimensions. I cannot imagine this space, and I doubt that the reader can either. That leaves us in good company: Several physicists who have published papers on curved spaces with three, four, and more dimensions have openly stated that they cannot imagine any curved space with more than two dimensions. But in two dimensions, they can visualize such spaces, and so can we. A plane is clearly a two-dimensional space without curvature; the surface of a sphere is, equally obviously, a two-dimensional space with curvature. Any reader who can imagine a two-dimensional bug that lives in a plane and has no concept of what lies outside that plane will also understand what we said about curved space. Our two-dimensional bugs cannot crawl in a direction at right angles to the plane in which they live; this direction does not exist for them. And in precisely the same fashion, there is no fourth spatial dimension for us inhabitants of three-dimensional space.

Now let's imagine bugs that live on the surface of a sphere (see fig. 17b); a point in the world of this bug is any point on the spherical surface; a straight line is a great circle on that surface. We call a *great circle* any circle on the surface of a sphere that has its origin in the center of that sphere. An example of such a great circle on the surface of our Earth is the equator; the meridians fall in the same category. Just as straight lines define the shortest distance between two points in a plane, so do great circles on a spherical surface. That's why air connections follow great circles rather than, say, lines of equal latitude: Vancouver, British Columbia, and Frankfurt, Germany, share the same latitude; the shortest distance between them is the great circle through both of them. Following that circle, Air Canada and Lufthansa will fly over Greenland rather than following the parallel (see fig. 17b).

The great circles are the shortest lines connecting two points on any spherical surface; they are the "straight lines" for the bugs living in this curved space. It is easy to convince ourselves that the great circles can be substituted for straight lines in all of Euclid's axioms, with the exception of the axiom of parallel lines. For this axiom, great circles won't do. After all, the great circle that connects Vancouver and Frankfurt will intersect every other great circle on the surface of Earth; there is no "parallel line," according to our definition, in this space.

That is exactly what figure 17c is meant to make clear. The figure shows three great circles on the surface of Earth. Two of these are meridians, and one is the equator. Let's look at the first meridian and at the point of intersection of the second meridian with the equator. Obviously, every great circle that goes through this point of intersection will also intersect the first meridian. Our second meridian does so at the North and South Poles.

The spherical surface that conforms to all of Euclid's axioms except the axiom of parallels is a curved two-dimensional space. The well-known eighteenth-century mathematician Carl Friedrich Gauss was the first to conjecture that we are living in a curved three-dimensional space. In that space, we would exist a bit like our two-dimensional bugs in figures 17b and 17c. Gauss is said to have tried to prove this point experimentally. To do so, he obviously could not leave his—our!—space in order to take a look from the outside; that would be impossible. But we can determine the possible curvature of a space from its inner characteristics—characteristics that become apparent to inhabitants of a space without forcing them to leave that space for a look from the outside. Suppose they want to find out whether they can draw one or several straight lines through a given point such that they are parallel to a given straight line outside that point. Gauss, of course, was not able to investigate straight lines beyond the surface of Earth. Instead, he is reported to have triangulated three easily identified mountain peaks in the Harz Mountains of northern Germany. He joined these points by light rays instead of following the surface of Earth from one to the other; he correctly surmised that light rays define the shortest distances in our three-dimensional space. Should these rays be curved, the sum of angles in triangular configuration will not amount to 180 degrees, as we learned in high school math, which, after all, deals with triangles only in flat Euclidean space. The sum of angles on a spherical surface is always larger than 180 degrees; in the specific triangle of figure 17c, it amounts to 270 degrees. Gauss found the sum of angles in the Harz Mountains to be 180 degrees. That means our space does not have enough curvature to make a deviation from flat space measurable with Gauss's instruments for a triangle the size of the one he observed.

The larger the radius of our sphere, the less curvature its surface will have. If we observe only small areas on the surface of a large sphere, we can safely ignore the curvature for all practical purposes. In the space we inhabit, this is exactly what happens. Space on a large scale may well be curved, but our everyday experiences take no note of that. Architects who build houses on Earth's surface need not worry about the spherical shape of Earth. They do quite well on the assumption that the surface is flat. On the other hand, it would be unthinkable to use a rectangular street pattern like Manhattan's for a very much larger city: It is impossible to divide up the surface of a sphere into rectangles bordered by straight lines that are the shortest connections between the corner points. Similarly, the spherical geometry of Earth influences the shape of long bridges, tunnels,

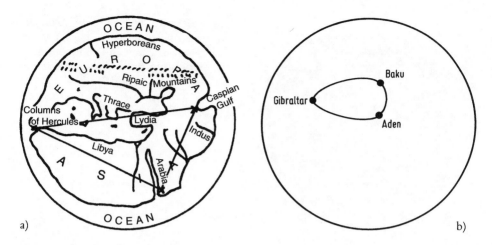

Figure 18 If Earth were flat, as in figure 18a, the shortest flight connection between three cities would form a triangle whose angles add up to 180 degrees. On our Earth, the sum of the three angles in any such pattern connecting three cities is larger than 180 degrees (18b).

and pipelines. And we know well that triangles made up of flight connections between three cities are a long way removed from triangular shape (see fig. 18).

The properties of space determine which path light rays will take. Light rays that form a triangle determine the sum of angles in that triangle; this fact alone means that space as such has properties. Pure logic tells us that a space containing light rays that are able to probe its properties is fundamentally different from a space without light rays. Quantitatively, true enough, space is incomparably less influenced by rays than it, in turn, influences them. Light rays should be seen as probes in an otherwise empty space—much like the magnetic needles and mass probes we used in the prologue of this book, in our gedankenexperiment for the measurement of the magnetic and gravitational fields of Earth.

Our space also determines the behavior of the probes we use to determine its properties. Let's again make them sufficiently small so that space acts on them but we can ignore their action on space. Let us release several such probes in different locations, with different velocities, and/or at different times; they will follow trajectories from which we can deduce the properties of that space. As long as Newton's mechanics are valid, our probes in empty space remain at rest or move at a constant speed in a straight line. However, if the probes are far apart, the expansion of space becomes an important factor. Given that there are these possibilities and many more, it is quite clear that space is anything but a "nothing" without properties.

One of the giants in the history of the topic of our book, Otto von Guericke,

Figure 19 Just as Poincaré's and Feynman's creatures (in figure 20), the angels and devils in Douglas Dunham's computer drawing diminish in size as they approach the confines of their world.

translated Aristotle's definition of the void as "a space that is not taken up by any extended object, but that is capable of being filled with such objects." We might speak of something like an empty box. Aristotle also thought that there could not be such a void—everything that can be filled has in fact been filled by nature. To the mathematician, space is nothing but an ensemble of points that are connected by some set of relations. In this interpretation, the laws of nature that apply to objects, including our probes, have nothing whatever to do with the properties of the space around them. The great French mathematician Henri Poincaré is one of those who contributed significantly to this mathematical view. To physicists, on the other hand, the properties of space can be read off the trajectories of the test probes they release in it.

But we have more evidence than those trajectories. The shortest path from one point to another is defined not only by the trajectory of the light ray that connects them but also by its very length, measured in whatever units—say, feet or meters. Take two points in an ordinary Euclidean plane and an arbitrary amount of one-meter-long measuring sticks (see fig. 20a). Assume that the length of these sticks, defined by the so-called ur-meter, which is carefully stored in Paris, is very small in comparison with that of the path. The length of that path, measured in meters, equals the number of measuring sticks we have to lay down end to end in order to join the two points of reference. The shortest path from one point to another is defined by the fact that fewer measuring sticks are needed on this path than on any other. Obviously, the straight line defined in this fashion is the one we drew in figure 20a, for the motion of a bug in a Euclidean plane.

For a curved space, we have to define the meaning of a 1-m measuring stick—the last paragraph showed that this is not easily done. Seen from the outside, lengths may well change when we move from one system to another.

Figure 20 This figure illustrates the constructs realized by the bug in a plane (20a)
and on a hot plate (20b), as described in the
text. The circular lines stand for constant
temperatures. We assume that (20b, 20c)
all distances shrink to zero as soon as a
certain lowest temperature—that is, absolute zero, or −273 degrees Celsius—has
been reached. This means the world of the
hot plate bugs has an impenetrable edge:
The circle with zero temperature cannot
be reached, because the bugs' steps become
infinitely short once they get arbitrarily
close to the edge.

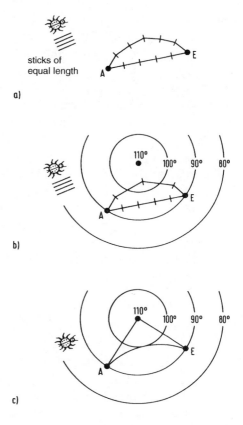

Poincaré gave a model for this, using spherical surfaces. Richard P. Feynman
gave another detailed two-dimensional model in terms of a hot plate. In his
famous Caltech lectures on introductory physics, he asked his listeners to imagine
a hot plate inhabited by two-dimensional bugs, such as we discussed in the case
of spherical surfaces. The temperature of the hot plate is not the same everywhere.
It may be cold at the perimeter and hot toward the center (see fig. 20b). These
bugs have measuring sticks that Feynman assumes change their lengths as a
function of the temperature: The higher the temperature, the longer the sticks.
This is borne out in reality by railroad tracks; spaces are left between adjoining
track lengths so that their expansion in summer will not lead to a deformation.
Feynman's bugs change their measurements as a function of temperature along
with the sticks.

Feynman uses his hot plate model to illustrate many properties of curved
spaces. Let's address only the topics of the shortest path between two points and
of triangles. To join two points on a straight line on the hot plate, the bugs will
need more sticks than for the curved line that deviates through the (hotter)

central part of the hot plate (see fig. 20b). Remember, the sticks lengthen with increasing temperature. But the bugs do not know that. They say, "The sticks are the same length; we know that from putting them down next to each other and comparing." The bugs are so obtuse that they now insist on calling their shortest connection a "straight line." Figure 20c shows that if they connect three points with what they call "straight lines" they will notice that the sum of angles in the triangle they just constructed is smaller than 180 degrees.

Our own space is in fact curved. This is certainly true for distances observable by us, in the presence of masses. Whether space as a whole is curved or not transcends our knowledge. Maybe the local curvatures due to distortions caused by individual masses are nothing but valleys and mountains in an overall flat universe—or, for that matter, in a spherical universe. We cannot exclude other forms of curvature beyond those mentioned. To make matters worse, we will have to include time as a fourth dimension in any complete description of the curvature of space as it is realized in our universe. But that goes beyond the ambition of this book.

SPACE AND BODIES

Whether we investigate the properties of space by means of light rays, mass probes, or measuring sticks, it is only by means of the objects that space contains that we learn about it. To probe it, we chose very small masses in order to avoid noticeable action of those masses on the space. Masses, after all, deform the space that surrounds them, just as additional heat sources on our hot plate would further deform the two-dimensional space in which the bugs of figure 20 live. Light rays that pass close by the Sun on their way from distant stars toward Earth will be deflected by the Sun. The experimental proof of this phenomenon made a huge splash in 1919 and served as pivotal evidence for the validity of Albert Einstein's general theory of relativity. We stick to our belief—as did Gauss and Einstein—that light rays follow the shortest path between two points. We have to interpret the fact that light, on its way from a star to Earth, deviates from a straight line as the effect of a deformation of space by the Sun.

As we mentioned above, galaxies move apart as though they had been propelled fifteen billion years ago from a common origin, and they move at velocities that increase with the distance between them. The interpretation of this finding, however, inverts the causal relation of "distance" and "velocity" implied by this correlation. The larger its initial momentum, the farther a galaxy has moved to this day, and the faster it is still moving.

It is improbable that our position in the universe is special in any way. That implies that all galaxies move apart with velocities that increase with the distances between them. We can convince ourselves of such a possibility by pulling evenly on each end of a rubber band. Were that rubber band infinitely long, it is quite evident that any two points would move apart with a velocity proportional to their distance on the band.

The space of that rubber band is one-dimensional. A frequently used two-dimensional model of an expanding space is a balloon. If we regard small markers on its skin as galaxies, the balloon's inflation will make these galaxies move apart at rates that increase with their distance (see fig. 84 in chapter 8). In a three-dimensional space, the galaxies move apart just like the raisins in a cake that expands in the oven. These images share one correct point: As the galaxies move apart, they will not expand themselves but rather maintain their sizes. We can treat them as probes that move along as space expands. The space between them increases—more rapidly in the early moments of the universe, more slowly today. The galaxies attract one another and thereby slow down the increase of the space between them. Gravity holds their own masses together and does not permit them to grow as the space around them does. We find analogous behavior in atoms, which are held together by electrical forces.

Bodies influence the space this surrounds them; they tell us whatever we can learn about that space. Does this mean that, possibly, there is no such thing as space by itself? Could it be that space is nothing but a theoretical construction invented for the purpose of giving the observed world an ordered framework? Could it be that what we perceive as the reality of space is nothing but the influence of abstract laws of nature on the behavior of massive objects or bodies? That space is wedded to these bodies and will vanish if they do? This is a respectable position to take. For more than two thousand years, it has coexisted with the view of space as the primary stage that permits material objects to make their appearance. Natural philosophers who theorized about space can easily be charted on a scale between these two extreme positions. On the left is Thales of Miletus, whose "space" is nothing but one shapeless fluid; on the other side, there is Democritus, with his empty space in which material objects whir around. Leibniz on the left, von Guericke and Newton on the right. Albert Einstein juxtaposed these two concepts of space as, on one hand, the positional qualities of the physical world (left) and, on the other, the container of all physical objects (right). In his left-hand case, there is no space without an object; on the right, such an object cannot be thought of except in conjunction with the space that surrounds it—thereby assigning to space a higher reality than that possessed by objects.

PRE-SOCRATIC IDEAS, MODERN IDEAS

The two concepts listed by Einstein may be seen to summarize the twenty-five hundred years' worth of discussions about space. When the pre-Socratic Greek philosophers were looking for unity behind diverse phenomena, they broke with the tradition of ascribing each phenomenon to an individual force of nature, an individual god, or the whims of a particular demon. Their novel idea was that the world can, in fact, be understood; that there are laws that run the mechanisms inherent in the workings of our world; that ur-matter is the basis for all substances observed.

The earliest pre-Socratic philosophers from Ionia—Thales, Anaximander, and Anaximenes—did not differentiate between substance and law. Their basic question was about being and its negation, nonbeing. Their train of thought had enormous influence: According to them, whatever truly exists, the "being" must be infinite and eternal and immutable. Were that not so, we would have to see the "being" before and after a transformation. That, however, would contradict the fact that there is only one basic substance. The transformation that we observe daily cannot concern the true being. The last consequence is that transformation doesn't exist; it is but a deceptive appearance.

This opinion, obviously, cannot possibly stand: After all, there is such a thing as change, transformation. The question must address the nature of that change: What does it leave untouched? What do we leave out when we deny change? It may sound paradoxical, but physics starts where that change is denied: Denying transformation is tantamount to the assertion that there is something that simply exists and is not subject to change. Take, for a random example, the dancer's two-step: The only permanence in its existence is its formula—one–two–change your step. Similarly, behind each phenomenon there are the laws of nature. Parmenides, a later pre-Socratic philosopher, put the laws of nature into a class of true existence all by themselves and denied true existence to anything that can change. This idea, as we said before, has been superseded. There is, after all, matter that is subject to change. Parmenides speaks of a single "being," thereby dividing our world into two categories—that of the laws of nature and that of random coincidences.

Champion for unity that he was, Parmenides did not intend that division. But modern natural science is based on the separation he defined. From our vantage point, the distinction between eternal laws beyond the range of our influence and coincidental phenomena that can change at random is the greatest achievement of Parmenides. It was, in fact, the seed of physics as a science.

There is no fixed demarcation line that separates the laws of nature from the remaining phenomena. Parmenides' tiny seeds have spawned a vast field of action, a field that keeps expanding as fundamental research grows. The ultimate dream of physics is still that of Parmenides: to understand everything—that is, to elucidate the realm of the laws of nature to the point where nothing remains beyond. But that is still a long way from where we are today. Ultimately, we would like to have the many laws of nature replaced by a single fundamental law governing the behavior of the single fundamental substance of the world. We call that one law we are seeking a *unified theory;* its basic substance is the ground state, or the vacuum.

Today, the world of physics can be divided into two areas. First, there are the laws of nature—timeless, immutable. We have no influence over them. Second, there are the *initial (or boundary) conditions:* Archers set the boundary conditions for the flight path of their arrows by choosing the location, the time, and the direction in which they are sent off. Once these conditions are fixed, the laws of nature determine the flight path.

Those laws needn't be fundamental. What we see as a given by natural law and by the initial conditions depends, as our example will show, on our position in the universe as much as on our understanding of those conditions. The reason why the arrow does not take off to infinity along a straight line is, simply, the presence of Earth below it. That is not a law of nature; it is simply a consequence of our specific location in the universe—a boundary condition, not a law.

Where is the demarcation line between the two areas? Our very existence on Earth is due to our specific position in time. Ten billion years ago, there was no clump of matter fit as a domicile for intelligent creatures like humankind, and ten billion years hence, no such clump may remain. If we go back further and further in time, we will reach the early universe in its initial conditions. Nobody can say what difference there may be between the consequences strictly due to those initial conditions and the consequences due only to eternally valid laws of nature. If there is only one universe, both apply, both are unique.

The regime of eternal laws of nature within the boundary condition of a given point in time lends the world a measure of uniqueness. The laws of nature apply to numerous natural phenomena that differentiate themselves by their initial conditions. We don't know whether there are laws of nature beyond these—laws that set the initial conditions of the universe.

The laws of nature are defined in a causal relationship to time: First, the state of some physical system is fixed at a given time; then this causal relationship fixes the state of that system for all time according to the laws of nature. (Here is an aside for chaos buffs: There is a difference between determinism and predictability.) As we said before, the logical order then defines the two realms of the universe, that of the initial conditions and that of the laws of nature. The division between them is not arbitrary; it is such that the mutable initial conditions fix whatever follows according to the immutable laws of nature. It is the job of the natural sciences to find, to define, this division. The importance of this point was clearly stated by Eugene P. Wigner, a 1963 Nobel laureate in physics:

> *The world is very complicated and it is clearly impossible for the human mind to understand it completely. Man has therefore devised an artifice which permits the complicated nature of the world to be blamed on something*

which is called accidental and thus permits him to abstract a domain in which simple laws can be found. The complications are called initial conditions; the domain of regularities, laws of nature. Unnatural as such adivision of the world's structure may appear from a very detached point of view, and probable though it is that the possibility of such a division has its own limits, the underlying abstraction is probably one of the most fruitful ones the human mind has made. It has made the natural sciences possible.

What we have discussed so far does not put us in a position to describe all of reality—only isolated systems that we have chosen for the purpose. In these systems, the methods of physics have enjoyed overwhelming success. Newton introduced them in his seventeenth-century treatise on mechanics. Before him, Ptolemy, Copernicus, Kepler, and Galileo looked at the skies and observed mainly the geometrical trajectories of planetary motion. Most important, prior to Newton, nobody conceived of the idea that the planetary system might function very much like a clock: If, like setting a clock, one sets the planetary system into motion at a certain time in a certain way, the laws of nature will determine its behavior for all times in the future. Initial conditions and the laws of nature are thus identified by Newton as the two very different types of causes, both of which are needed to predict the behavior of a planetary system in the course of time.

Newton's idea of separating the two regimes—initial conditions and the laws of nature—has survived to this day. It certainly is a brilliant concept, but, more important, it is successful. I am, however, not sure if Newton ever explicitly mentioned this separation at the basis of his work, or if he even was aware of it. Maybe he would not have been interested. What he did is this: He discovered laws on the basis of which he was able to understand the orbits of the planets. Finding these laws and applying them—that was the bulk of his epochal work. He realized that every planet moves, to a good approximation, as though there were nothing in the universe except itself and the Sun. The smallest physical unit that lets us understand the planetary system consists of one planet and the Sun. Kepler, on the other hand, thought the planetary system could be understood only as a whole—that everything was interdependent. To this day we have no proper understanding of all the observations on which Kepler based his interpretation of the planetary system: the number of the planets and the interrelations among the radii of their orbits. Some aspects of these observations are today understood in the framework of the theory of chaotic phenomena; on the whole, however, we have to invoke random coincidences that occurred during the formation of our solar system to explain the number of the planets and the radii of their orbits. Newton derived, from the laws of nature he thereby discovered, which orbits are possible for planetary motion—they are ellipses, and the Sun is located in one of their focal points. He had nothing to say about which ones out of all the possible orbits around the Sun the planets would occupy.

UR-MATTER

The historical development of thinking about empty space largely begins with the pre-Socratic philosophers' idea that all of the physical world issues from one and only one basic matter element—what we call ur-matter. This ur-matter cannot be broken down or analyzed any further. Later pre-Socratic philosophers, in a logical leap, redefined flat negations of the "there is no such thing" variety into such (non-)concepts as the existence of nonexistent matter, only to lapse immediately into the denial that there could actually be any such thing as nonexisting existence.

The first to consider all aspects of nonexistence was Parmenides. He showed that Thales and his successors' thinking about "being" left plenty of room for what is not—for the nonexistent. Consider, for example, motion: Something that exists is by necessity fully defined—it cannot change, it cannot move. One of Thales' followers, Anaximander, is not willing to assign the distinction of being *the* ur-matter to any one of the known elements. According to him, there is only one thing we know for certain. Ur-matter has no limit, it may even be—and this is not quite the same—infinite. Everything we can define originates in ur-matter; by inference, we cannot define that original form of matter. Everyday elements such as water, air, and earth come into being and then vanish again because of internal motions of ur-matter. Their emergence is due to change, which marks them as not really existing.

The existence of one thing means, to these ancient philosophers, the nonexistence of something else at the same location and at the same time. For its existence at the expense of something else's chance to exist, this one thing has to pay a price; Anaximander borrows the term "has to atone for" from judicial thinking. It does so by giving up a bit of existence, and that gives a chance to the nonexistent. In the process, the limitless ur-matter remains the same. It secretes matter according to its inherent rules and absorbs it again. True existence is limited to ur-matter.

We are to understand that Anaximander's ur-matter enjoys real existence. Plato, in his dialogue *Timaeus,* introduced a related substance that is essentially like putty, out of which the world as we know it was formed. Anaximander's limitless ur-matter and Plato's putty are forerunners of Aristotle's *materia prima.* The difference is that Plato's substance cannot become apparent in its true form. It is a purely spiritual object. Natural scientists may not have to take cognizance of its specific properties, but they cannot rob matter of those properties.

The idea that ur-matter has to be infinite is shared by the philosophy of

Anaximander's follower Anaximenes. But the latter doesn't go as far as ignoring everything except the infinite: For his ur-matter he chooses air. It makes sense to him to consider air as extending to infinity—something that cannot be postulated for Thales' water. On the other hand, the conditions known on Earth—temperature, pressure—permit only the substance water to appear in three phases: It can be a rigid body (ice), it can be fluid, it can be gaseous (steam). Because of this quality, water is particularly successful as a substance that shows the unity of existence in those different phases. We are now pointing again toward Parmenides: For him, ur-matter has to be infinite; if it were not so, there would be room for an additional real existence.

TO BE OR TO BECOME

In pre-Socratic philosophy, things being are opposed not only by things nonexistent but also by things developing. From the vantage point of things existent, both things nonexistent and things developing (into existence) touch on the question of whether or not a being can grow out of a nonbeing. Xenophanes and Heraclitus, two immediate predecessors of Parmenides, introduced very different models of thinking about change into pre-Socratic philosophy. According to Xenophanes, "the universe remains immutable." His ideas tend to deny all change; he claims that "the universe is unique, of spherical shape; it is not infinite. It did not originate, but has always been there, always remaining at rest." Heraclitus, on the other hand, sees change everywhere, even though our senses may suggest everlasting constancy. "You cannot step into the same river twice," he says; "the river's flow makes it change rapidly, its currents moving apart and reuniting perpetually." He wants to unmask the seemingly immutable as never in fact remaining the same, and this view made him enormously popular. He did not theorize about the possibility that there is an existence of the nonexistent. Rather, his opinion makes all existence real in appearance only. Therefore, he wants to eliminate this question from our description of the world.

Heraclitus, like all of Thales' followers, wants to reduce the world to one guiding principle, and for this principle he chooses change. Depending from where we look, the principles of change and constancy have their merits and their problems. Take a river from afar: It looks constant—a ribbon that appears at rest all the way from the mountains to the ocean. But then take a look at close range: Now it is a current of water in motion. If I am a follower of the principle of constancy, I will take the ribbon as describing the true nature of the river, as its natural law. If, on the other hand, I am partial to the principle of change, I have to explain this appearance of constancy in terms of the river's flow. I will elevate the river to a paradigm and use my eyes to detect all along its ribbon the flow it hides. Indeed, the Greek expression *panta rhei*—everything flows—is closer to reality. Ever since Leucippus and Democritus, physicists have seen the dance of the molecules in all matter, even in the most rigid bodies, such as rocks or metals. Or take a magnet and put it on a tabletop next to a plastic comb that we have rubbed so that there is an electrical charge on its surface: There will be energy flows on closed paths through the air above the tabletop. In both cases, we see the laws of nature running the show: One–two–change your step; the rules of the game determine which motion is possible.

Heraclitus did not imagine immutable laws of nature behind all that change: "Change alone makes not a law." But the idea that we can create the appearance of constancy from substrates in perpetual motion is by itself a most useful one. A number of different aspects of a river can be described in terms of static qualities, but the flow of water is best interpreted in terms of a dynamic description. It alone can interrelate width, depth, angle of flow, and the overall shape of the river, and thereby facilitate our understanding of the whole system. For another example, and maybe a more impressive one, consider a cyclone: Seen from outer space, it will change its position and its form slowly; both position and form can be described without reference to the violent motion inside it. But to understand even the external perimeters of the cyclone, we have to study that internal violence; indeed, we have to start with that study.

PARMENIDES: TO BE, TO THINK, TO SPEAK

Although the modern concept of the laws of nature might be ultimately due to Parmenides' teaching, it constituted, scientifically speaking, a step backward. He does not make any attempt to discover unity in diverse phenomena. To do so, he would need detailed insight and he would have to propose concrete mechanisms. Instead, he gives up and announces that there is no such thing as mere phenomena: What truly *is*, the truly existent, is immutable, in contrast to merely apparent phenomena. Thus, he divides the world into two areas: First, there is the area of thought, which alone is related to the truly existent; then there are the mere appearances, or phenomena. The latter do not really exist, and are deceptive at that. The school of Elea assumed that there were two substances: "One of these has true existence and can be grasped only by the mind; the other is mutable, developing, and our senses can perceive it. This second one they do not recognize as being truly existent; its existence is only seemingly true." "That is why," Parmenides believes, "we can recognize truth in the truly existent, while we can form only opinions about the changeable."

While Heraclitus sees change everywhere, Parmenides recognizes none. To him, the worlds of thought and of true existence are identical. There is no path that connects the latter to mere appearances—not even a one-way road in either direction. This is exactly what distinguishes the philosophy of Parmenides from Wigner's division of the world into the realm of natural law and initial conditions. Wigner sees the laws of nature as immutable, just like the truly existent of

Parmenides. In contrast to the latter, however, the laws of nature have their influence on the world of phenomena. It is the job of the natural sciences to gain insight into them. Wigner describes the natural sciences "such as they are." Greek philosophy didn't have the necessary knowledge and could therefore only develop a rough copy of a true science of nature. That was within its scope, but Parmenides' philosophy also falls short of this aim.

According to Parmenides, only the being—the truly existent—actually *is*. Its inverse, the nonbeing, does not exist, which bears heavily on the topic of this book. Let us repeat this statement: "Only the being has reality. It may well be tangible in front of us—whereas the nonbeing cannot possibly have any reality." This is tantamount to a negation of the world's having sprung from the void, the *creatio ex nihilo*.

> *How would you invent the origin of true existence? Whence did it come, how did it grow? I will permit you to think or say it came straight out of nonexistence. We must neither think nor say that it does not actually exist! What possible compulsion could lead to its starting from nothing and growing into something? In this fashion, we can extinguish all growth, get rid of all decay. The same goes for thought and for the object of thought. You will not find thought outside the realm of existence, where it finds its expression. After all, there is nothing outside that existence, and there never will be. Therefore, everything which has been put into language by mere mortals is nothing but empty names; they may well believe that there is some reality at the very basis: "growth" and "decay," "being" and "nonbeing," "change of position" and "change of brilliant color."*

NOTHINGNESS AND EMPTY SPACE

While Parmenides assigns the "being" to the realm of thought, he astonishes us by considering it, at the same time, as material, pervasive in space. This gives a completely new aspect to our topic, the "nothing," the opposite of the "being"—an aspect that makes it the same as empty space! By denying reality to the "nothing," Parmenides also negates that of empty space. Since space is an object of scientific study, he therefore elevates the nothing to an object of science, not just of philosophical speculation.

CHANGE BY WAY OF UR-MATTER

The fundamental ideas of the pre-Socratic philosophers were carried too far by Parmenides and Heraclitus. Both claimed that only their own view of the world could be true; in the process, they accepted astonishing contradictions to everyday experience. The philosophy of their successors Empedocles, Anaxagoras, and the atomists tried to join their points of view, and in the process dispose of latent contradictions to experience.

Change is a part of everyday experience. To account for it, Empedocles replaces one unique ur-matter with four ur-elements: fire, air, water, and earth. Since these four are existent, in Parmenides' definition they are immutable. The change they were introduced to explain is nothing but the external appearance of their internal intertwinings.

This means that something is happening in the world of Empedocles. The forces he sees at the basis of what is happening are "love and discord"—attraction and repulsion, in today's language. He interprets these forces as independent phenomena to the point of according them, at least on occasion, material existence on a level with the elements. "Love" and "discord" are supposed to bring about, for instance, the formation of a sphere out of the elements, and its subsequent disintegration. This concept, obviously, is wrong; there are no such forces. But the *idea* of independent forces is important: It transcends the concept of immutable laws, since it awards concrete shapes to them. The laws of nature now define independent forces and thereby imply phenomena that can be observed. This is where progress sets in.

Independently existing forces determine the behavior of matter, but they are not matter themselves. Their characteristics can be detected by experience. It is not the work of the rock that attracts it to Earth; rather, gravity, the gravitational force, is at the basis of this observed phenomenon. Anybody who has ever fallen from the branch of a tree is witness to the independent existence of gravity.

Empedocles was the first to distinguish between matter and forces. There is no way to exaggerate the importance of this distinction for the development of natural science. It resurfaced in Newton's distinction between initial conditions and the laws of nature.

There are three details of Empedocles' thinking that bear on the topic of this book. First, he denies the existence of empty space, which had taken the place of the metaphysical "nothing" as a physical concept of natural philosophy. Empedocles, however, maintains that there is no such thing as empty space. In his own words, "The universe has no space that is empty nor space that is overcrowded." True, his denial of empty space belongs to the tradition he grew up in—but the idea as such does cause him trouble. He therefore resorts to his version of an experiment on the subject, which we will discuss below.

Second, he denies the formation and disintegration of elements. Again in his own words, "There is no empty space in the universe. How then should something be added to nothing?" And: "It is impossible that something grows out of nothing; and equally impossible that something present disintegrates into nothing at all." And: "There is a coming together and a growing apart of existing elements."

Third, Empedocles tries to define the hidden mechanisms by which nature accomplishes the physical happenings we observe. It is impossible to overstate the importance of this element of his thinking for the development of the natural sciences. Empedocles furthermore attributes the intermingling and separation of substances to what he calls excretions of solid matter; this concept implies that there must be pores penetrating all bodies. For our purposes, this theory of pores penetrating matter has one difficult aspect: How is it possible that there are these hollow channels of emptiness if there is no such thing as empty space? The originators of the idea of pores denied the existence of empty space; the pores must therefore differ from that concept—they must be filled with some invisible substance. This is where we find the first mention of a fluid medium that will dominate discussions of empty space for the next twenty-five centuries. This medium is the least massive of all matter, later on chosen by Aristotle as a fifth element. He called it *ether*. The ether, as we see, was invented to ensure that there could be no such thing as a true vacuum, a truly empty space. Out of Empedocles' four elements, only air might qualify for the role of ether. As Aristotle puts it: "The physicists who see pores penetrating solid matter don't suppose them to be empty, but rather filled with the very light substance, such as air."

THE HYDRA MONSTER AND AN EXPERIMENT

It is significant that Empedocles did not limit himself to idle speculation on the question of whether the air is a substance or nothing but empty space. He was the first to introduce a detailed experimental observation into that philosophical discussion. In a highly poetical fashion, he uses the example of the mythological nine-headed Hydra, a contraption that was known to raise the level of water in a controllable fashion, to explain that air is, in fact, a substance and not just empty space. (See fig. 21.) This is frequently recognized as his most important contribution. We can divide his observations into four steps. First, take the contraption shown in figure 21a as his version of a hydra dipping into water. If the top of the neck is closed, no water will penetrate through the holes at its bottom. Second, open the top so air can escape (see fig. 21b). Water now enters from the bottom, and air escapes at the top in a tight flow. This shows that water and air cannot coexist in one and the same space—the water cannot enter the hydra before the air leaves it.

It is the third in his series of observations that is the most spectacular and has brought the hydra experiment fame for over two millennia: If we lift the hydra, now filled with water, out of the tank while keeping its neck tightly sealed, the water will not escape through the holes in the bottom (see fig. 21c). We have to uncover the top to let the water flow out of the bottom (see fig. 21d), while,

Figure 21 Torricelli's experiment (fig. 2) is basically nothing but a repetition of this experiment using mercury rather than water—and with one decisive difference: Unlike the column of water in this experiment, Torricelli's column of mercury could not be supported by the pressure of ambient air.

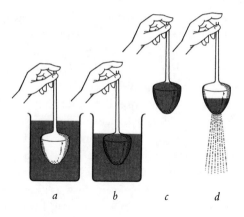

a   b   c   d

according to Empedocles, "air wildly replaces it in a turbulent stream from the top."

The poetry here does not contribute to clarity. We can, however, convince ourselves from Empedocles' original description that he correctly understood the first two steps of the hydra experiment. The spectacular third and fourth ones he still explains correctly as showing that both air and water are forms of matter—"substances." He avoids the ancient *horror vacui*. But the details of his description are a bit muddled; they cannot pass for a correct interpretation of the effect in physics terms—to wit, that it is the pressure of the surrounding air that provides the definitive interpretation. It remained for Torricelli, Pascal, and their contemporaries, who were able to observe the level of water (or mercury) inside their versions of the hydra, made out of glass, to supply that interpretation, a couple of thousand years later. Our figure 21 gives the erroneous impression that Empedocles might have similarly seen the level of water inside his hydra; but his was made out of "brilliantly shining iron ore," so he could not.

ANAXAGORAS

Anaxagoras was active at about the same time as Empedocles. In a number of ways, he was quite modern. He is, however, famous for introducing the "Mind" as the primal mover: "It seemed to me in a certain way right that Mind should cause all things, and I reflected that, if this were so, then Mind in ordering all things must order and arrange them in the best possible way." What is modern about his philosophy is his *not* taking recourse to such an ill-defined concept as long as he could find concrete causes for the phenomena he described. Plato later reprimanded him on this point: "My magnificent hopes were shattered, my friend, when, as my reading progressed, I found a man making no use whatever of Mind and ascribing to it no causal action in the ordering of things, but assigning such causes as air and aether and water and many other strange things."

Anaxagoras, just like Empedocles, dismisses the existence of empty space and the formation of "something" out of "nothing." He even formulates a conservation law: "We have to realize that when we take all things together, they will neither increase nor decrease; their totality will remain the same." He differs from Empedocles by assuming the existence of endlessly many elements, which he calls seeds: "Let us consider matter of a given kind as elemental—let's say, flesh, bones, and the like. . . . In any one of these elements of living beings there are the seeds of hair, nails, veins and arteries, sinews and bones; they may be too small to be seen, tiny as they are. But let them grow, and you can tell them from each other. After all, how could hair grow out of matter that is not hair? Flesh out of nonflesh?"

This, of course, makes no sense at all. Still, it puts Anaxagoras on the trail of two important ideas: the concepts of what we call today *the continuum* and *entropy*. About the first of these, he says: "Among things small there is no smallest—there always will be something still smaller; it is unthinkable that something existent can be divided until it no longer exists." He adds: "And it is the same with things large: there is always something larger."

We do not know exactly how Anaxagoras meant to deal with the endless divisibility of all things being. Obviously, this makes no sense in terms of the seeds, where there is no infinite dividing without change. We have long known that we can break ice molecules into atoms, atoms into nuclei and electrons, and so on. But we don't know whether the physical space that surrounds the atoms and the constituents is infinitely divisible—remaining the same but becoming smaller and smaller in measure. Mathematically, space is, in fact, infinitely divisible, which means that the idea, at least, makes sense. To illustrate, let's take a ruler. Between any two points on a ruler you can always place another point. The distance between two neighboring points can be arbitrarily small—an inch, a thousandth of an inch, a billionth, and so on to infinity, to infinitely small distances. Mathematically, we call that a continuum of possible lengths: There can be arbitrarily short rulers. Physically, however, that doesn't work. There is no way of having a ruler shorter than the diameter of a single atom, about 10^{-8} cm (one-hundred-millionth of one centimeter). This is not to say that distances smaller than this one cannot be measured; there are other ways to do the measurement, without a ruler. But we do not know whether there is a limit beyond which the divisibility of lengths ceases to make sense. We cannot exclude the possibility that, at some level, space becomes discrete. It is thinkable that some elementary length—say, for instance, 10^{-33} cm—is the smallest possible separation of two points. We can imagine the smallest division possible along a straight line like the individual keys on the keyboard of a piano, in contrast to the positions of a violinist's finger along the fingerboard of the instrument. The length of the keyboard is divided into discrete smallest distances, the violin string into a continuum of possible finger positions.

Anaxagoras's second important idea concerns the relation between order and disorder, or chaos. He does not doubt the difference between "something" and "nothing," which he takes for granted. He believes that everything has been in existence from the beginning of the world, as some structureless medium. How then, he asks, has structured matter risen from this chaotic state? By gradual ordering, or unmixing, obviously. But how could that have happened? He was the first to recognize the problematic nature of this unmixing, or increasing order, and concluded that "spirit" played a role in it. This spirit, he opines, set the initial chaos into a turbulent motion that initiated the process of unmixing, of ordering.

Today, we know very well that the concept of unmixing presents a considerable problem in many ways. Increasing the order in one place implies destroying it somewhere else—but overall, disorder carries the day. That is what the law of entropy says: The degree of disorder either remains unchanged or increases. When we apply this law to the universe as a whole, we have the difficulty of not really knowing how to define the "overall" of entropy increase. What does "overall" mean in an expanding universe? We don't know all of that universe, and in particular we don't know the degree of order it had immediately after the Big Bang. But we maintain that the order in all kinds of structures here on Earth—be they living creatures, solar panels, cathedrals, accelerators for elementary particles, books about the void—originated only at the expense of undoing order someplace else.

Anaxagoras, of course, couldn't have known about any of this. He felt that unmixing presented a problem, and he was right. But he was far from any possible solution—and so are we to this day, especially when we include the universe as a whole. Even on a small scale, it is not obvious that order deteriorates in all processes. This is because energy in the form of heat contributes considerably to the balance of order and disorder. If a system gives off heat energy, it may be raising its own state of order. We have already mentioned water: When water freezes, it passes from a disordered to an ordered state, while at the same time giving off melting heat. This heat can be used to thaw ice someplace else. Should that happen, it will transform the water molecules in that faraway place from a state of higher order (ice crystals) to one of lower order (the ordinary liquid state). Ordering a substance in this case obviously causes disorder to grow in a substance someplace else. And overall, disorder has the upper hand.

Here are a few more examples, taken from everyday life. Let's say all the ancestors of the Salisbury family lived in and around Salisbury, all the Salazars around Salazar Castle in Portugal, and that all the men in the Cook family were cooks. In the course of time, this order was destroyed without necessarily being replaced by any new order. Or take the books in your library: If you rearrange all of them according to subject matter from a previous order based on book size, you destroy one order at the expense of another. Or consider bricks being unloaded from a truck in front of a building site: Masons pick them up and arrange them in precise patterns for the construction of a wall, a house. They appear to increase order, but in fact overall order decreases. Just think of all the gasoline or diesel oil that had to be burned to transport the bricks, and the resulting fumes that were released into the atmosphere. And think of where the bricks came from: maybe some old palace had to be taken apart to yield them, or beautifully undulating sand dunes were dug up to yield the sand out of which the bricks were then baked. And think in particular of the energy dissipated in the form of heat by that very baking.

We see that the creation of material objects cannot stand by itself; an ordering of matter, a building of structure, is also needed in the process. We can interpret Anaxagoras's arguments in this way: He considers the act of creation as equivalent to the formation of order out of chaos. And millennia had to pass before the juxtaposition of order and disorder came back into focus.

In Anaxagoras's world, filled to the brim with continuous matter where Empedocles spoke only of poorly defined pores, there is no room for empty space: "There is no such thing as empty space," he says. Still, he investigates the nature of air experimentally. Is air a substance? Is it just empty space? In his own version of Empedocles' experiment with the hydra, he uses air-filled wine-skins. With these he demonstrates the pressure exerted by the air inside them; it is proof that air is not the same as empty space, and he infers that there is no experimental indication for the existence of empty space at all.

With the appearance of Empedocles' *four elements* and Anaxagoras's *infinity of seeds,* the unity of all things being that had governed the natural philosophy of Thales and the school of Elea was relegated to the dust bin. Parmenides showed that we can insist on the immutability of all things being only by declaring that all change and all motion are real only in appearance. But there is no denying change or motion in our world; in fact, Heraclitus saw change and nothing but change all over. It was all a bit hard to take: Empedocles saw his agents of change in terms of love and discord, which was still supposed to leave no trace on the sum of things existent. Just like Anaxagoras's "Mind" of the universe, this does not really make sense and should probably be seen as a consequence of one fundamental reluctance of early Greek philosophy: It was loath to give up its bastion built on the unshakable idea that there can be no such thing as empty space.

MOTION AND EMPTY SPACE

Motion brings about change. Couldn't it be that all change is based on motion? The principal tenet of the atomists is that matter consists of invisible tiny particles, and their motion brings about all change. This motion happens in empty space. If we accept the existence both of the particles and of empty space, it is easy to explain condensation and dilution of matter in terms of an assemblage and a dispersion of particles. The mixing of all substances is then nothing but the random motion of the particles they consist of, in the otherwise empty space they occupy.

The notion of the existence of empty space was first advanced by the atomist Leucippus. With this idea he undid a web of tangled thought as unwieldy as the Gordian knot, in the process providing for a mechanistic view of the world that is free of obvious contradictions. Atoms move in empty space; and that is all there is: The space the atoms occupy is filled, and the rest remains empty space. The changes we observe in the world do not imply changes in the atoms. They move, but they carry all the qualities of true existence as defined by the Eleatics— eternal, unchanged, immutable. They didn't spring from the void; that is impossible: Nothing can spring into existence from the void; nothing can vanish into the void from existence.

Atoms are compact and indivisible. It is the atomists' contention that matter cannot be divided below the atomic level into the infinitesimally small. (The Greek word atom, *a-tomos,* means "without parts.") Their indivisibility is fundamental; it is not just a practical property. The atoms have mass; they are not part of the void.

THE EARLY ATOMISTS

Leucippus and his student Democritus adopted the concepts of "what is" and "what is not" in terms of matter and void, of the "being" and the "nonbeing," from their predecessors. Their scientific intuition told them that they had to provide explanations for their observations of existence and change. If there were only one kind of atom, there would be no way to explain their observations: There must be various kinds of atoms. But how to tell them apart? They postulate that all atoms are made up of the same basic substance and differ only in their shapes, as pieces of gold might. Still, they do away with Parmenides' unity of being by situating the atoms, the "being," inside empty space, the "nonbeing." As a result, all being is made up out of the same substance; the atoms are merely the smallest units of Parmenides' "Sphere of the Being."

The atomists, of course, could not come up with a reason for atoms taking on only "certain shapes in preference to others, so they assumed that all shapes must be possible." Consequently, there is an infinity of possible shapes, with no restriction imaginable. But how big are atoms? That is where Leucippus and Democritus differ: The former takes them to be so small that our senses cannot discern them; the latter believes there might be at least some atoms of macroscopic size. It is not clear that Democritus followed this notion to its conclusion: He never addressed the question of why those visible units of matter would have to be indivisible.

Epicurus later interpreted the teachings of the atomists: Atoms, according to him, cannot take on arbitrary shapes and sizes, certainly not sizes large enough to be visible. To match the atomists' theory to our observations, Epicurus waters down their strict perceptions. He does not accept their rule, later called the "rule of sufficient cause" and which for the atomists implied that there cannot be any restriction on the possible shapes and sizes of the atoms.

The atomists also included the concept of empty space in the framework set by their predecessors. We know what they mean with their tenet: There is no more existence to "what is" than to "what is not," just as empty space is as real in its existence as matter (which equals nonempty space); but there is no need for the paradoxical expression "the existence of the nonexistent" for the expression of their thought. The main use to which they put empty space is to make room for the atoms and allow motion without resistance. Just as the atoms are too small to see, there is no way to perceive the empty space between them.

ATOMS AND EMPTY SPACE

The atomists recognize that change in the physical world and perception by our senses are real, but this reality is qualitatively different from that which they grant to the atoms and to empty space. Sensory perception is by no means deceptive, but it can be seen as just one aspect of ultimate reality: The atoms move, they form larger structures, they drift apart. As Democritus says: "Only seemingly does a thing have color, only seemingly is it sweet or bitter; in reality, we are dealing only with atoms and empty space."

Detailed criticism aside, the atomists' perception of reality marks an important step in the direction of scientific understanding in the modern sense. The fundamental questions they address have not been answered to this day. I'm not even thinking of the philosophical questions of "what is" and "what is not"; rather, here are important physical problems to be solved: Are there units of matter that are indivisible? At that level, are they all the same, or do they come in different varieties? Is there meaning to the idea of Anaxagoras that space forms one geometrical continuum?

Maybe some of the atomists' fundamental ideas were correct; if so, that is more or less coincidental. Considering what was known to pre-Socratic natural philosophers, there was no way to decide between the continuum theory of matter and the atomic model. Both are important constructs of their understanding of nature—no more, no less.

For the moment, let's leave the fundamental questions posed by Leucippus and Democritus aside, while admitting that they proved to be amazingly fertile ground for thought. After all, they hinted at elements of the theory of what

Ludwig Boltzmann in the nineteenth century called the "ideal gas": The theory says that any gas of low density and of temperature that is neither very high nor very low can be considered as an assemblage of indestructible solid spheres that fly through an otherwise empty space and bounce off each other elastically. That's how the atomists thought of fire, where spherical atoms were engaged in something like a three-dimensional billiard game. Of course, this theory of the ideal gas is an approximate one; its validity is limited to a relatively narrow range of temperature and pressure. Once the temperature increases beyond that range, the collisions between the molecules or atoms of the gas become violent enough to destroy them: Their indestructibility no longer holds. At the other end of the temperature scale, the gas will condense to become a fluid and finally jell into a rigid body. The atomists thought up a detailed but incorrect image for this phenomenon: They supposed that atoms hook up to each other by means of mechanical hooks and eyes.

These hooks and eyes, in our modern understanding, are not mechanical: They are electromagnetic forces, governed by the laws of quantum mechanics. What is more important in the atomists' thinking for the topic of this book is their concept of motion in empty space. They correctly imagined that motion in empty space, once started, continues forever unless there is a force acting on it. To the best of our knowledge, nobody else had such clear notions on the subject prior to Newton or Galileo. Wisely, they abstained from speculation about the initial cause for the atoms' motion in empty space; never mind that Aristotle berated them for that reason.

MOTION IN THE ABSENCE AND PRESENCE OF MATTER

The tenet that motion, once started, keeps going forever holds true beyond what we usually call empty space and extends into the abstract vacuum defined by physics. This vacuum is better described in terms of a medium. The natural philosophers of antiquity developed their ideas of motion in it along the lines of the stirrings of a spoon in honey or of a feather in the wind. In fact, however, the notions of filling up a space and of displacing its contents cannot be applied to the ideas of motion in the empty space of physics.

As we understand physics today, these notions are not fundamental and don't lend themselves to detailed observation. The physical size of atoms, nuclei, and elementary particles is, to the physicist, an extremely complex and derived concept. Today's physics sees the interaction between the particles of matter in very complex terms—only rarely, and in loose approximation, do they coincide with what the atomists saw as direct hits leading to displacement. To give an example: There is nothing wrong with having an elementary particle in a given location interact with an electric field while, at the same time, gravity pulls on it. It is therefore impossible to categorize gravitational forces and electrical forces as the ancient natural philosophers did. These forces *act,* which means they are "something," not "nothing." They cannot be bodies, because they are present in the same point in space. And since they are present there, that space cannot be empty.

The laws of physics will not admit the existence of a completely empty space. They say, at the same time, that bodies left to their own devices in the emptiest space possible will move as though this space were in fact empty. Their motion, once started, will keep going forever. Moreover, it took the instruments of modern physics to permit the observations that prove there can be no such thing as empty space. The motions that the ancient natural philosophers wanted to describe are still most accurately seen as happening either in empty space proper or in a medium that fills empty space, such as honey, water, or air.

In analogy, we can discuss a thickening or thinning of matter in space. By using concepts developed by the early atomists, we today not only understand condensation and dilution in empty space—which, as we know, is not really empty—but also why the bicycle pump heats up while pumping air into an inner tube. When we push on the piston, it will accelerate the air molecules inside the pump. They move about faster in random disorder; the temperature has risen. As with billiard balls, the air molecules kick against each other and are deflected by the walls of the pump. Otherwise they move as though there were nothing around them but empty space.

The atomists did not understand the concept of heat. What they did understand very well is the equivalence of condensation and dilution to atoms in a gas or liquid moving closer together or farther apart. Their insight challenged the accepted view that all motion must be described as that of a spoon in honey or a feather in the wind. Aristotle then developed the notion that there are several, sometimes mutually contradictory, reasons for declaring the impossibility of having something move in a true void: There cannot be a void, after all, where there is motion. The atomists thought that motion in a medium that fills all space cannot be possible; as a result, there must be empty space everywhere.

Within the framework that the atomists chose, this conclusion holds. We will cite Epicurus, and Lucretius, later followers of the atomists, on the impossibility of motion inside a filled space. They assume—in the language of physics—that no effect can propagate with arbitrarily large velocity. Take a fish swimming through water: Where does the water go when the head of the fish pushes it away? If there is no empty space—remember, the atomists want to prove that it does exist!—the water cannot be compressed. This means that the displaced water must in turn displace other water, and so on; finally, the displacement will reach the space vacated by the tail of the fish. This looks like an infinitely rapid propagation of displacement in water. But the atomists don't believe it is possible.

The forms of motion in media such as honey, water, or air, disclaimed by the atomists, really do not exist. All substances are compressible, and no effect propagates with infinite velocity. For many actually existing forms of motion, this is almost irrelevant—bodies will move through media as though they were incompressible and permitted the effects to be transmitted with infinite velocity. This makes a certain amount of sense as long as the velocities of motion are small with respect to the velocity of propagation of the signals in water. When these conditions are met, the physics world speaks of *dry water*. In reality, there is no such thing; but real water can approximate its properties closely.

Upon patient observation, the observer will always notice that real media—air, "wet" water, honey, and so on—put up resistance against all motion of bodies inside their volume, and that includes resistance against stirring. Any motion not fed by outside energy will die off with time in a real medium. The moving body will lose energy to the surrounding medium through friction; it loses motion energy, and the medium absorbs the same energy in the form of heat. Real media are, in fact, compressible. In the process, they heat up, as does the air that we compressed in the bicycle pump.

The atomists needed their empty space so that the atoms would not stop moving. To date, we know only two real media that offer (almost) no resistance to moving bodies. One such medium was discovered in the 1930s: If we cool down helium, a noble gas, to temperatures close to absolute zero, it will flow through the thinnest of tubes with almost no friction. This phenomenon is called superfluidity. Another substance that shows superfluidity is a rare isotope of the same element, called helium-3. It takes experience to explore all the facets nature offers us. No form of logic can replace it.

DEMOCRITUS'S WORLDS, AND WHAT THEY MEAN

The atomists took a stab at cosmology, too. They started from the notion of an infinitely large empty space. Unending existence and infinite extent coincide for them. Hence the notion of limitless space: "Democritus took the universe to be infinite on the grounds that nobody could possibly have created it." This creation would not just mean that "something" replaces "nothing"; there must also be motion and, as Anaxagoras and the creation myths imply, ordering of shapes. "Democritus was one of the philosophers who took the cause of the heavens and all of the universe to be self-explanatory, a 'matter of course.' Turbulence, motion, they thought, started on their own; they brought order to matter, shape to the universe." Unlike Anaxagoras, Democritus goes beyond some of the creation myths in assuming that order might have sprung up spontaneously.

In his opinion, there are countless worlds in our universe. He has detailed views on their sizes and distances. As we saw previously, some of the opinions he passes on to us sound fairly modern. But he fails to invent mechanisms that could have contributed to his assumed structure of the cosmos; for that reason, his cosmological ideas are less illuminating than his musings about atoms in empty space. About those worlds in our universe, he has this to say: "Some worlds are still growing, others are in the prime of their existence; others again are on the wane. In one place, worlds originate, in other places, they disappear." Replace what he called "worlds" with the word *stars,* and the concept makes sense to the modern scientist.

The atomists developed detailed ideas about bodies in space, but not so about space proper. They never conceived of the idea that the properties of those bodies are based on the properties of space. Recall: To them, atoms were "something," and empty space was "nothing." This pragmatic differentiation was basically elevated to a fundamental tenet by their successors: As Grant wrote, "The atomist identification of real, albeit empty, space with 'nothing' guaranteed that the history of spatial concepts would, from its inception, be rooted in paradox and enigma."

PYTHAGORAS

It is Pythagoras who raised the notion of empty space to the level of an object of scientific inquiry. All statements about objects or bodies in space can be read as saying something about those objects just as much as about that space. Consider, as an example, the famous theorem on the sum of squares of the sides in a triangle, which takes its name from Pythagoras: Our space is made such that all objects it contains will be subject to this theorem. Pythagoras held that all can be expressed in terms of numbers—only to be shocked by the realization that he could not succeed in numbering all the points contained in a straight line. His proof for this observation I will describe; the formulation I choose is strictly modern. Since they cannot be numbered, those countless points between the numbers were declared by the Pythagoreans as their version of the void, the nothing. Thus, they had to believe in a nothing.

Pythagoras and his followers—unlike Thales—were interested in all possible forms of matter rather than in matter itself. They were fascinated that visible, tangible properties of matter can be formulated mathematically. In this way, properties can be transferred from one object to another; mathematical rules of geometry acquire their own abstract existence.

They were convinced that the essentials of the visible world can be registered only through abstraction. Later, Plato would emphasize that the essentials are not rooted in matter but in the shapes, the laws, the symmetries, the ideas. For the Pythagoreans, the numbers are what really count: Theirs is a real existence, unchangeable in space and time. All ideas that deal with the world of physical matter can and must be formulated mathematically.

Modern physicists would agree to the extent of finding my description astonishingly up to date. We do, after all, believe that the laws of nature determine which form or shape matter can adopt. Maybe there is only one *ur-matter,* and all that differentiates the matter we experience is the form those laws permit it to assume. That is today's scientific opinion—but we do not yet know that for sure. We think we know the basic stuff out of which all matter is made. But we know that there is positive and negative energy; this means that some amount of energy in one place may well cancel an equivalent amount of opposite sign in another place. If that is so, we do not need, in the final sum, any basic stuff at all. In that case the universe may be just one manifestation of empty space or, if we accept the definition of space in terms of a distribution of energies, of the void proper.

I will be more specific about such speculations as we go on. They are mentioned here simply because they appear as a new and different perspective on the thinking of the Pythagoreans and Plato. Their two schools of thought arrive at opposite conclusions on our topic—the void, nothing, or empty space. The Pythagoreans were the first occidental thinkers to concentrate on the relationship between numbers and space. They used numbers to enumerate points along straight lines or sides of triangles in order to reduce distances to integer numbers. However they approached that goal, there always remained a finite distance that separated one point from the next. But their philosophy, based as it was on numbers, had no use for a space between the points that were marked by those numbers. For them, the unavoidable space between points was identical with the void, which therefore exists: "The Pythagoreans said that void exists. . . . It is the void which keeps things distinct, being a kind of separation and division of things that are next to each other. This is true first and foremost of numbers; for the void keeps them distinct." This "void" or medium might be air or ether or any other dilute medium, as long as it leaves the numbers or the bodies that those numbers denote as discrete units.

The basic idea of the Pythagoreans, reducing ratios of specific distances to relationships between integer numbers, was ultimately unworkable for a very simple reason. We can demonstrate this easily by looking at the diagonal of a square: Every line can be subdivided into a number of short sections of equal length. Let's do this for two different straight lines, and those short sections will, in general, have different lengths. In figure 22a, a 20-cm-long ruler is divided into four sections of 5 cm each, and an 18-cm-long ruler is divided into three sections of 6 cm each. It is equally possible to subdivide both rulers into sections of equal length—say, into the centimeter scale also shown in the figure. The Pythagoreans believed that any straight lines of arbitrary length can be subdivided into shorter but equal units of length; if this doesn't work with inches or centimeters, we might have to resort to smaller scales such as millimeters or fractions thereof.

This, however, is not true: There is no way to divide the sides and the diagonal of a square into sections of equal length. This is shown in figures 22b and 22c. Take the first of these: The three sides of the right-angled triangle shown there measure 3, 4, and 5 cm, respectively. The existence of triangles with these properties follows easily from the theorem whose discovery is at the basis of

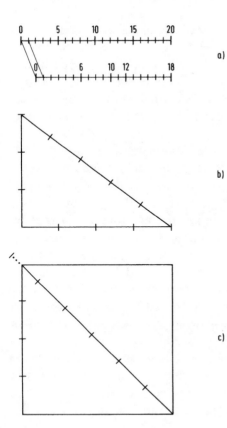

a)

b)

c)

Figure 22 The two rulers in figure 22a, of lengths 20 and 18 cm, respectively, can be divided into length units of 1 cm. If we want to divide them into sections of 5 or 6 cm length, we can do so only for one of these rulers each. Similarly, the sides of the right-angled triangle with lengths of 3, 4, and 5 cm (22b) can equally be divided into 1-cm sections. This is not true of the triangles that are formed by a diagonal division of a square (22c). As the Pythagoreans proved in antiquity, there is no way to subdivide the sides and the diagonal of a square into sections of equal length. This is illustrated in figure 22c.

Pythagoras's fame. The lengths of the three sides correspond to the expectations of the Pythagoreans—their lengths can be subdivided into sections of identical extent.

But the same does not work for either of the two triangles in figure 22c. The Pythagoreans proved that there is no way to subdivide the diagonal and a side of a square into sections of equal lengths. No matter what length we choose for these smaller unit sections, our figure illustrates that at least one of the lines always has a length different from a multiple of these units.

The Pythagoreans were deeply shocked by this result. They had divided straight lines into sections; they had counted these sections in the expectation that any straight line would correspond to an integer multiple of an appropriately chosen fraction of the length of any other line. That was the basis of their belief that numbers, either integers or fractions thereof, would suffice for the description of space. They took this abstraction seriously, only to find out that it was mistaken. The properties of integer numbers imply, instead, that it is *not* possible to express the lengths of arbitrary distances in space, and the relations among those lengths, in terms of integers.

The Pythagoreans held on to their fictitious ideas that integer numbers are all it takes to describe space. As a result, they had to correct the properties of space; their space, as a consequence, is not a continuum but a kind of lattice built up out of points at discrete distances from one another. Contrary to their initial intent, they demonstrated in this fashion that there is no such thing as a true continuum that can be completely described by integer numbers and fractions thereof: Every straight line whose length can be described in terms of these numbers contains infinitely many smaller sections for which this is not possible.

PLATO

The world is "artfully shaped such that it acts and suffers everything by its own devices." This is how—if we follow Plato's *Timaeus*—the Creator God, or Demiurge, arranged it. There is one central point, equally distant from every point at the world's edge; in short, the world is spherical and of finite extent.

Plato's Demiurge found the stuff from which he shaped his world in random, chaotic motion. That made him the agent of the ordering of the universe. There is no mention of his own origin, presumably predating all times. His world differs by its ordering from that of the atomists, who saw all atoms moving about in arbitrary patterns.

Timaeus gives a detailed image of the world as a whole: It realizes Plato's views of perfection. Starting with some hesitation, he develops a notion of having the Demiurge create the properties of the universe following his wish to shape perfection. Thus Plato arrives at four elements and their mixtures, at the spherical shape of the world, and at the orbits of the celestial bodies, from which he derives the notion of time. The universe is, in this fashion, the living realization of a preconceived idea: The Demiurge "intended the world to resemble the most beautiful of all ideas, as close as possible to perfection in all respects, and thus he created this visible, living world."

The Demiurge also created time—as a "movable image of immortality." What Plato's predecessors had called the *one* and the *empty,* the *being* and the *becoming,* we find in his writings as the *idea* and the *image.* The *many* corresponds to the world of physical phenomena, the *one* to that of ideas. Astonishingly, Plato dedicates only a short and rather obscure section of the *Timaeus* to the formidable problem of how time can be seen as an image of some eternal unchanging existence.

For the Pythagoreans, the numbers count as ideal building blocks. This idea is transferred by Plato from arithmetic to geometry, even physics. He considers ideas, abstract mathematical forms, to be the true objects of physical perception.

Observed phenomena are of interest only as long as they partake in these ideas. This is well exemplified by the astronomer, whose most significant task is the reduction of the seemingly irregular motions of the planets to circles or ellipses, and thereby to ideal and "true" orbits of celestial bodies.

In this fashion, Plato formulates an important aspect of the inquiry into physical phenomena. In complete analogy to Plato's ideas, fundamental physics research is concerned with the laws at the basis of observed phenomena. The objects of research are chosen accordingly: It makes sense for the physicist to investigate the structure of crystals rather than the structure of, say, liverwurst. It is, in fact, one of the most important and difficult tasks of the experimentalist to create pure systems, those whose structure and behavior are determined solely by the natural laws to be investigated. But we differ from Plato in one significant way: To us, the laws of nature mean *questions that we ask of nature.* It was Plato's tenet that all statements concerning the physical world that conform to his criteria of logical consistency and aesthetic beauty must count among the laws of nature. We cannot agree with him on this point. To us, natural law must be validated by experience—by observation or experimentation. In philosophy, this is called a contingency condition: The laws of nature are not true because of their logical deduction; they are contingent on verification. Things *could* be otherwise: The laws of nature, such as we see them, make statements about our world that could conceivably be found to be invalid by observation. We might even say that every so-called verification of a law of nature is tantamount to a failed attempt at falsifying it. There is no such thing as definite verification.

Of all conceivable reasons for some sequence of physical events, Plato recognizes only its ultimate aim as worthy of attention. The world is as it is simply because the Demiurge wanted to create something perfect. With this background, Plato's viewpoint on matter and empty space becomes understandable. Mathematical possibilities appear as ur-shapes of matter, giving structure to space. Matter and space are mathematical forms that distinguish themselves through symmetry and beauty. These forms, or shapes, belong to the realm of ideas; the ultimate aim of matter and space is the realization of these forms. Once they are perfect, they *must be realized*—that is the logic of the proof advanced in Plato's *Timaeus.*

To the four elements enumerated by Empedocles, Plato assigns four out of his total of five regular bodies, as shown in figure 23a: to Earth he assigns the cube, to the water he assigns the icosahedron, to the air he assigns the octahedron, and to fire he assigns the tetrahedron. Just like the atoms of the atomists, Plato's regular bodies, which stand for the elements, are surrounded by empty space. They differ from the atoms in the hollowness of their interior. They are ideal geometrical structures with infinitely thin surfaces; they have no reality that would give them mass, weight. They divide up space; and how they accomplish that reflects the implicate order of Plato's universe. They are not immutable and

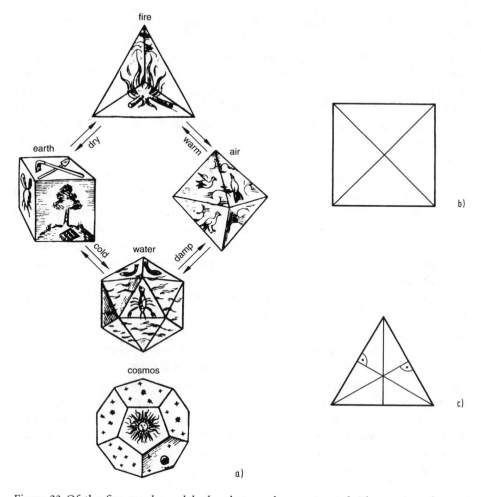

Figure 23 Of the five regular polyhedra that can be constructed, Plato assigns four to Parmenides' elements: water, air, fire, and earth; the fifth one he assigns to the cosmos as a whole. Aristotle stressed the connections among the four elements as indicated in figure 23a. Following a drawing by Kepler, a square can be subdivided into four identical triangles (b); similarly, an equilateral triangle consists of six smaller ones (c). The surfaces of the polyhedra assigned to water, air, and fire consist of identical equilateral triangles; this implies that the elements can be transformed irrespective of the subdivisions shown in 23c.

indivisible, like atoms—their surfaces, and hence their shapes, can be built up out of two types of triangles.

A cube has six square surface areas. Each square is subdivided by its diagonal lines into four equal triangles (see fig. 23b). The surface of the cube thus can be divided into twenty-four of these triangles. The surface of the tetrahedron, consisting of four equilateral triangles, can be made up out of twenty-four right-angled triangles, as shown in figure 23c. Similarly, the octahedron's surface contains forty-eight such triangles; the icosahedron's, one hundred and twenty.

The triangles are the most basic of Plato's units. They are infinitely thin; therefore, they occupy only two out of three spatial dimensions. They do not appear individually, but only as the parts of overall surfaces of regular bodies—for example, twenty-four of them in the case of the cube representing "earth" in figures 23a and 23b. The modern reader might be reminded of the idea of quarks, since just like Plato's triangles the quarks do not appear individually. Rather, they are always contained in larger configurations, which we know as elementary particles. Three quarks make up a proton; the proton, in turn, is a building block of the atomic nucleus. Protons appear individually; quarks do not.

The quarks were originally introduced as mathematical symbols—the carriers of certain properties that would facilitate mathematical understanding of elementary particles. This very role was assigned by Plato to his infinitely thin triangles. But there is a difference: Plato's composite "elementary particles," the polyhedra that we discussed, are models of thought, not physical entities. To Plato, it made no difference whether their existence was ideal or real. Triangles and the structures built out of them, to him, served only to show that nature could be described in terms of such mathematical entities.

ARISTOTLE

With his notion that a body is equivalent to the space its surfaces surround, Plato meant to reduce matter to space, physics to geometry. He did not care whether or not empty space could exist on its own; he never addressed the question. As an object of geometry, Plato's space is closer to this idea than bodies are; to him, therefore, space is more "real." In an everyday sense, however, space "exists" only when surrounded by surfaces—that is, in terms of a body.

Plato's student Aristotle, on the other hand, did have very specific ideas about empty space: It cannot exist. He piles argument upon argument in order to prove this. These are best understood in terms of his preoccupation with the larger connotations that will help to establish his system rather than with any individual features. He wants to convince his readership by all means available to him that there cannot possibly be any such thing as empty space.

SPACE AND MOTION

Whatever does not fit into his system, his view of the world, Aristotle classifies as absurd. In the process, he comes up with concepts that reappear some two thousand years later as scientific truths, freed from the label of absurdity. Aristotle's formulation of Newton's first axiom may serve as an example: "Every body perseveres in its state of being at rest or of moving uniformly straight forward, except insofar as it is compelled to change its state by forces impressed upon it." Those were Newton's words. Aristotle's were: "Nobody can give a reason why a body that has been put into motion in empty space should stop on its own account. Why should it stop in one place rather than in another? Thus it will either remain at rest, or it will of necessity keep moving ad infinitum unless it is hindered from doing so." This theorem yields one of Aristotle's absurdities of empty space: His physics states that all motion will come to a standstill unless some external force keeps driving it—continual motion in empty space is therefore absurd; there cannot be empty space.

This vacuum or empty space of Aristotle's is isotropic; it contains no preferred direction. Today, we might say there is no gravity acting in or on it. But he also discusses gravity as the agent of free fall inside an otherwise empty space. His physics claims that heavier bodies fall more quickly than lighter ones, and that they do so in relation to their sizes, or masses. The same would have to be true for really empty space. But that cannot be. After all, for what reason should one body move more rapidly than another one? Inside a medium, that would be necessarily so: The "larger body" breaks through the medium more rapidly, because of its greater momentum. But in a void, all bodies will move with the same velocity. This is clearly not correct, however. Result: There cannot be a void, an empty space.

Here again, Aristotle speaks a physical truth, then proceeds to deny it. I fail to understand why he would insist on the dependency of the velocity of free fall in empty space on weights or masses—after having just argued that this observed dependency is naturally explained by the presence of a medium.

According to Aristotle, the world is finite; it is contained inside a large sphere defined by the firmament of fixed stars (see fig. 24). That sphere is also its limit: There *is* nothing outside it; there *cannot be anything* outside it, not even empty space. As he says: "At the same time, it is clear that beyond the heavens there is no place, no void, no time. A place is a location where a body might exist"; we might call it empty if there *is* no body there, but there *could* be one. We can put this idea in context with a different wording in Aristotle's famous definition

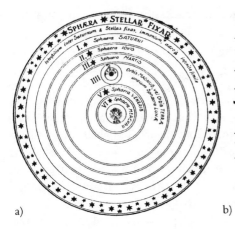

a)

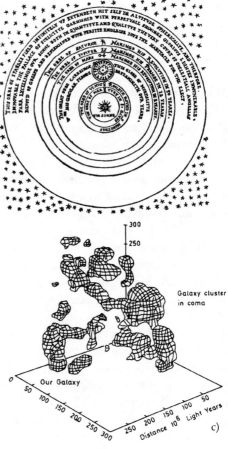

b)

Galaxy cluster
in coma

c)

Figure 24 Aristotle believed the universe to
consist of ten concentric spherical shells surrounding Earth. The elements water, air,
and fire made up the three "terrestrial
spheres" delimited by concentric spherical
shells below the shell inhabited by the Moon.
The spheres beyond the Moon sphere are
made up of his fifth element, the ether.
Those spheres are governed by celestial laws
that permit no change: The only motion
permitted there is rotation of spherical shells that do not change in the process. The
motions below the sphere of the Moon, having a beginning and an end, however, do
mean change. Contrary to this, ever since Newton, science has insisted that there is no
subdivision of the world into regions where different laws of nature reign—the laws of
nature must be the same everywhere. It was one of the foundations of Aristotle's view of
the universe that it is finite. A physical argument in favor of this feature was this: The
sphere of the fixed stars revolves once a day about Earth; an infinite universe would mean
that there are fixed stars that move at infinite velocity. Figure 24a is from Kepler's *Prodromus*
(1596). It shows Aristotle's finite world with three important modifications: The Sun has
a fixed position at the center of the sphere of the fixed stars. Earth belongs to the planets
and is orbited by the Moon. The skies do not revolve; Earth does. This means that the
fixed stars, which are at an infinite distance, do not have to move with infinite velocity.
Figure 24b is from a 1576 book by Thomas Digges. The universe may be infinite in
extent—whether it actually is or not is beyond our knowledge; but it is certainly very
large. That is stressed in a modern "scale-drawing" in figure 24c, where galactic superclusters
are shown to be removed from our Milky Way Galaxy by up to three hundred million
light-years. The Milky Way, on this scale, is nothing but a point of invisibly small size.

of empty space, which he calls "a place that contains no body, but that could contain one." Clearly, then, there cannot be empty space beyond the outer confines of the world. If there were, his definition would require the possibility of bodies out there—and that he denies.

A world without limit cannot be reconciled with Aristotle's view. It is only through its finiteness that we can define a center or origin, that we can distinguish directions like "up" and "down." We can associate bodies with given locations in space only when they are permitted to reach those points. Aristotle defines the location for heavier bodies as "below," of higher ones as "above." This means bodies will sink like rocks or rise like smoke in order to reach the locations destined for them. We can speak of "free fall," and similarly of "free rise." These motions are the natural ones. In addition, there is unnatural motion—like that of a cart being pulled or an arrow being shot.

The forces that cause these "unnatural motions" issue from animate sources, such as the archer who shoots off an arrow; Aristotle feels they need not be explained beyond the animate prime mover who put the world as a whole into motion. In the absence of such a mover, earthly bodies consisting of the four elements—earth, water, air, and fire—assigned to them will move along straight lines. They will fall down (earth, water), they will rise straight up (air, fire), or they will move ahead along a straight line, only to fall or rise eventually. All this cannot exist in space that is really empty. If the space is not even the carrier of information of what is "up" and what is "down," no stone knows in which direction it should fall. Aristotle considers this consequence absurd; and again, he concludes that there cannot be such a thing as empty space. At the very least, the difference of the proportions that mark the various directions in space should be imprinted everywhere inside it. Once that is given, space will not be empty: "There is no getting around the alternative—either there is no natural motion or, if that doesn't hold, there can be no void, no empty space."

The consequence is that the occurrence of natural motions precludes the existence of empty space. Motions that do not distinguish particular directions do not belong in this category, as defined by Aristotle; they are therefore compatible with the existence of empty space. This is the kind of motion that the atomists postulated for the smallest particles: They interact only with one another when they come into contact, when they scatter. As a result, they move about in random motion—the kind of motion that Aristotle also considers compatible with empty space. If the existence of empty space is admitted and if there is only one body in that space, it cannot change its form of motion, which includes remaining at rest. Freed from Aristotle's absurdity, this statement will, some two thousand years later, reemerge as one of the foundations of Newton's mechanics.

Aristotle does not tire of arguing against the existence of empty space; his worldview would collapse otherwise. The directions "up" and "down," which play a vital role in his natural philosophy, have no meaning in empty space.

Were there such a thing as empty space, Aristotle would have to admit that the finite material world of his thinking could be embedded in it. But in that case, there would be the question of *where* exactly our world is situated—why in one place rather than in another? An infinite space cannot be built up around some center of forces toward which Earth, like all matter, would have to gravitate. But more than that, just as he denies the existence of empty space, Aristotle argues that there cannot be limitless straight-line motion, the only motion permitted in the void; it would, after all, necessarily lead to points beyond a finite world.

Aristotle argues that circular motion is the only motion that can persist unchanged forever without reaching unlimited space, which does not exist. Thus the fixed stars move in circles. More precisely, we can consider them to be attached to a rotating sphere. We recall that nothing exists beyond that spherical limit—not even empty space. Aristotle defines motion in terms of his concept of rest and location: Earth is at rest in the center of the universe; it is one of the massive bodies that are attracted toward this center. There is no law of nature that implies the identity of the center of Earth and the center of the universe. Rather, that is the way things must have arranged themselves in the course of time; otherwise, a massive Earth would have dropped to a different location. But Aristotle also argued in the opposite direction, stating that the center of Earth defines the center of the universe. He defines the location of a body as the "internal limit" of an extended physical entity. In this sense, he argues in only two dimensions: The location of wine, say, for him is the internal surface of the barrel; motion is equivalent to a change of this location; in this sense, a boat well anchored while a current of water moves by, according to Aristotle's definition, is in motion.

To Descartes two thousand years later, *every* motion is relative motion. He does not know of anything that is absolutely at rest; as a consequence, he can equally well speak of the boat at rest or the boat in motion. For Aristotle, the consequence of his definition of motion—that bodies move although they are at rest with respect to Earth—was an absurdity. He therefore specified his definition: "A fixed location is defined as the first immobile limit of an extended space or body"—that is, the location of the boat at anchor is defined by the banks of the river.

SPACE AND LOCATION

This definition is not quite clear; it has to be adapted to individual cases. In this spirit, Aristotle's thinking found many different interpretations. It is instructive to consider what the term *location* does *not* stand for, according to Aristotle: It is neither "shape nor form, nor the matter of a body, nor the extension between the bounding surfaces of a containing body." This *space* cannot exist independently; if it did, it would be the nonexistent empty space in Aristotle's definition. Take a barrel: Its inner surface can define the location of wine, of water, of air—the location of something that is material. Location is not defined by one kind of body rather than another but is a property specific to bodies only: The world is full of matter that fills all space, that touches all inner surfaces. Matter borders on matter, always and everywhere; there is nowhere and never a gap in between.

Space, in this sense, is equivalent to the full set of all locations; and as such, it exists. Empty space, however, does not. If it did, it would be another body. By equating a hypothetical empty space with just another body, Aristotle means to prove that it cannot exist. He argues that if it did, two bodies could occupy the same location: empty space and, let's say, a wooden cube. That is, of course, just as absurd as the idea that a wooden cube in water was not displacing some of that water but rather coexisting with it in the same location.

In addition, Aristotle feels no *need* for the existence of empty space. Even if it were there, there would not be any way of telling it apart from the volume taken up by bodies. Volume is one of the properties of bodies, regardless of where they are. But bodies are present all over. As a result, there is nothing that could be explained by empty space that is not already explained by volumes of bodies.

This reasoning is firmly grounded in Aristotle's teaching. Today, we might simply say that he differentiates between matter and shape. They cannot exist independently of each other—there is no matter without shape, no shape without matter. Matter necessarily has shape and thus becomes what Aristotle calls a *substance.* This substance is at the beginning of every natural philosophy. It is as concrete as can be; it simply is, say, a statue of Socrates that we might be looking at in a given museum at a given time. Proceeding from here, we may start to disregard details of the shaped matter we are looking at: the here, the now, the shaped image of Socrates, the marble it is made of.

The searching mind, according to Aristotle, can also disregard the rigid matter that makes up the statue without at the same time endowing it with another property—say, liquidity. He even believes in the possibility of abstractions that forgo all properties and forms, so that ultimately nothing but pure matter with no property or form remains.

This takes us back to ur-matter, *materia prima.* It is clearly nothing but an abstraction, not truly observable. It is impossible to remove all properties, all specificity, from a substance so that only pure, shapeless *materia prima* remains. This imaginary or idealized matter carries the potential of all forms and shapes imaginable; it possesses no individual property but is capable of assuming any one of them.

In other words, it may show up in arbitrary form. Aristotle denies the existence of matter without shape and, even more so, of shape without matter. Shape owes its existence to the fact that it may be assumed by matter; but as an idea, abstract and without ties to physical bodies, it is an impossibility. In this argument, Aristotle draws a line between himself and his teacher Plato. As the searching mind progresses from a given substance to *materia prima,* specific shapes recede into possible shapes. If there were an independent extended volume that existed without reference to any body, we would have empty space before us; but then again, empty space, according to Aristotle, cannot exist.

The *materia prima* will become a substance by realizing one of the forms or properties out of the totality that it contains "virtually." This is the only process admitted by Aristotle: Potential properties can be transformed into actual properties and vice versa; one shape may replace another; there is no such thing as a completely new property or shape. This includes motion as something that changes a potential new position into a real one: The potential property of the arrow to be "there" is realized at the expense of the previous property of being "here."

There is, according to Aristotle, no actual motion involved. The arrow, rather, replaces a label it carries that is marked "here" by another label marked "there." If real motion were nothing more than that, it would not need any space independent of bodies in order to occur. Aristotle, in his completely occupied universe, can always tie a "here," a "there," to substances. Once this is done, the world looks like one immense three-dimensional pointillist image. Its points are made up by individual material objects that can be resolved individually, even without reference to their location. Note that, to make matters easier, we have been assuming a world composed of almost pointlike elements that have a finite size; it is easy to make a transition to the continuum by reducing that finite size further and further.

We can describe the world at a specific moment by identifying every pointlike element of matter with the location it holds at that time. Now let's turn to motion to change this picture: We observe the way all our points move in space to change the picture; distinguishable material points will change their positions. Aristotle describes it differently: He has the material points remain unmoved. It is their location that changes as one of their properties, just as a color might change.

Because Aristotle's ur-matter contains everything possible among its potential properties, it becomes closely related to the vacuum of quantum mechanics. We said so in the prologue and can now be more specific. But in contrast to the vacuum, the *materia prima* does not really exist; it is only a concept, a model of thought. It is an abstraction, fashioned after what Thales, Anaximander, and Parmenides imagined as some basic matter that filled the universe in its entirety.

The finite world of Aristotle is everything that exists: It is not an island inside some larger empty space. In this sense, our world has no location that could be defined as the "inner limit of a larger body surrounding it." In particular, if the world were infinite, this argument would hold; a surrounding body for an infinite world is not even thinkable. But if there is no surrounding body, then there is no definable location. "And thus we can speak of Earth as being surrounded by water, the ocean being surrounded by air, the atmosphere being engulfed in some ether, the ether filling our world. But that is where it stops: Our world is not further surrounded." And given that there is no location outside our world, even questions about the location of our world or its potential motion are an absurdity as far as Aristotle is concerned.

ARISTOTLE'S SYSTEM AS A STANDARD MODEL

Aristotle's system was for two thousand years the standard model of philosophy and natural science. Just as we have our doubts about the ultimate validity of the modern standard models of elementary particle physics and cosmology, Aristotle's system has not been impervious to doubts. But those doubts notwithstanding, all scientific thinking up to the seventeenth century took as its reference this model—contradicting it, arguing against it, adding to it. It served as the basis for communication; today we would call it a paradigm. With two or three important exceptions, Aristotle discussed all the ingredients of later theories in the formulation of his own system—rejecting them (the concept of atoms) or accepting them (the *horror vacui*). We might think of the offerings a restaurant puts on its menu, where the variety of dishes is actually due to inventive combinations of very few basic foodstuffs: The philosophical and scientific theories developed in the two millennia after Aristotle are similarly based on a few original ideas formulated by him.

The main reason that his system proved so solid is not that it was correct. No, today we know that his system is fundamentally flawed. Soon after Aristotle's time, it became evident that his system could not be completely correct. But its success persisted simply for the reason that it is a self-contained system. You cannot replace an entire system by hauling in a variety of detailed new information; you have to bring in a completely new system. And that did not happen until Newton.

VARIATIONS OF THE STANDARD MODEL

Aristotle did not admit the possibility of empty space, or a vacuum, either inside our world or beyond it. An inner vacuum might manifest itself in two varieties. It might either be a "microvacuum"—in the version of the atomists separating minuscule units of matter—or a "macrovacuum," a macroscopic empty space. Today, we would have to call the latter simply a space without atoms and without radiation. The world of Aristotle is a continuum filled with matter, with no mention of atoms. It does not date back to some creation process; it will never cease to exist.

One early variant of the standard model was formulated by the Stoics. Their main preoccupation was a philosophy of life represented notably by Seneca and

Marcus Aurelius; they also developed a scientific view of the world. This was done notably by Zeno of Citium, Chrysippus, and Poseidonius. Their universe includes an external vacuum but not an inner one. Their world has a finite volume; it forms one and only one island of matter inside an infinitely large empty space, the external vacuum. The Stoics follow Aristotle in his belief that "empty space" would have to be capable of accepting the presence of bodies. To his decree that outside the world there can be no further bodies, they add the question of why that should be so. Of course, it is a trivial truth that there are no bodies beyond the confines of our world. But to what extent does Aristotle's decree extend beyond this simple statement? What precisely are those confines? Why can't we extend a hand, a projectile, beyond them?

TO THE EDGE OF THE WORLD

This question goes back to Archytas of Tarentum, a contemporary of Plato: "If I am at the extremity of the heaven of the fixed stars, can I stretch outwards my hand or staff? It is absurd to suppose that I could not; and if I can, what is outside must be either body or space. We may then in the same way get to the outside of that again, and so on; and if there is always a new place to which the staff may be held out, this clearly involves extension without limit."

The reader may compare Archytas and his staff to the bugs and sticks of Feynman's imagination on a hot plate (see figs. 19 and 20b): None of them reaches the outer edge of its world because they all diminish more and more in size the farther out they get. Similarly, a bug-size Archytas on a spherical surface (see figs. 17b and 17c) will never hit an edge; rather, he will find himself moving through the same space again and again.

These kinds of solutions that explain the riddle of the edge of our world in terms of non-Euclidian geometry have become accessible only in the middle of the last century. Discussions on the question of whether or not the space of our universe has all the properties that our immediate observations suggest were moot before that time; a mathematical model other than Euclid's had not been developed. The latter-day atomist Lucretius inquired about edges of the world just as Archytas had done. His answer was not dissimilar:

Let's assume, for now, space is limited, and somebody throws a javelin beyond its outer edge. What do you think the projectile that has been sent off at great speed toward the edge will do? Will it continue in that direction beyond the edge, or will it stop dead in its track when it hits that edge? However you might move or displace that outer edge, I will ask you again: "What about the javelin?" From this you should learn that our universe is limitless in all directions.

The same answer about the limits of the universe is given by the Stoics: There are none, no matter what direction you choose. But the Stoics start from a finite world occupied by matter, or bodies. The space outside that finite world, however, is capable of accepting bodies, thus expanding the world; where previously there had not been bodies, and hence no material world, Archytas's staff created a new one. In Aristotle's definition, this means there must be empty space beyond the material world of bodies—after all, the space beyond the material world does not contain such bodies in the first place, but is capable of accepting them. And this empty space extends without any limits to infinity. There cannot be a limit to how far the edge is extended by Archytas's staff to create an ever larger universe.

The idea that the world can expand simply because bodies it contains are moving farther and farther apart is astonishingly modern. The formulation I will use is, of course, different from that of the ancient natural philosophers. They could not possibly know what we have learned since about the makeup of the universe; their means of observation did not permit that. Their concepts of the qualities of the universe—continuous versus discrete, empty versus full, infinite versus limited—were developed into a consistent theory through many generations. But only experiment would be able to decide on the correct choices. Experiments—expensive ones at that, with telescopes, research satellites, particle accelerators—answer the questions that make up our cultural inheritance.

Let's return to Archytas's staff and to the expanding universe. A javelin that is thrown straight upward will drop back down a long time before we can sensibly ask the question about the edge of the universe. The javelin cannot escape the gravity of Earth. A spacecraft that runs out of propellant only after it reaches the velocity that permits its escape from the gravitational pull of our solar system will continue on its path, but it will remain inside our galaxy. The Milky Way's gravitational field holds on to it. Proceeding like this, we pass from a practical problem to a fundamental question. Can there be objects that may escape our group of galaxies, and that will ultimately leave our universe altogether? The answer must be no. The universe, after all, keeps expanding. Light that reaches us today was emitted shortly after the birth of the universe; its "headwaters" are moving away from us with a velocity close to that of light. There is no light older than what originated at the time of the Big Bang. Since no information travels at a speed greater than that of light, the observable universe is contained within the surface of a sphere with a radius defined by the distance traveled by light that was emitted at the time of the Big Bang.

Of all the material objects we know, quasars are closest to Lucretius's javelin. Quasars are galaxies in an early state of evolution. The light that tells us about them today has traveled for billions of years. They move away from us at a speed close to that of light; in that way, they mirror the expansion of the universe.

Galaxies, quasars among them, are interpreted by astrophysicists in terms of probes that are attached to points in space and that move with these points. The increase of distances of the galaxies is seen as the growth of space. Space itself expands when galaxies move farther apart. The fact that quasars move away from us indicates that new space originates in between. Conversely, the masses of the universe help to determine both time and space. We might say that space expands *because* of the behavior of the masses; the masses of the universe, in their joint action, take the place of Archytas's staff. The expansion is caused by the initial momentum that was imparted to matter by the Big Bang.

THE STOICS' PNEUMA AND ETHER

The Stoics, like the atomist Lucretius, do not know about an edge of space, but they do know about an edge that confines the world of bodies. This edge opens up like a deep moat across which Lucretius might toss his javelin: The world of bodies with its given size could, at any time, increase when the javelin is tossed. It is embedded in empty space, but it does not contain empty space, neither microvacuum nor macrovacuum. The Stoics believe that the world is filled with what they call *pneuma*. This pneuma, in their imagination, is an elastic substance, a mixture of fire and air. It holds the world together; it keeps the world from diffusing out into infinite space. This pneuma is not a passive substance like Aristotle's ether, eternally at rest. It has "tension," which can change with location and time and transfer various effects from one place to a neighboring one. In modern terms, we would call the dynamic theory of the Stoics a theory of close interactions: If one body acts on another at a different location, this action is not instantaneous. It propagates with a given velocity from one location to a neighboring one. The pneuma "vibrates." This vibration starts at a given point; it expands into neighboring territory—just as a wave expands around a pebble we throw into a pond.

The Stoics noticed with remarkable scientific insight that sound is the same thing as vibrations of air. Their interpretation of the propagation of observable phenomena anticipates important aspects of field theory. To each point in space we assign a number; this number marks the strength of the field at that point. The field is then a map made up of all numbers assigned. One such example of a field is the distribution of temperatures on Earth: 98 degrees Fahrenheit in Los Angeles, 105 degrees F in Las Vegas, 61 degrees F in New York, and so on. This field, the temperature distribution, derives its physical reality from external sources: The molecules in air and ground are in a higher state of motion in Las Vegas than in Los Angeles.

Sound can equally be described in terms of a field. This field is another totality of numbers—air pressure as a function of location. The phenomenon of sound is based on a periodic change of pressure in a medium such as air. Pressure, of course, is another word for the density of molecules in a gas (in this case, air). Air molecules oscillate when a bell strikes. This is no different, in general terms, from the wake of a boat moving through the ocean: It causes a moving deformation of the water surface. We call this deformation "waves," and the waves are ultimately nothing but oscillations of water molecules. Whoever is familiar with the definition of fields will find examples in many places—in the distribution of velocities in the river flow from one bank to the other; in the inclination of rye stalks when wind passes over a field.

Once it became clear that the Stoics' interpretation of sound in terms of air vibrations was correct, it was only a small step to interpret light as an oscillation phenomenon. But of what? That is where the term *ether* came in. The theory that light is nothing but vibrations of a hypothetical, omnipresent ether dominated physics in the late nineteenth century. Its triumphant advance started with Maxwell's equations, which describe electrical and magnetic phenomena jointly. In this context, light is easily seen as vibration—but Maxwell's equations have nothing to say about what precisely it is that vibrates. In fact, there is no such substance at all—something present in the dark that lights up once it starts oscillating. Maxwell's carrier substance of light, his "ether," was supposed to be capable of elastic oscillation. In that sense, it had more in common with the pneuma of the Stoics than with Aristotle's ether.

THE ATOMISTS EPICURUS AND LUCRETIUS VERSUS ARISTOTLE

The atomists needed the empty space between the atoms to permit atomic motion. They did not make a distinction between such concepts as inner, outer, microvacuum or macrovacuum. The world had to be infinitely large; otherwise there would not be enough room for the infinite number of atoms. Epicurus, successor and disciple of the early atomists Leucippus and Democritus, contrasts their worldview with Aristotle's in the following way:

> *Furthermore, the universe consists of bodies and the void. That bodies do exist, our senses attest to; they also permit us to reason about what is beyond our perception. . . . If there were nothing like the void, the empty space, a realm beyond what we can touch, then there would not be any room to locate bodies, to have them move the way we see them moving. . . . Also, the universe does not limit the amount of atoms it contains on the expanse of the void. . . . And what is more, there is an infinite number of worlds, whether they resemble ours or not.*

The Roman philosopher and poet Lucretius wrote an instructive poem, "On the Nature of Things," in which he summarized the teachings of the atomists. To show that motion cannot happen in space that is occupied, he refers to fish swimming in water:

> First, let me make sure that you will not be taken in by misleading arguments advanced by many. They maintain that the water opens up for the swimming fish, closing up behind the fish once they have passed. In this manner, other objects may well move and exchange their positions though all space be filled a priori. This argument, of course, is completely wrong. Where should the fish go if the water did not open up space to start with? But how can the water recede when the fish cannot move? We have to conclude that either there is no motion of bodies or that there is a void at the basis of all motion of objects through a medium.

This view of motion, with all its problems, is certainly superior to Aristotle's reducing motion to the exchange of labels. Prior to this, Lucretius had explained his interpretation of directional motion:

> We cannot say that bodily objects fill space all over; there is empty space inside matter—empty space that we cannot touch. Were it not there, the objects could not move. After all, resistance against motion is a natural property of all bodies, and it would have to show everywhere. Nothing could move because nothing would be yielding. But we do see a great deal of motion by land, water, or through the skies. If there were no void, all this motion would be impossible; and nothing could originate because of a lack of space—after all, matter would be densely packed, immobile everywhere.

The physics question concerning the void fuses for Lucretius and his contemporaries with the metaphysical question of the nothing. From this nothing, Lucretius says, nothing new can originate. His ideas are therefore incompatible with the Christian teaching about the creation of the world out of nothing. (It will take many centuries before Galileo's contemporary the French priest Pierre Gassendi will heal this rift.) "In the beginning of all rational observation of matter, there must be the recognition that nothing can be generated out of nothing; there cannot be a creation *ex nihilo* by, say, some divine creator." As Lucretius said, the world was created all right, but not out of the nothing. Conversely, matter cannot vanish into the nothing. For the topic of this book, it is interesting to see him restate Democritus's argument that only two components make up the world: the atoms and the void. He formulates it like this: "Let me add that we cannot find anything beyond bodies and the void; there is no third component of the world." Anything that acts is a body. Anything that does not act is empty space, "but space cannot be if there is no void. And thus the void and the bodies do not leave an opening for any third component that could manifest itself to our senses or to our minds."

TWO CONCEPTS OF SPACE

The two concepts of space considered by Aristotle also determined the philosophies of his disciples Theophrastus and Strato. The former defined space via *the positioning qualities of the world of bodies;* the latter assumed space to be *the container of all bodily objects.* We have previously seen these concepts contrasted in Einstein's thinking. Two millennia after Aristotle, Leibniz and the English philosopher Samuel Clarke, the latter inspired by Newton, squabbled bitterly, and not always on a scientific level, over these ideas.

Theophrastus replaces Aristotle's concrete space, which is defined by the bodies it contains, with an abstract space. By this he means the spatial relation among the bodies, their "positioning qualities": "Maybe space itself is not real; rather, it may be determined by the position and the ordering of the bodies with respect to their properties. This is just what we are accustomed to from animals and plants and other inhomogeneous bodies, bodies with structure. After all, these bodies have a well-determined ordering and positioning, by which their parts make up the total." The continuation of this citation of Theophrastus shows his Aristotelian schooling; he does not implement his own concept of space in the direction later taken by Leibniz, Mach, and Einstein. Instead, he says: "Every body that takes its rightful position has a specific ordering. Every part of a body seeks the location and the correlations for which it is destined." Space, in this reading, is not dependent on the actual positions of the bodies; it is to be abstracted from the locations that Aristotle's system assigns to them. According to our previous definition of fields, we can interpret space as a field that directs the motions of all bodies toward the locations assigned to them.

Theophrastus ascribes reality only to bodies; in the absence of bodies, there can be no space. Therefore he agrees with Aristotle that there can be no vacuum—neither microvacuum nor macrovacuum, neither inside the world of bodies nor outside.

After Theophrastus, his follower Strato headed the school of philosophy founded by Aristotle. He adopted many of the atomists' ideas: Objects move in empty space without displacing a continuum; in dense matter, the atoms are closer to each other than in more dilute matter. He imagines that there is empty space between the building blocks of matter. His space is a container for bodies, and as such an absolute quantity in itself—its existence does not depend on whether it contains bodies or not. If not, it is empty space, a void or vacuum. He differentiates between what actually is and what could be. As his tradition passes on to Philo, then to Hero, these ideas develop so as not to permit the existence of a macrovacuum in nature; but they allow for the possibility that it could be produced by artificial means.

Strato, however, maintains that under normal circumstances "there cannot be a continuous vacuum; there may well be small vacua scattered throughout the air, the water, the fire, and all bodies. We want to adopt these ideas and we will seek experimental proof that this is indeed so."

STRATO'S EXPERIMENTS ON THE VACUUM

Strato then thinks it possible that, perforce, there may be an expanded empty space, a "macrovacuum"; but this would have to be under abnormal circumstances. And he does mean to prove it experimentally. As a method, this must be seen as revolutionary. He starts out with a sober lab protocol devoid of all poetic frills, describing the observations we already know from Empedocles' experiment with the hydra (see fig. 21). He summarizes his result: "We have thus proven that air is, in fact, a physical substance." He then addresses all those who categorically deny the existence of vacuum; he means to prove to them, again by experiment, that there is in fact a phenomenon that can be called a continuous vacuum, even if it does not normally occur in nature; and then again, that this vacuum may be normally present—but in small amounts and widely dispersed. These scattered microvacua will fill up with bodies under pressure. Our demonstration, he says, will make it impossible for all those idle wordsmiths to talk their way out of these observations.

With pedantic precision, he then describes an experiment in which air is blown through a pipe into a metal sphere—a space normally filled with air. His explanation for the phenomenon that more air can be added to the space already filled with air still makes sense to us: The air molecules move closer together. That goes against the grain of the logicians, those finaglers of words; they argue that since there is no empty space, there is no way to add more air without taking some of the already present air out of the vessel. We don't know how these critics interpreted Strato's experiment. They probably abstained from comment. Experiments designed to address a particular point did not, after all, belong in their arsenal.

PLOTINUS AND AN INTERPRETATION OF TODAY'S PHYSICS

We now have assembled all the concepts that were to dominate scientific discussion on empty space to the present day; physicists, of course, are interested in solutions and detailed answers rather than in mere questions and inquiries. But many of these questions, the answers to which they keep discussing, date back to Leucippus, Plato, Aristotle, Theophrastus, and Strato.

Plato holds a special place among them. His detailed interpretation of matter as a hollow space surrounded by regularly shaped bodies was too specific to hold up when scientific development advanced. On the other hand, his concept, taken over from the schools of the Eleatics and of Pythagoras—that only ideas possess reality—has persisted to this day. To him and his successors, matter really exists only by virtue of its form. It is only through its form, after all, that matter is tied to the ideas. It takes mathematics to describe form precisely. This means that the true laws of nature are expressed mathematically. We might say, the laws pin down the quantities that make up real existence.

Among Plato's successors, the most remarkable thinker to pick up this strain of argument is Plotinus. He offers little to those who would pin him down to rational argument. His intricately defined system combines ideas owed to both Plato and Aristotle. He categorizes our world in terms of four hierarchical steps, each of which is subtended between two quantities that have no existence of their own. At the upper end, there is the abstract quantity he calls the One, or the Good. It cannot be described by any sequence of positive terms. Every other existence can be so described, but the One, the Good, cannot; we'll leave that to Plotinus. Now to the lower end of our range of observation: That is where we find matter miserable, contemptible, poor to the degree that it, again, has no existence of its own. But it contains a potential for all physical phenomena. We might say that it is fighting for its existence: "Its essence can be described in some measure by such images as utter poverty, constant want, perennial longing for making its appearance in the realm of reality."

In and by itself, matter is shapeless; as Plotinus says: "The very idea of matter implies absence of form." As a result, it will not make its appearance in the real world. If, however, it does show up, that must be due to its having taken on a specific form, to its having changed. This feature of Plotinus's matter—its ability to take on shapes notwithstanding its own lack of all shape—reminds us of Aristotle's *materia prima*. Matter, according to Plotinus, is the carrier of all properties of bodies. This includes all physical extent: "Absolute matter must take its magnitude, as every other property, from outside itself."

Plotinus also mentions a different form of matter, a form tied to the spiritual world—but I will not discuss this here. "Matter," he says, "is understood to be a certain base, a recipient of Form-Ideas. . . . There is, therefore, a Matter accepting the shape, a permanent substratum. . . . The Matter must be . . . ready to become anything. . . . Matter, not delimited, having in its own nature no stability, swept into any or every form by turns, ready to go here, there, and everywhere, becomes a thing of multiplicity: driven into all shapes, becoming all things. . . . The distinctive character of Matter is unshape, the lack of qualification and form. . . . Matter is therefore nonexistent." The concept of existence is rooted deep in Plotinus's mystical thinking. It can be rationally approached only to the extent that things immutable were distinguished by the ancient philosophers from things that are subject to transformation, change, passing.

Figure 25 Can a cat vanish, but not its grin? Can structure exist without matter? This may be possible in the abstract interpretation of Plato and Plotinus; in a concrete sense, it is impossible. Structured matter can be interpreted as an excitation of unstructured matter, just as the cat with the grin may be seen as an excitation of the nongrinning cat.

Plotinus may choose more poetically elevated terms about what he thinks of matter; but in essence, it is not much different from Aristotle's *materia prima.* The main difference is that form, for Aristotle, has no reality of its own but is inextricably tied to *materia prima,* while Plotinus accords an independent existence to form or shape. Maybe we are talking about a mostly semantic notion, somewhat akin to the grinning of Lewis Carroll's Cheshire Cat (see fig. 25) in *Alice's Adventures in Wonderland:* Does that grin have an independent existence of its own? The cat vanishes from the tree, disappearing piece by piece from our observation, "ending with the grin, which remained some time after the rest of it had gone. 'Well! I've often seen a cat without a grin,' thought Alice; 'but a grin without a cat! It's the most curious thing I ever saw in all my life!'"

I think Aristotle and Plotinus would have agreed that there cannot be any reality to a grin that is not tied to a creature that produces it. On the other hand, there is evidently the abstract concept of a grin; it is a form that can be assumed by all kinds of bodies—by candy animals, by flower arrangements, by all types of matter. In Plato's succession, Plotinus accords an existence of itself to form, a feature lacking in Aristotle.

Plotinus's ideas exceeded those of Plato's school in creating a psychological climate for disregarding—or, worse, looking down upon empirical investigation of—natural phenomena. I would not have dwelt on them in such detail were it not for the fact that Ugo Amaldi, the illustrious contemporary Italian experimental physicist, recently used them to illustrate a number of contemporary concepts:

Figure 26 Chaotic or disordered distributions are symmetrical on the average. An observer cannot conclude anything about his location from looking around him. An example is the chaotic disposition of sand grains at the beach, or of flecks of light on a television screen when the station closes down. In the chaotic distribution, no location or direction is preferred. Only when there is structure appearing in a chaotic distribution does orientation become possible, as the overall symmetry is broken. Our figure shows a computer simulation of the formation of structures in the universe.

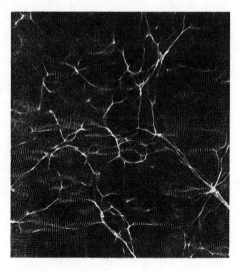

What Plotinus called the abstract One, or the Good, Amaldi interprets as Energy. Plotinus's "nonbeing" Amaldi compares "with the all-pervading fields of quantum theory; indeed the fields are not matter, but their presence determines potentially the nature of the matter-particles which we directly observe. For Plotinus," according to Amaldi, "the overall movement from the One to matter gives rise to actual being, which matter by itself seeks in vain to attain." Energy and the fields of quantum theory are concepts to be exactly defined; there is no way of tying them one by one to Plotinus's ideas. Later on, after having considered fluctuations of fields and particles in the vacuum (chapter 6), where only the lack of energy keeps particles from making their appearance in reality, the reader may well reread the present section. At this point, I think we will agree that today's quantized vacuum and its chaotic fluctuations are but a modern version of Plotinus's image of matter in its "perennial longing to enter into reality" (see, for example, fig. 26).

IS THERE ROOM FOR GOD IN THE UNIVERSE?

After this excursion into the poetry of physics, let us turn back toward "space as container for all material objects and as an expression for the positional quality of the physical world." Those, after all, were the concepts for which Strato and Theophrastus gave a clear formulation. True, through centuries they were modified and imbued with philosophical trappings of later interpreters. In particular, the rise of Christendom posed the question of where God might fit into the universe, into its spatial structure. I will stay away from this question as much as I can. To set the tone, we might listen to Philo of Alexandria, the Jewish religious thinker (born about 25 B.C.) so popular and influential with early Christian theologians: "Space has a threefold significance. First, it is the volume that contains all bodily objects. Second, it is the repository of divine order. Third, space is as God himself—comprising all, comprised by nothing." From this quote we can deduce that Philo of Alexandria understood space as a receptacle in the same way as Strato did. Beyond that, his contributions were in the realm of theology.

ARISTOTLE AND CHRISTIAN TEACHINGS

In the thirteenth century, Robert Grosseteste, Albertus Magnus, and Thomas Aquinas modified Aristotle's system to the extent that it became the dominant teaching of the Christian Church. There was one problem: For Aristotle, the world was not created—it had always been and would remain forever. Christian faith, however, held that the world had been created from nothing and would cease to be on Judgment Day. Moreover, Aristotle's determinism left no room for God's intervention in the affairs of this world. In these two cases, the church declared Aristotle to be in error; but it adopted his teaching that there could be no vacuum—as well as his view of Earth's standing still, with the Sun orbiting around it. Thus, the church opened itself up to the perils of being proved wrong by dint of declaring scientifically observable notions as matters of faith. This was bound to cause conflict in scientists' consciences and danger to those among them who arrived at conclusions that differed from church teachings.

All this is well known. The denial by this amalgam of Christian and Aristotelian teaching that there was no way a vacuum could exist was so complete that it had to be seen as the expression of a consensus already existing. According to this teaching of the church, God—or, more specifically, God's immeasurability—must pervade all space, which therefore could not be empty. Not that this made God an object that could be spatially divided; the doctrine implied that God is fully present at all times and at every point in space.

An external sign that Aristotle's teaching on the impossibility of a vacuum had pervaded Christendom came in 1325: In that year, the church revoked a decree with which, in the year 1277, the bishop of Paris, Étienne Tempier, had intended to counteract the spreading of Aristotelian ideas in the church. He forbade the defense of 219 articles in theology and natural philosophy. For our discussion, article 49 is the relevant one. It was directed against an idea that might be due to a Christian Aristotle: God will not move the world along a straight line, because that would create a vacuum behind it. This did not sit well with Tempier, who clearly did not appreciate Aristotle's arguments against the vacuum: God in his omnipotence can obviously do what to Aristotle was absurd— move the finite cosmos inside an otherwise "empty" space. (The argument was revived centuries later in an exchange of letters between Clarke and Leibniz.)

The experts in the field are not agreed on the consequences of Bishop Tempier's edict. I tend to believe that it eased the tension on the point: He meant to do away with very specific statements that could hinder the unfolding of scientific activity in the wake of narrow adherence to Aristotle's views, such as the one denying the possible existence of the vacuum. His message was that if some phenomenon *does not exist,* this does not necessarily imply that *it cannot exist;* God's omnipotence could make it happen. Whether the powers of God Almighty could be replaced by experimental human endeavor is left an open question. Philo and Hero, two successors of Strato's of whom more will be said in the next chapter, assumed that while a vacuum does not exist, it can be built up.

As of 1325, Bishop Tempier's decree was declared null and void: Aristotle's teachings prevailed; moreover, the decree contradicted the positions of Thomas Aquinas, the main champion of Christianity's adoption of Aristotle's ideas.

SCHOLASTIC TEACHING ON EMPTY SPACE

The philosophy of the Scholastics was a wild mixture of preconceptions and logic. It was governed by conventions and compulsions in the wake of Aristotle's teachings and those of the church. It did not search for new, hitherto unheralded truths; rather, it was content with nailing down well-known revelations. Its evidence came in the form of indirect proof, which might appear either as contradiction or as an absurd consequence: The method was to make an assumption, then reject it if it led to a contradiction or to some nonsensical conclusion. To the latter category the Scholastics relegated everything that did not sit well with their conventions and with the narrow set of their beliefs.

One principle of Scholastic philosophy was the tenet that three-dimensional extension implies bodily existence. Similarly, it held that two bodies cannot coexist in the same location. As a consequence, three-dimensional space would have to be seen as a body; no other body could additionally occupy the same volume. Therefore, there cannot be three-dimensional space; alternatively, it would have to be seen as a physical extension of bodies, simply parading under a different name.

That is not very different from some of Aristotle's teachings. The Scholastics adopted from him the definition of location as the concave inner shell of surrounding space. The Scholastics did not share the audacity of the Neoplatonists, the followers of Plotinus, who were willing to scuttle some of Aristotle's principles. This attitude might have led to trouble in the fourteenth century, given that the church had adopted many of Aristotle's ideas. Rather, the imminent task of the Scholastics at that time was a reformulation of Aristotle's teaching such that it would include the notion of the omnipotence of God.

All the Scholastics agreed on one point: There is no practical way to produce a vacuum. To prove it, they did gedankenexperiments that involved a bellows and a hydra—we will discuss them in the next chapter. Then there was a question that carried religious significance: Does logic by itself forbid the existence of a vacuum? Clearly, God Almighty can do anything that is logically possible. But even God will not transcend logic: He will not draw a square circle. By implication, if logic itself forbids the existence of a vacuum, even God will not be able to create one, though he might do so if it were hindered only by practical considerations. Let's recall the Parisian decree: Given that God can move a finite world such that the space it vacates remains void, it may be that an empty space, a vacuum, is impossible simply for practical reasons; but logic alone permits its existence. It took the lifting of that decree to make the church adopt fully the Aristotelian notion: The concept of a vacuum is contradictory in itself; therefore, logic suffices to exclude its existence.

In the discussion of the Scholastics, this topic did not die down. Next, I will describe the ultimately unsuccessful attempt of Albert of Saxony (1316–1390) to establish harmony between the potential existence of empty space and the remaining teachings of Aristotle.

First, Albert defines the vacuum as a place "that is not filled by bodies." He adds immediately that we cannot conclude the existence of a void simply from the fact that it can be defined. Just as location can be understood in two different ways, so can the vacuum: "Some have imagined that 'place' is equal to all the dimensions of what is placed [or situated] in it; others, however, as Aristotle and his followers, have imagined that place is not such a space but [rather] a body external to the body placed [or situated and] equal to it in two dimensions, namely width and length." And "vacuum must also be imagined in this twofold way: in one way as a separate space in which there is no other body conjointly; in the second way as body between whose mutually closest sides there is nothing." Albert theorizes that God might annihilate all bodies within the concave shell of the Moon; that would have created a vacuum—but what would it signify? He equates this vacuum with the nothing pure and simple—the nothing that is certainly not a body, that certainly has no extent. He concludes that there cannot be a vacuum in the first definition: Space has three dimensions; therefore it is "something," not "nothing." It is in the same category as the bodies that have to be removed to create a real void.

And so it is the encapsulating body—here, the concave spherical shell of the Moon—that makes up empty space. But even this leads to a contradiction: "Either the sides of the sky would be distant [that is, separated] or not. If they are not separated by a distance, they would be conjoined as two leaves of a book or an empty purse; but then it ought not be conceded that the sky is a vacuum, because in the aforesaid case the sides of the sky would be immediate [that is, in contact]. If, however, it be stated in this case that the sides of the sky are yet distant [or separated] there would then be some dimension between them by which they would be distant [or separated]; hence there would be no vacuum."

Only a logical hara-kiri appears to remain: Opposite sides of the skies are indeed separated, but it is not a straight line that keeps them apart; rather, the separation is curvilinear, like a line that follows the inner shell of the Moon. Taking it all together, Aristotle's system has proved extremely rigid: The existence of a vacuum is unthinkable in its context. As the historian of science Edward Grant observes: "Arguments demonstrating that space could be neither substance nor accident were of great utility, because they could infer from them the absolute nonexistence of space." And even Newton had to deal with this reasoning, in his inimitable temperamental way: "As for physical extent, maybe you would expect me to define it as an accident or as nothing at all. But I will not do anything of that kind. It has its own particular existence, devoid of either substance or accident. . . . Even less can space be seen as a nothing: It is more than just an accident, it comes close in its nature to that of a substance."

MATTER IN THE VOID?

All those who, in the wake of the Stoics, held that space extends to infinity in all directions also filled this space with matter, with substances; a true vacuum was beyond their imagination. Thus Nicholas Oresme opined in the fourteenth century that space is unlimited even beyond the material world observable to us, and that God himself filled it in some fashion. Nicholas of Cusa, another believer in unlimited space, thought it is filled with both matter and spirit. Giordano Bruno held that space infinite is replete with ether, and the English philosopher Henry More simply put the presence of God himself forward as ubiquitous all over space.

This listing may serve to show that the notion of *empty* space as an actual vacuum—as Gassendi, Torricelli, Pascal, Guericke, and Boyle would develop the idea in the seventeenth century—was not only new but sensational. Their immediate precursor, Francesco Patrizzi, still held on to the idea of the Stoics, filling his empty space with light. He argued that light is the substance that comes closest to being "nothing." René Descartes, confronted with the experimental evidence created by Torricelli and his followers, admitted that a macrovacuum could obviously be prepared in our world but denied the existence of truly empty space, for philosophical reasons, in a categorical fashion. To do so, he took refuge in the idea of an omnipresent extremely diluted matter.

THE VOID—REAL OR IMAGINED?

Robert Grosseteste, one of the fathers of the Christian version of Aristotle's philosophy, admits to the theoretical construction of empty space, but stops right there. According to his definition, a vacuum is nothing but a mathematical space. Whatever is introduced into this vacuum will not replace another something. In this sense, a vacuum clearly cannot be compared with water or air. And for that very reason, a true vacuum cannot exist; a mathematician may well imagine a space infinite and empty, but reality knows no such space. Grosseteste has a fairly modern concept of this nonexistent empty space of infinite extent, symmetric under translation, and therefore identical wherever we might look. He contends that "if space thus imagined existed it would contain no local differences, especially if the space were infinite."

Roger Bacon, a student of his, joins his master in holding empty space impossible. He argues that light cannot penetrate a vacuum. Just as sound needs air to propagate, light needs a medium in which to move; and that medium fills all seemingly empty space. He held that light was "not a flow of body like water but a kind of pulse propagated from part to part." In this, light was analogous to sound, about which Bacon held realistic views, as did the Stoics.

Prior to the mechanization of our view of the world by Gassendi in the seventeenth century, the only Christian philosopher who thought that a true vacuum could actually exist in the real world was Nicholas of Autrecourt (active around A.D. 1350): "There exists space that contains no bodies, that is capable of accepting any and all bodies." He held that we have to choose the most likely of all different explanations, and that may or may not be the same as that chosen by Aristotle. Autrecourt believed that the concentration and dilution of matter are better explained in the system of the Greek atomists than in the one proposed by Aristotle. To him, concentration means that the particles—that is, atoms—of a body move closer together in an otherwise empty space, thus implying the existence both of atoms and of empty space, and that " 'dense' and 'rare' do not come to be except by the local motion of the parts." When it comes to motion and change, he also disagrees with Aristotle. He says that "motion in vacuum" is a simpler concept than "motion inside matter"—and that is the way we see it today.

HORROR VACUI

While agreeing with all the Scholastic philosophers on the impossibility of an extended truly empty space, a "macrovacuum," Nicholas of Autrecourt did believe that the real world might hold microvacua. No force in this world, he felt, is strong enough to increase the distances between the atoms of matter sufficiently to make room for a macrovacuum. By his time, the question of whether air is a bodily substance had long been answered: Air, in fact, *is a body*. Until Torricelli's experiments proved the existence of a space that was empty, at least to the extent of not holding any air, there had been one principle that governed all experiences and theories on the question of the macrovacuum—the so-called *horror vacui*. Nature, according to this principle, does not admit the vacuum's existence: Nature is afraid of a true void. In the first century A.D., Hero of Alexandria opined that one can generate a macrovacuum, but he did not manage to do so. His experiments left the guiding principle of the *horror vacui* intact.

Today we might smile at the attempt to explain nature in terms of this principle. But that would be a mistake: Just as the natural philosophers up to Galileo or Torricelli were convinced of the unconditional validity of the principle of the *horror vacui*, we hold to the belief that energy can neither originate nor vanish, that the sum of all energies is a constant. The law of the conservation of energy agrees with all our experiences. Its validity can be proved mathematically on the basis of the symmetry of the laws of nature under time displacement. This principle simply says that the laws of nature don't change with time. But we have no absolute proof that this law holds completely. Just like the symmetry laws, we have elevated the conservation law of energy to a guiding principle. Physicists believe in it so firmly that they will not even stop to look for the mistake in any purported plan to build a *perpetuum mobile*. There must be a flaw in every such plan; the law of energy conservation says so, and they will not give the matter any further thought.

This law of energy conservation, we believe, will hold also for processes that are governed by principles beyond our present understanding—in analogy to our scientific ancestors' holding on to the idea of the *horror vacui* up to Torricelli's experiment. The great physicist Wolfgang Pauli, who received the Nobel Prize in 1945, went as far as postulating the existence of a particle that had never been seen, just in order to be able to hold on to the law of energy conservation when this law appeared to be violated in certain radioactive decays. His 1930 conclusion, originally opposed by one of the guiding lights of quantum mechanics, Niels Bohr of Copenhagen, was triumphantly confirmed many years later, and the Nobel Prize for the experimental discovery of the conjectured particle called a *neutrino* was not conferred before 1995.

Still, this does not provide absolute proof that the law of energy conservation will hold up to every challenge; it is correct to the limits of our knowledge. Clearly, it's a good idea to hold on to it. But let us not forget that, quite similarly, it was reasonable to believe in the principle of the *horror vacui* until Torricelli's experiment proved otherwise. The understanding of the effects of external air pressure, previously mistaken for the evidence in favor of the *horror vacui*, did not precede Torricelli's experiment; but once the experiment was there, it followed easily.

CHAPTER 3

PROBLEMS WITH
NOTHINGNESS:

•

HOW TO MAKE IT A
PHYSICAL REALITY

•

IN HIS HIGHLY INFLUENTIAL TREATISE *DE MAGNETE,* WHICH APPEARED in A.D. 1600, the British medical doctor and natural scientist William Gilbert correctly describes Earth as a giant magnet with its north and south poles rotating about its own axis. According to Gilbert, Earth's magnetism keeps matter on the surface of Earth from being ejected into space; this also holds for Earth's atmosphere. Consequently, the space between the planets cannot contain air.

EFFECTS OF EMPTY SPACE

Is there a vacuum between the planets? Along with Gilbert, who ascribes electrical phenomena to what he calls "effluvium"—a sort of ether—there were many others who, around A.D. 1600, directed their curiosity at empty space. Galileo (1564–1642) is one of the first, after Hero, to deem it possible that a vacuum could be produced. He believes, however, that nature does oppose, in a quantifiable way, the existence of a vacuum. He has learned that suction pumps will not raise water higher than about 10 m (see fig. 27). He knows that two flat plates lying on top of one another will stick together. He shares the viewpoint of his contemporaries who hold that those parallel plates will not separate because, if they did, a vacuum would have to emerge between them before they part. It would be filled by air, which streams from the edges toward the center, over a certain time interval. And since nature abhors the vacuum, it is impossible—or, as Galileo says, very hard—to pull the plates apart.

Considering the famous hydra experiment of Empedocles, Galileo compares

Figure 27 A suction pump is unable to lift water higher than 10 m. Therefore, in order to drain a deep well, several pumps have to be used in a series.

a water column suspended inside a tube (fig. 21c) to a dangling wire. Both of them have to carry their own weight. Wires will tear when they are long enough (and therefore heavy enough) so that their tensile strength is no longer sufficient to withstand their weight. But what is this tensile strength due to? According to Galileo, there are two answers, only one of which is related to the vacuum. Let us imagine a wire as a stack of a great many small circular plates; they hold together because if they did not, a vacuum would have to appear between them. The strength of their resistance, according to Galileo, will amount to a given magnitude.

Similarly, he assumes that the columns of water inside the closed tubes of suction pumps hold together by these very forces. He devised an experiment (see fig. 28) in order to measure this effect, but it was probably never performed. Still, the resistance with which nature opposes the formation of a vacuum can be estimated in terms of the weight of the longest water column it will permit; this amounts to about 10 m. Galileo had the perspicacity to ask himself, and his student Vincenzo Viviani, how tall a column of mercury—which is a good deal heavier than water—a suction pump could produce.

Figure 28 This is the apparatus Galileo proposed to investigate the cohesive forces of a water column. The corresponding experiment, which is equivalent to Torricelli's, was probably never performed.

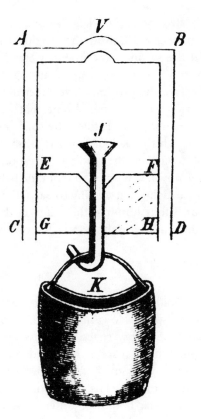

Galileo was the first to suggest quantitative experiments on the *horror vacui*. He stipulates that this principle exists in nature to a certain extent—but to *what* extent? What exactly is the force, if any, that makes the column of water tear? If that column were to resist the strongest forces the experiment can realize, the principle of the *horror vacui* would be found to hold at least to that quantifiable extent.

THE WEIGHT OF AIR

Galileo proves experimentally that, just like a rock, air has weight. To prove that, he determines the weight of an open bottle, then heats the bottle before weighing it again. It turns out that the bottle has lost some weight. It was generally known that air escapes out of an open bottle when it is heated. Galileo's conclusion was, therefore, that air does have weight and does not tend to rise—because a smaller amount of air will produce less upward momentum. In his *Dialogue Concerning*

Two New Sciences, he lets Salviati, his stand-in, describe two experiments to determine quantitatively the weight of a given amount of air.

In the second experiment, water is being pressed into a balloon "without permitting the air to escape, causing it to compress." After we have added as much water as possible, "we determine the weight of the balloon on a scale. Next, we pierce the membrane that kept the balloon airtight, permitting it to release as much air as filled the space now taken up by water. Subsequently, we again weigh the balloon, only to observe a decrease of the reading on the scale. The difference in the reading corresponds to the weight of the air that filled the volume now taken up by water in the balloon." The result of the other experiment by which Galileo, according to Salviati, measured the weight of air—determining that air is four hundred times lighter than water—differs from the correct value of eight hundred only by a factor of two.

FREE FALL IN A VACUUM

Galileo holds on to the belief that nature is opposed to the formation of an evacuated space. He is not worried about the philosophical implications of the existence of empty space. The less air a cavity contains, the closer it is to a vacuum. Empty space is not fundamentally impossible; it is just very hard to realize. The vacuum that he is not experimentally able to generate is of great importance to him: His theory of free fall, after all, makes specific predictions about how fast bodies of different weight will fall in it. Galileo thinks that a feather and a lead ball fall at different rates in air simply because of the resistance of the air; in a vacuum, they would drop at equal rates. That is what Aristotle had thought—with the one difference that Aristotle considered this result absurd and concluded that there could not be a vacuum.

GALILEO EXTRAPOLATES

Galileo was not able to verify his ideas as easily as we can today (see fig. 29). Because of his inability to create a vacuum, he was dependent on the method of extrapolation. Today's physicists use this when investigating the properties of empty space by means of probes that they make successively smaller (see chapter 2). Let's listen to Galileo, as he has Salviati say that

> *we want to study the motion of very different bodies in a nonresistant medium, such that all differences between two cases will be due to those bodies themselves. Given that we could only demonstrate that motion in a space that contains neither air nor any other matter, however diluted and nonresisting, and also given that we cannot possibly create such a space, let us proceed as follows: We will examine what happens to motion in a*

Figure 29 Galileo was the first to ascertain that all objects fall at an equal rate in evacuated space. This was experimentally proved by Robert Boyle, who gleaned the design of air pumps from Otto von Guericke. The first experimental investigation of this question was done by J. T. Desaguliers, on October 24, 1717, in Newton's presence. He worked at the London Royal Society, which reported in its journal:

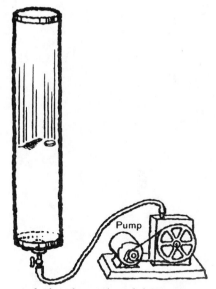

> *Mr. Desaguliers shew'd the experiment of letting fall a bitt of Paper and a Guinea from the height of about 7 foot in a vacuum he had contrived with four glasses set over one another, the junctures being lined with Leather liquored with Oyle so as to exclude the Air with great exactness. It was found that the paper fell very nearly with the same Velocity as the guinea so that it was concluded that if so great a Capacity could have been perfectly exhausted, and the Vacuum preserv'd, there would have been no difference in the time of their fall.*

Today, the demonstration comparing the drop of a feather and of a coin has become routine in high school. More recently, precision experiments have tried hard to find deviations from the velocities of different materials dropping in an evacuated volume. None of them has succeeded.

> *highly diluted and less resistant medium in comparison to less diluted and, consequently, more resistant medium. Now suppose that different bodies actually move less and less differently from each other as we make the media less and less resistant, and that the decreasing differences are independent of the nature of the different bodies. Then we should be confident to include that in the most diluted medium those differences become minimal—and that they will vanish in an actual vacuum.*

Galileo's vacuum is an actual vacuum in the real world, not merely the theoretical empty space of the Scholastics. In Galileo's day, as long as no claim of reality was made, anything whatever—empty space, a moving Earth—could be discussed with the blessings of the church. But to claim that these things were real was quite another matter.

REALIZING A VACUUM

Galileo was able to show, through experiments with wooden and lead balls that he rolled down an inclined plane, that bodies will fall at approximately the same rate, independent of their weight and of the material they are made of.

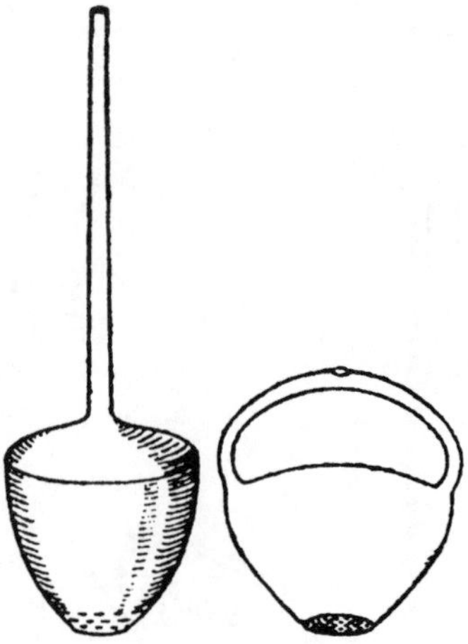

Figure 30 Two water lifts, or clepsydrae, as Empedocles knew them.

But the experiments he undertook to clarify the nature of the vacuum are less important than the one that he suggested to his pupil Torricelli. Following Galileo's guidance, Torricelli was the first to succeed in actually evacuating space, in 1644. His experiment made it impossible to hold on to the unconditional principle of the *horror vacui*, which was not supposed to allow any exceptions. Anybody who wanted to maintain it would have to deny that a space devoid of air is a vacuum. This, as we know, is in fact true, but in a way that is far beyond any possible experimental verification or theoretical understanding at the time of Galileo.

The Torricelli experiment initiated a new era in the experimental and theoretical research on the vacuum. No natural scientist who wished to be taken seriously could ignore the experimental results that were produced from 1644 on. In order to characterize their great significance, I will go back and describe some of the preceding experiments and chance observations that the *horror vacui* principle was in a position to explain.

VACUUM EXPERIMENTS PRIOR TO TORRICELLI

First, let's go back to Empedocles' hydra experiment, which actually was based on a chance observation. It has been described in the previous chapter. For everybody who knows that air is a substance such that an air-filled space cannot also accommodate water, the interest of this experiment is limited to the fact that water will not gush out of the container as long as it is shut at the top (see fig. 21). So why is water not following its natural inclination (to use Aristotle's reasoning) to flow downward? Aristotle's reason was that this would create a vacuum, while nature abhors it. When nature's choice lies between the evil of an emerging vacuum and some other evil, the abhorrence of the vacuum will prevail.

I repeat that, in effect, the pressure of the outside air is responsible for all effects that have been ascribed to the *horror vacui*. I will discuss details in their correct historical context. Modern scientists accept the *horror vacui* as a guiding principle—though not the mechanisms that were invoked to explain how nature chooses to follow that principle. One unacceptable argument is this: Nature chooses the easiest available method to prevent the appearance of a vacuum. To avoid a vacuum in a drinking straw, a column of water rises, quite contrary to what nature would normally prescribe; similarly, water will remain in the hydra instead of flowing out its open bottom just to avoid the creation of a vacuum on top; in both cases, nature chooses the easiest way—as opposed to more violent means, like breaking the straw or the hydra vessel.

Whatever else might happen, the principle of *horror vacui* will absolutely and under all circumstances hold up; the strongest possible pipe would break rather than yield to forces that otherwise would create empty space. This absolute principle might remind us of similar principles of present-day physics such as the principle of conservation of energy. There is an important difference, however: Conservation of energy was postulated not only to summarize experimental observations of energy conservation but also to better explain a host of processes to such an extent that we know *how* energy conservation comes about. There was nothing of the kind in case of *horror vacui:* All mechanisms that were invoked to explain how nature manages to avoid a vacuum were invented for this purpose alone. They had no other scientific basis. The prime postulate, inherited from Aristotle, was this: All space in nature is filled to the brim with something, always and everywhere. After this first principle of "nature in general" came lesser ones invoked by "nature such as it is." The idea was that "nature in general" will not admit the emergence of a vacuum. To respect this principle, the specific tendency of water to flow downward cannot assert itself. A vacuum, after all, will not transmit any force. Thus, if a (macro)vacuum existed in nature, the world would break up into distinct, unconnected parts: The vacuum would not permit the transmission of dynamic forces from "outside" to "inside"—or, in Aristotle's scenario, from the immobile mover to the moving stars in the skies.

When water freezes, its volume increases; that's why a closed vessel of water will be destroyed in the freezing process. This phenomenon had been observed, but was interpreted, ever since Aristotle, incorrectly. The assumption was that water decreased its volume when changing to ice. If it did so in a closed container that it had completely filled in the liquid state, a vacuum would ensue unless nature came to the rescue by shattering the container. Nature could also opt to keep the water from turning to ice—but that would have been "costlier"—that is, harder to achieve.

Unlike ice, air will expand when heated. How can it be that a continuum

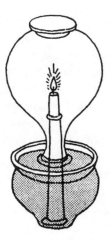

Figure 31 The burning candle in a closed container is one of the classic experiments on the vacuum. It was known even in antiquity.

expands? Let's take a look at artificial leeches, "cupping glasses," which are used to suck blood from animal bodies. When heated and applied to a body, the cupping glass will draw blood up as it cools. If it didn't do so, a vacuum would be generated. The cup does its sucking to prevent this from happening.

Similarly, we might use a bellows as a vacuum pump, only to encounter the now familiar problem in a more spectacular form. The Scholastics take to it wholeheartedly. For example, the fourteenth-century French philosopher Jean Buridan declared that when the opening of a compressed bellows is sealed, "we could never separate their surfaces. . . . Not even twenty horses could do it if ten were to pull on one side and ten on the other." But once we separate the parallel enclosures of the bellows, *another body* will fill the space between them. Buridan's gedankenexperiment reminds us of Otto von Guericke's famous demonstration of the effects of air pressure using two sets of eight horses at the Reichstag (the Imperial German Diet) in Regensburg in 1653–1654.

Many of the early vacuum experiments are based on two phenomena: the expansion of air when it is heated, and the fact that gases generated as a candle burns oxygen occupy less space at room temperature than the oxygen lost.

In the second century B.C., Philo of Byzantium described an experiment involving a burning candle—an experiment that would be discussed over and over in subsequent times (see fig. 31): Assume we have a burning candle standing in a bowl of water. Place an empty cup over it with its mouth immersed in the water. The water will rise inside the cup until the candle finally stops burning.

Why does the water rise? Philo believed—almost correctly—that the flame destroys the air and the water rises in order to replace it. Were that not so, a vacuum would form. A mechanism that nature might have used to make the

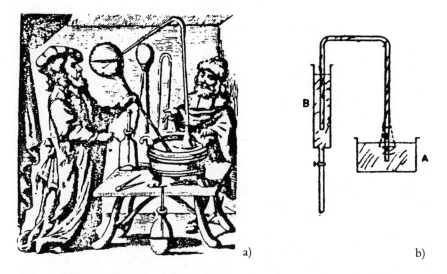

a)   b)

Figure 32 The two clueless gentlemen in (a) are attempting to do Philon's experiment, schematically shown in (b). The containers A and B and the tube connecting them are initially filled with water; the valve is closed. When it is opened, water flows through the tube from container A to container B. A simpler version of the same experiment is the water lift in figure 33a. In Philon's original experiment, only the horn-shaped connecting tube (drawn in dashed lines) was transparent. The two gentlemen in our picture use glass tubes throughout, but hold on to the horn.

water rise is the tendency of all substances to cling together. One might think that the surfaces of both, water and air, hold fast as though connected with magic glue.

Philo knew that air expands upon heating. After all, his experiment showed that an "empty cup," filled only with air, when cooling—open top down—inside a water-filled bowl, will have water rise inside it. A vessel such as Philo might have used for a more detailed experiment can be seen in figure 32a. The clueless gentlemen in the illustration appear to be attempting an experiment as described by Philo (see fig. 32b).

Just like Hero of Alexandria three centuries later, Philo was an engineer as well as a scientist. Both summarized in their writings what was known in their time about pneumatics and hydraulics—the study of air and fluids, respectively. As it was, Philo could have used his experimental setups for quantitative experimental investigation of issues vitally related to the vacuum. It would certainly appear to us today that Philo should have studied the dependence of the amount of water rising in the bottle of figure 32a on the ambient temperature. But that would have put him ahead of his time.

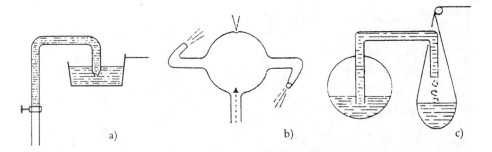

Figure 33 Both the vessel and the tube of the liquid lift of figure 33a are filled with water. When the valve is opened, water from the vessel escapes through the tube. The ancient explanation starts from the notion that the water column cannot rupture. If it did, there would be a vacuum, which is unthinkable because of the *horror vacui*. In truth, the water column cannot break because the ambient air pressure supports it on both ends. A water column that cannot break behaves like a rope: If both hang down from one point of support, the longer part will pull the shorter part after it. The principle of the *horror vacui* says that the water column cannot break under any circumstances; the ambient pressure forbids such a break only as long as the columns are not too long. By "too long," we mean vertical lengths of 10 m or more. It has been experimentally proved by Pascal and others that it is not a universal principle at the bottom of these phenomena; it is just the variable pressure of air. The so-called Hero's ball in figure 33b is driven by steam, just as a lawn sprinkler is driven by water. When we heat the vessel on the left in figure 33c, the air expands and pushes the water out. The vessel on the right, into which it flows, gains weight and is therefore able to activate various mechanical tasks.

EARLY VACUUM TECHNOLOGY

The lifting device for liquids, as we show it in figure 33a, is a simplified version of the experiment illustrated in figure 32b. "Hero's ball" (see fig. 33b) will rotate when steam flows through it. The third and best-known device in this category is the one sketched in figure 33c, also due to Hero and variously used as a door opener, as in figure 34: As we heat the sphere in figure 33c, the air inside it will expand and push water into the receptacle on its right. The latter, having gained weight, actuates a mechanism like that of figure 34, by pulling on the string from which it is suspended. This device opened the doors of the temple, once offerings to the gods had been deposited in a bowl and set on fire. Figure 35 shows a baroque toy that functions in the same way: Water is being pushed out of a heated volume by expanding air, only to appear as a fountain emanating from the mouth of the crocodile of figure 35.

MICRO- AND MACROVACUA

The Dutch historian of science E. J. Dijksterhuis notes: "We take it for granted that technical feats that can be understood in terms of specific scientific insights

Figure 34 The fire on the altar to the right is capable of opening the temple's doors by making use of Hero's mechanism (figure 33c).

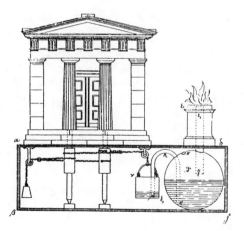

Figure 35 Water stemming from steam produced over pressure in another volume, as in figure 33c, feeds this toy baroque fountain.

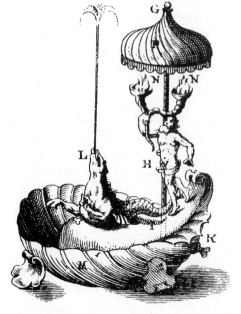

imply that these insights have actually been held." But quite to the contrary, Hero had no notion that the pressure of the ambient air was partly responsible for the functioning of his devices, especially of his lifting mechanism for fluids. Along with Philo and the atomists, he was of the opinion that air is made up of particles embedded in empty space. The space available to these particles can be increased or decreased—but both of these processes are "unnatural" and will therefore meet with resistance. Hero observes: "If a light vessel with a narrow mouth be taken and applied to the lips, and the air be sucked out and discharged, the vessel will be suspended from the lips, the vacuum drawing the flesh towards it that the exhausted space may be filled." In his opinion, nature's resistance against dilution or evacuation, later elevated to an absolute principle in terms of the *horror vacui*, is the origin of this and related effects; he was not aware of the true cause—the pressure of ambient air.

He also imagined a detailed arrangement of the air particles in space: "The particles of the air are in contact with each other, yet they do not fit closely in every part, but void spaces are left between them, as in the sands on the seashore: the grains of sand must be imagined to correspond to the particles of air, and the air between the grains of sand to the void spaces between the particles of air."

Just like Leucippus some five hundred years earlier, Hero interprets concentration and dilution as external signs of the clumping together and flying apart of atoms. They can occur only if there are microvacua between the atoms. The same applies to the mixing of substances. His image might be due to Leucippus: "When wine is poured into water, it is seen to spread itself through every part of the water, which it would not do if there were no vacua in the water." Hero was also correct in his observation that empty space will not offer resistance to the motion of bodies: "Bodies will have a rapid motion through a vacuum where there is nothing to obstruct or repel them, until they are in contact." Hero incorrectly invokes the unopposed mobility of air particles in empty space to explain the resistance of air against dilution. He claims correctly that under pressure the air particles move closer together; if the pressure were removed, they would return to their original position.

MECHANISMS AND . . .

According to Aristotle, there has to be room somewhere else for objects supplanted by other objects; how else could there be motion? If the upper hole of the water-filled hydra is open, the water can flow out, since the air supplanting it can then set other air in motion and allow the space vacated by the water to be filled. If the hole is closed, the water will remain in the hydra; the previously removed air has no way to take up the space that the water would have to vacate. In other words, the ambient air has the effect of stopping up the lower openings of the hydra; in effect, the water is resting on the outside air. This air has nowhere else to go, and therefore will not permit the water to escape from the hydra.

. . . ABSTRACTIONS

The Scholastics tested their acumen within limits that we are not familiar with. Nothingness, empty space, vacuum—these concepts could be characterized only by negatives. It was therefore impossible according to generally recognized doctrine that a vacuum could initiate any action; a vacuum cannot show up either as cause or as effect. But the *horror vacui* describes the *avoidance of a vacuum* as a goal: Water climbs upward so that no vacuum will form. The necessity of maintaining the world in a state of being filled with matter supplies the final positive reason for all the processes by which nature manages to avoid creation of a vacuum. To this end, surfaces cling together—such as the surfaces of water and air in a drinking straw. This may even serve as a causal mechanism for the lifting of water by air inside the straw.

GASSENDI AND MECHANISTIC PHILOSOPHY

The way the Scholastics saw and theorized about nature, it was nothing but a set of abstract phenomena, subject to the physics and logic defined by Aristotle; this view had little in common with actual nature. Their reasoning finally collapsed in the seventeenth century, when mechanistic philosophy took over.

One of the most important intellectual pioneers of this triumphant new philosophy was Pierre Gassendi. He is often recognized, together with Thomas Aquinas, as pivotal for the acceptance of Greek natural philosophy by the church. What Thomas Aquinas had done for the teachings of Aristotle, Gassendi did for the teachings of the atomists Leucippus, Democritus, and their followers. Just as Aristotle had held that the world had always been there and was not a part of creation, the atomists attributed these same properties to the atoms. In denying these particular features—which are not, after all, subject to experimental verification—Aquinas and Gassendi managed to revive these teachings of ancient philosophy and make them acceptable to the church. For his part, Gassendi argued:

> We must give up the view that atoms exist from eternity, to eternity; and in numbers infinite, even while they can assume any form whatever. We can admit that they make up the materia prima *that God created in the beginning, in finite amounts. If we improve our thinking in this way, it can stand up to critical discussion just as well as that of Aristotle and others who originally took the* materia prima *to be eternal and not created, infinite and not limited in number or space.*

The terms *eternal, not created,* and *infinite* raise red flags for the priest Gassendi. For one thing, Scholastic teaching reserves the adjective *infinite* for God alone. Whatever quantity might be called infinite will have to be seen in its relation to God. He effortlessly refutes the notion that an infinity of atoms, each of finite size, might be needed to build up a finite world. He does not resort to numbers; instead, he amusingly limits himself to counting the digits of the numbers necessary—in the process anticipating notation in powers of ten.

Gassendi accepts the existence of empty space—both of microvacua between the atoms of our world and the limitless vacuum in which our finite world (which the Stoics bequeathed to him) is embedded. The former he needs so that the atoms can move. It also forms a necessary part of his "corpuscular" philosophy, which holds that there are only two realities—space and material bodies. These transmit mechanical effects by moving and coming into direct contact with other bodies. The motion of the atoms is mandated by God's will instead of being arbitrary, as the atomists had assumed.

Gassendi knows about Torricelli's experiment. Although he knows all the tricks of the Scholastic trade, he recognizes what that experiment implies: It means that a macrovacuum can exist in the world, and that humans are able to create it. That implies empty space may exist, although only passively. As Gassendi describes it, "space cannot act or suffer anything to happen to it, but merely lacks resistance . . . , which allows other things to occupy it or pass through it." Its passivity distinguishes Gassendi's space from Newton's absolute space. Newton's space does act, but it cannot be acted upon. This is what Einstein will object to in Newton's interpretation of space, some two hundred fifty years later.

Gassendi's scientific opinions clear the path for the insights of Guericke, Pascal, and Newton: "Place and time do not depend upon bodies . . . for even if there were no bodies, there would still remain both an unchanging place and an evolving time." And "Space and time must be considered real things, or actual entities, for . . . they do not depend on the mind . . . since space endures steadfastly and time flows on whether the mind thinks of them or not." While the world, made up of atoms, was created, it is surrounded by space infinite and eternal, and that space beyond the confines of our world is truly empty.

Gassendi's conclusions thus conflict with his own Scholastic tradition. How can something that was not created exist? According to Gassendi, this is so because space can be characterized only negatively—just as the Scholastics had said. Even the real existence of space differs from the existence of created objects; space itself is independent of God, whose creations are characterized by positive properties. And yet Gassendi's space disputes God's singular property of existing uncreated for all eternity, past and future, and of being infinitely extended. Even though Gassendi's space has all the properties that he disputes in the case of the atoms, he still manages to keep God unique—and thus to make his argument acceptable to the church. He brings together Scholastic arguments and scientific notions; and we can see him at the crossroads of the scientific revolution that began during his lifetime—which he cannot fully accept.

TORRICELLI'S EXPERIMENT

Evangelista Torricelli's experiment in 1644 showed for the first time that we can, in fact, evacuate a well-defined space (see fig. 2): In the tube above the mercury mirror, the experiment produces a space without air. A heated discussion soon

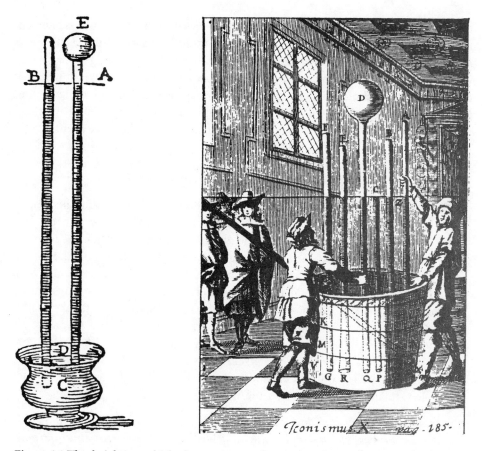

Figure 36 The height to which the mercury column rises in Torricelli's experiment is independent of the vacuum volume (36a). It also doesn't depend on shape, length, or inclination of this column (36b).

ensued about the origin of this "empty space," and about its possible identity as a vacuum. Let us, for the time being, stick to this experiment. It was quite a modern notion to invent and carry out experiments in order to answer theoretical questions. First, it was noted that the level of mercury inside the tube was independent of the volume of the airless space enclosed (see fig. 36a). Consequently, the mercury cannot have been suspended from above as though by a thread, due to the *horror vacui;* otherwise the increased vacuum would have to exert an increased force. Second, it was possible to fill the entire evacuated space with water. This led some to the conjecture that the space was not only evacuated but empty—a true vacuum. Others argued that fine substances such as the ether penetrated the evacuated space through pores in the walls of the tube and filled it; the substance would be pushed out again through the pores as the water entered. Third, it was seen that if the tube is inclined, the level of the fluid inside remains the same: This implies that mercury, as the water before, fills some of the airless space (see fig. 36b).

Interpreted in an unbiased fashion, these experiments show two things: that a "force from without" acts to have the mercury rise to some given level, and that the space above that level is at least approximately empty. Both of these inferences are true—although we may have to qualify the last statement a bit.

AIR PRESSURE AND THE *HORROR VACUI*

Without being able to prove it, Torricelli was convinced that the aforementioned force from without is exerted by the weight of the air. This was proved four years later by Blaise Pascal: It is only the weight of air that prevents the forming of a space devoid of air. In fact, according to Torricelli's calculations, the weight of air should create an even greater resistance to the formation of a vacuum than what was actually measured in the 10-m water column or a 760-mm mercury column. We need not go into the details of what was wrong with Torricelli's calculations. His lack of knowledge of the atmosphere, and the incorrect air density inferred by Galileo, were at the bottom of his error, and he found that out by himself. In a letter to Michelangelo Ricci dated June 11, 1644—which we might call Torricelli's only publication on the controversial topic of the vacuum—he explains his thoughts on the matter in some detail:

Many people have said that it is impossible to create a vacuum; others think it must be possible, but only with difficulty, and after overcoming some natural resistance. I don't know whether anybody maintains that it can be done easily, without having to overcome any natural resistance. My argument has been the following: If there is somebody who finds an obvious reason for the resistance against the production of a vacuum, then it doesn't make sense to make the vacuum the cause for these effects. They obviously must depend on external circumstances. I did some simple calculations, and found that the reason I have mentioned, namely the weight of air, offers more resistance to the creation of a vacuum than what we actually measure. I say so to make sure that certain philosophers, who feel obliged to agree that it is the weight of air that hinders the formation of a vacuum, acknowledge the importance of that weight; this may keep them from insisting on "nature herself" resisting the formation of a vacuum. We exist on the bottom of an ocean composed of the element air; beyond doubt, that air does possess weight. In fact, on the surface of Earth, air weighs about four hundred times less than water. The authors of "Crespusculi" report that visible, steam-loaded air rises to a height of some fifty or fifty-four miles. I am not convinced this is quite true. I was able to show that if it were, it would be

Figure 37 A schematic view of Earth's atmosphere: Its overall height is hard to define. At an elevation of 250 km, its pressure is a factor 10^{10} smaller than at the sea level. This factor grows to 10^{12} at 400 km elevation, and to 10^{19} in interstellar space. The concentration of matter close to Earth's surface is enormous when compared with that in the universe. If we started a rocket from Earth and sent it to the ends of the observable cosmos, it would traverse more matter in the first 10 km of its journey than over the entire remaining distance. The air pressure on Mount Everest, at an elevation of about 9 km, has already dropped by a factor of 3 with respect to sea level. Essentially all of our air is confined to a shell around Earth that has a thickness of only one-half of a thousandth of its radius.

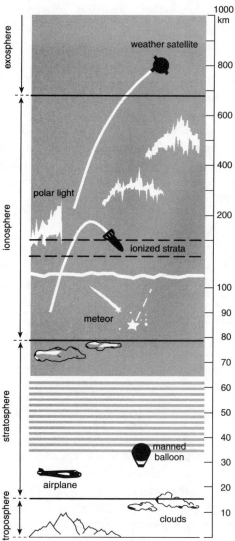

even harder to create a vacuum than what the evidence warrants. Someone might refer to the argument that the weight of air such as determined by Galileo is correct for the altitudes commonly inhabited by man and animals, but not high above the mountain peaks; up there, air is extremely pure and much lighter than the four hundredth part of the weight of water.

Figure 37 gives an image of Earth's atmosphere. As Torricelli suspected, the density decreases with altitude. Its weight at sea level under normal weather conditions—that is, halfway between what we call high and low atmospheric pressures—is that of a mercury column 76 cm high. This is equivalent to having roughly 1 kg's mass rest on each square cm of Earth. We know that at high atmospheric pressure the "weight" of air is higher, and the mercury column rises. When there is an atmospheric low, the column will sink. In fact, the weight of air is at the bottom of all that has been ascribed to the *horror vacui*. If the air were massless, the levels of mercury in the glass tube and in the bowl would be the same. The weight of the air alone keeps the water from flowing out of the hydra when the upper opening is shut. Actually, we should use the term *pressure* instead of *weight*. This "weight," after all, presses upward, which makes no sense if taken literally. Air pressure, however, is a *consequence* of the weight of air; it acts in all directions—upward, downward, sideways. The same applies to pressure in fluids. The discovery and clarification of these facts constitutes one of Pascal's greatest contributions.

Humans and other animals retain their form due to a complicated system of pressure and counterpressure. If there were no atmospheric pressure, our blood would escape through the pores of our circulatory system and our skin. This is how cupping glasses function: They relieve sections of the skin from atmospheric pressure; this permits blood to escape by the action of its own pressure. Along with Empedocles, we can compare the blood in the veins with the water in the hydra. That water would similarly flow outward, were it not for the ambient pressure of the air.

LIQUID LEVELS, LIQUID WEIGHT

Torricelli was unable to make any absolute prediction about the height of the water or mercury columns in his experiment; still, his idea of air pressure as the cause of the *horror vacui* yielded the correct proportion of the relative heights of the columns. Clearly, the "weight of air" will support the same weight of the fluid column in the cases of both water and mercury. We know that mercury is 13.5 times heavier than water. As a consequence, the liquid levels in the water and mercury tubes should differ by this very factor. If the water level measures 10 m, mercury should rise to 10 m divided by 13.5, or about 74 cm—and that's pretty close to the experimental value of about 76 cm. In addition, the water level was not measured precisely, and was not measured simultaneously with the mercury level, so that the agreement must be seen as very good indeed. Torricelli's theory of air pressure passed a quantitative test with flying colors! It is such quantitative tests that mark the difference between success and failure of physical theories.

HOW EMPTY IS TORRICELLI'S VOID?

As Torricelli's experiments became known, discussions arose immediately on whether the space above the mercury column was actually empty or filled with ether—but to no avail. None of the experiments that could be performed at that time would have discovered the vacuum effects that are the ultimate subject of this book; they were not sensitive enough.

Shortly after 1644, Blaise Pascal learned about Torricelli's experiments. He repeated them and improved on them. From the outset, he carefully approached the question about what actually causes the effects that had been ascribed to the *horror vacui.* Some of the experiments he carried out in his investigation of this question were spectacular and became quite famous. Even though Pascal knew ahead of time what the results of his later experiments would have to be—in fact, he was so certain about some that he did not even bother to perform them—there was a quantitative result in one of his first experiments that he probably did not expect.

WATER AND WINE

This experiment was a reenactment of Torricelli's, but the mercury was replaced with water and wine. To make it work, Pascal had to utilize tubes at least 13 m long. He carried out the experiment in front of an audience of five hundred, with the help of ship masts that could be tilted. He needed the latter in order to raise the tubes filled with their liquid to a vertical position. He then invited the audience to guess whether the wine or the water level would wind up being lower. Their answer was that the wine level winds up lower: Since a vacuum was not assumed to form, the vapors of the fluids must be able to fill the seemingly empty space, such that the size of this space is determined by their ability to do so. Because wine is more volatile than water, the vapor of the wine will be able to fill more space above the fluid's surface, and the level of the wine is going to sink more than that of the water. No mention is made in this argument of the weight of the fluid—the more volatile it is, the more space its vapors will be able to fill and the lower it will sink.

Pascal knew that there would be no fluid column without the pressure of the ambient atmosphere. If this pressure alone determined the height of the column, the lighter-weight wine would wind up with a taller column in the tube than the heavier water, just as a column of water would wind up higher than that of mercury. The proportion of the heights of the columns of wine and water would then be the same as that of their respective weights—again as with water and mercury. Undeniably the fluids would also evaporate into the vacuum—but would that be observable? To what extent was the space above the fluid level really the universal vacuum that doesn't depend on the fluid?

Before the experiment showed clearly that the column of the "lighter" wine didn't sink as much as that of the "heavier" water, Pascal had not decided whether the seemingly empty space above the fluid column was actually empty. Only after the experiment he notices: "Above the mercury, above the wine, above the water, there is really a space devoid of air." Today we would say that, to explain the reported result of the experiment, we can neglect the vapor pressure of the fluids in comparison with the ambient atmospheric pressure.

TWO QUESTIONS

This experiment convinced Pascal that all the phenomena ascribed to the *horror vacui* could actually be explained as effects of the ambient air pressure. He also believed that empty space was created in the Torricelli experiment. Barring minor details, there are two questions that in view of this experiment need to be answered with either a clear *yes* or a *no:* First, is the space above the liquid in the tube actually empty? Second, is it the weight of the ambient air that sets the level of the liquid inside the tube?

All four possible combinations of yes or no to the above two questions have, in fact, been advanced. Pascal's answer was *yes* and *yes:* The apparently empty space is truly empty, and the level of the fluid column is due to the weight of the ambient air alone. Descartes's answers, *no* and *yes,* agree with Pascal's with respect to the outer pressure and the height of the fluid column. But Descartes did not believe there can be empty space. To him, "empty space is as absurd as happiness without a sentient being who is happy." *No* and *no* were the answers advocated by those physicists still beholden to Scholastic teaching at the time. *Yes* and *no,* finally, in diametric opposition to Descartes, was the judgment of his adversary Gilles Personne de Roberval, mathematician at the Collège de France.

THE WEIGHT OF THE VACUUM

Pascal dips the nozzle of a long syringe into mercury to pull the liquid metal upward. He knows, of course, that he can reach a liquid level of no more than 76 cm, just as suction pumps (see fig. 27) will not exceed a water level of 10 m. If he pulls on the plunger to raise the level, the mercury remains at the same level, but the apparently empty space above it increases. Galileo had weighed the air, and now Pascal weighs the vacuum. He finds it to have no weight; once a "vacuum" appears inside his syringe, its weight does not depend on the volume of that vacuum. When he pulls on the plunger while leaving the nozzle in the syringe, this does not change its weight (see fig. 38). If his balance were sufficiently precise, he would have to assign a *negative* weight. The weight as measured, after all, is the actual weight of the object in question minus the buoyancy of that object, which is identical with the weight of the air it replaces in the ambient atmosphere (see fig. 39). This buoyancy obviously increases with the volume. Thus when the vacuum inside the syringe increased in volume as Pascal pulled on the plunger, so did the syringe itself; with its greater volume, it replaces more air and therefore becomes lighter.

Figure 38 If the balance in our picture were more precise than it actually is, the right-hand arm would be at a higher level than the left-hand one: The piston on the right has been pulled out, and therefore displaces more of the ambient air than that on the left, leading to a stronger lift on the right. The vacuum below the lifted piston doesn't contribute to the weight measured on the balance. Had Pascal been able to measure this lift on his balance, his experiment would have forced his opponents to assign *negative* weight to the vacuum.

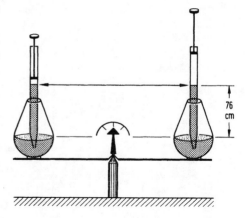

Figure 39 The real weight of an object in empty space is its weight measured in air plus that of the air it displaces. This is nothing but Archimedes' principle transferred from water to air: The buoyancy of an object in water—and in air—is equivalent to the weight of the water—or of the air—it displaces. A balloon filled with air will drift weightlessly. Filled with hydrogen, which is lighter than air, it will rise; filled with carbon dioxide, which is heavier than air, it will sink. All of the above is true only as long as we assume that the balloon itself has no weight and that it does not compress the gas it surrounds; neither of these assumptions is quantitatively correct.

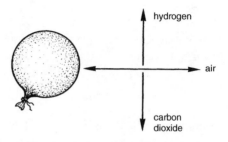

Whatever substance might fill Torricelli's vacuum, it has no weight; there is no experimentally detectable quantity associated with it. Two further experiments will, however, crown Pascal's achievement and serve to prove Torricelli correct. They verify that the pressure of the surrounding air is all that determines the level of the mercury column. As Pascal sees it, this pressure acts on the mercury in the bowl (see fig. 2); it is the same in all directions, and pushes the mercury up into the tube. But since the tube is closed on top, there is no ambient air to reciprocate the pressure there. As a result, the column will settle at a level determined by the condition that its weight equals that of the air pressure

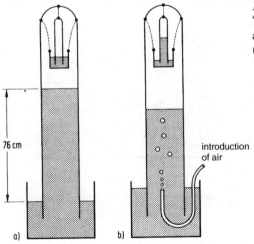

Figure 40 Pascal's experiment on the "vacuum inside a vacuum" (a) before admitting air into the outer vacuum, (b) after the introduction of air.

76 cm

introduction of air

a)

b)

multiplied by the area on which the column rests. The mercury column will not and cannot rise beyond that level—the space above it is empty.

VOID INSIDE A VOID

The first of Pascal's famous experiments deals with the "void inside a void." If it is the ambient air pressure in Torricelli's experiment that does not permit the mercury column to sink to the level of the mercury in the bowl, then the column will have to sink to the level of the liquid in the bowl when there is no air pressure in the surrounding space. Pascal needed to introduce a second Torricelli barometer into his first one. I will not go into the technical details of how he managed this task. I will just show the (expected) result in figure 40a: The small barometer placed inside the Torricelli vacuum indicates zero air pressure; the mercury levels are the same in the little bowl and in the tube protruding from it. As Pascal raised the level of the outer tube while leaving its bottom immersed in mercury, he increased the vacuum inside the outer barometer. That implies that the inner barometer still gives a reading of zero pressure. Pascal already knew that the height of the mercury level in the outer barometer would remain at 76 cm. And it did!

When Pascal let air bubbles rise into the vacuum, the first—the outer—mercury column sank and the second one rose (see fig. 40b). This is to be expected if we accept the air-pressure theory: The air inside the first vacuum exerts pressure on its surroundings—that is, onto the first mercury column and the second bowl. Pascal could also have added water instead of air into the first vacuum. This would have made a difference only because of its weight—the vapor pressure would not have made a difference: The water would have depressed the first mercury column while leaving the second one unchanged.

After air has penetrated the first vacuum, both mercury columns will show a difference if we lift the outer tube in its bowl: The outer one rises, the inner one sinks. While different from the previous case, this is still easily understood. If a given amount of air can occupy a larger space, its pressure will decrease. Altogether, this experiment not only shows that air pressure is responsible for the level of the mercury column but demonstrates that Torricelli's setup can also be used to *measure* the air pressure. In fact, it is a useful barometer.

In principle, the theory of the *horror vacui* can explain this and other experimental phenomena. Ever since Torricelli's experiment, however, the theory based on it has to take second place to the far simpler explanations in terms of air pressure. As more experimental details become known, it takes more and more imagination to use this theory in order to keep the *horror vacui* the culprit. It has never been refuted in a logical sense; it just became obsolete after 1644. Once it was shown to be deficient, nobody felt like taking the trouble to give it the final blow. Today we are content to say that it has been passed by.

DOES THE *HORROR VACUI* DEPEND ON ELEVATION?

Back to Pascal: He was amazingly serious about the possibility of having his experimental results also explained by the theory of the *horror vacui*. Maybe, he worried, nature's distaste for the vacuum was satisfied by the first, outer vacuum in figure 40b, so that there was no further effect on the inner vacuum. He may also have found the experiment on one void inside another too difficult, too involved—not sufficiently direct, and requiring too much sophisticated rationalization. He was much more inclined to dwell on the importance of his spectacular second experiment, performed in the year 1648, which proved that the fluid level measured in Torricelli's barometer depends on the elevation where it is placed. He recognized this result as a proof for the theory of air pressure.

Pascal asked his brother-in-law Florin Périer to have Torricelli's experiment performed at the lowest level of his hometown of Clermont-Ferrand, and at the same time also on the neighboring mountain, the Puy-de-Dôme, at 850 m elevation. If the theory of air pressure was correct, then the level of Torricelli's mercury column should decrease as the elevation of the experiment increases. It is, after all, only the weight of the air *above* the experiment that presses onto the mercury surface in the bowl; and that pressure decreases with elevation.

This conclusion—as innocent as it may be—is controversial: Does it not imply that the air shell has a finite thickness and that there is a vacuum outside, as already noted by William Gilbert? The Jesuit Noel, speaking for Descartes, opposed the opinion that there might be a vacuum beyond the skies; he argued

Figure 41 An illustration of the experiment that permitted Pascal's brother-in-law Pér-ier to measure atmospheric pressure on top of the Puy-de-Dôme. When mercury columns at the foot and top of a mountain were measured simultaneously, the col-umn at the bottom measured 8 cm more than the one on the top.

that, knowing how useless empty space is, we would be forced to believe in useless creations of God in nature.

Such arguments—be they scientific, political, or theological—keep losing significance. The experiment that was done both on the summit of Puy-de-Dôme and at its base on September 19, 1648 (see fig. 41) showed the mercury column to be 8 cm taller in the lower location.

This result (which marked the invention of the altimeter) fostered great excitement. Pascal foresaw this in the letter to his brother-in-law asking him to do the experiment: "This experiment will probably show that the weight and the pressure of air are the sole cause of the relative levels of the mercury. It is not due to the *horror vacui*—we know that there is much more air pressing down on the base of the mountain than on its summit. It makes no sense to argue that nature abhors the vacuum more by the base of the mountain than on its top." In a posthumously published description of his scientific investigations, he formulated the same argument as a rhetorical question: "Does nature abhor the vacuum more on top of a mountain than in the valley, and even more so in wet weather than in sunshine? Does she not detest it equally on a church steeple, in our living room, in the courtyard?" And he concluded: "Nature does nothing to avoid the vacuum; rather, the weight of the air masses is the true reason for all these phenomena which we have been ascribing to an imaginary cause."

OTTO VON GUERICKE AND THE FIRST VACUUM PUMPS

Simultaneously with Pascal, Otto von Guericke investigated vacuum problems in Germany. The two scientists did not know of each other. Both had a flair for spectacular demonstrations, but otherwise they differed greatly. Guericke was fascinated by experimental equipment; he was caught up in romantic speculations about the vacuum, almost glorifying it. But his results equaled those of Pascal, who arrived at them after detailed experimentation and a comprehensive analysis.

Otto von Guericke, the inventor of the pneumatic pump, was well aware of the Scholastics' discussions about the vacuum. Within their philosophical reference frame, he develops his own system. He differentiates between phenomena created and not created:

> Everything in existence is either created or not created; there is no third possibility. Whether created or not, whatever exists is not a nothing. The question "What was there prior to the creation of the world?" admits two valid answers: "Everything that was uncreated" or "the nothing." And since all things were created from nothing, they are naturally comprised in these two answers. This means everything has its place in "the nothing." And if it pleased God to annihilate all that he created, everything would revert to the nothing, to the "noncreated" that has been there from the beginning of the world. Besides our world, there is only the nothing; that nothing is everywhere. Empty space is called a nothing and so is a space that exists only in our imagination. In fact, empty space is assumed to be a nothing.

This is quite different from the strict, abstract logic of the Scholastics; there are poetic connotations that cannot be missed. Guericke is quite taken with the concepts of his nothing, of empty space: His nothing is

> more precious than gold; it is beyond growth and decay; it is more refreshing than the light of grace, more noble than the blood of kings; it has celestial splendor, higher than the stars, brighter than the flash of lightning; it is perfection itself, universally blessed. The nothing is full of wisdom. Wherever

it exists, there is no power left to kings. The nothing knows no misfortune.
As Job said, by the nothing the world is suspended.

Guericke is clearly fascinated by the sheer size of the universe. He sees countless worlds inhabiting it. In the tradition of the Scholastics, he identifies infinite space with God himself. To avoid shrinking God down to three-dimensional existence, he differentiates between two kinds of space—one is the three-dimensional space we observe around us; the other is divine and knows no dimension. These details are shared with Scholastic tradition. "Empty space" is, however, according to Guericke and against the Scholastic tradition, truly empty. It is not filled with ether of any kind. Space is a "container" that is capable of accepting objects in its volume.

It is not clear how exactly Guericke imagined these things; and his greatness is in no way based on his views of the world. His importance derives mostly from his conviction that an experiment can decide what is true in all this rapturous poetical speculation—and from his having actually done these decisive experiments:

> *Scholars whose conclusions are based on their reasoning, without recourse to experience, are unable to make definite statements about the makeup of this world. Human thought must be supported by experiment; if it is not, it will distance itself from the truth farther than the Sun is from Earth. Scholars have been arguing about the void for a long time—its existence or nonexistence and all its qualities. They have been fighting a stationary war, and . . . defended their preconceived notions [as] a soldier defends a fortress against the onslaught of the enemy. As far as I am concerned, I could no longer curb my burning desire to find the truth to this questionable "something" by initiating an experiment as soon as I found the time to do it. I have approached this in various ways and my effort was not wasted: I have been able to build some equipment to prove the existence of the much-maligned void.*

Guericke was convinced that removing the air from a container was enough to create a vacuum. And so he did just that—he pumped air from a suitable container. Of course, he held the ambient air pressure responsible for all the practical difficulties he encountered in realizing his simple goal. There were no serious objections to his *yes* and *yes* answers to the questions we framed with respect to Pascal's experiment. Quite the contrary: Everybody was fascinated with Guericke's experiments; scholars, princes, even the Kaiser showed great interest.

The air pumps Guericke built were expensive technical masterpieces; they were the seventeenth-century version of today's major research enterprises. Guericke was not only the mayor of the city of Magdeburg; he was also a brewer, and

Figure 42a The two valves of the fire extinguisher that Otto von Guericke transformed into an air pump.

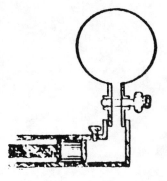

Figure 42b Guericke's ill-fated attempt to evacuate a barrel filled with water.

so the first vessel he pumped out was a beer barrel filled with water rather than air. By changing two valves (see fig. 42a) he transformed a fire extinguisher into a suction pump, attached it to the bottom of a beer barrel filled with water, and had two strong men pump on it (see fig. 42b). The attempt failed: The pump attachment broke and let air into the barrel.

In the second attempt, Guericke locks the barrel that he wants to evacuate into a second, larger barrel. He fills both with water. He pumps water out of the inner barrel, only to find water from the outer one seeping into the space he is freeing up. Undaunted, he replaces the beer barrel with a metal sphere. This time he forgoes the water and pumps directly on the air inside this sphere. This attempt

Figure 42c Guericke's successful attempt to pump the air out of a sphere made up of two hemispheres.

appears to work smoothly—but after initial success, his piston, which is supposed to pull the air out, can "barely be moved by two powerful helpers." Finally the metal sphere collapses—signaling that the pumping action has actually succeeded.

Now Guericke orders a stronger metal sphere to be built, with thicker walls. This one resists external air pressure (see fig. 42c) and can actually be pumped empty. Now Guericke has his vacuum pump and can start experimenting with empty space. And so he does, to his heart's content. On the technical side, he improves the pump. There has been a problem with the seals not being airtight. This he solves by immersing parts of the pump in water, which does not penetrate the seals as easily as air. The result of his efforts is the pump shown in figure 42d. In a further improvement that proved important for his later experiments, he succeeds in evacuating glass spheres. This permits him to look into the "empty space" and observe whatever happens there. The simple fact that he can look in there proves that light can penetrate empty space—an observation that Torricelli made before him on *his* empty space.

THE VACUUM/VOID, EXPERIMENTALLY

After pumping the air out of the glass sphere, Guericke closes the sphere with a valve and reopens this valve underwater. The water gushes forcefully into the empty sphere—and its motion is directed upward, contrary to its usual tendency. After eliminating a few errors in the procedure, he convinces himself that water

Figure 42d Guericke's air pump. He used water instead of air pressure gaskets.

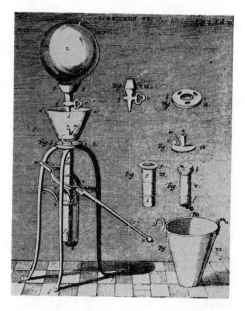

fills the empty sphere to the top; he concludes that his pump has actually removed all the air from his sphere. The fact that there is a tiny remainder left is beyond his power of observation. He weighs the air by placing an evacuated receptacle on a balance. When he opens the valve slightly, air enters slowly. And as it does, the receptacle gains weight. His experiment is sufficiently precise to show that the measured weight depends on the air pressure prevailing; as a result, he cannot determine a fixed ratio for the weight of air and water. In another experiment, Guericke connects a small glass sphere that is filled with air with a larger, evacuated one. As he opens the valve between the two, air from the small sphere penetrates the large, empty volume; in the process, water droplets appear and sink to the bottom. This effect—the formation of droplets during rapid expansion—has been used routinely in our century as a detection method for elementary particle tracks in an instrument called a cloud chamber. When Guericke puts small pebbles or nuts inside the large, otherwise empty, sphere, the turbulent airflow whirls them about upon entering through the valve, like a miniature hurricane.

Guericke notes that animals die in evacuated space. He reports that pikes spit out smaller fish they have recently swallowed (and I will spare you the more gruesome details he mentions). Beer is seen to froth; candles are extinguished. He supposes—correctly—that all of them are dependent on air for their "nourishment." He suspends a clock with a bell in a container by a wire. As he evacuates the container, he stops hearing the sound of the bell: Evacuated space does not conduct sound.

Figure 42e Water rises through a tube into an evacuated volume.

In an attempt at true scientific showmanship, Guericke has an evacuated receptacle suck water upward from a bucket at ground level (see fig. 42e). Curious onlookers want to know up to what level he can raise water in this fashion. Guericke doesn't know, so he tries it out. He finds that it is about 10 m—the height known already to Torricelli and Pascal. He replaces the tube in figure 42e with one of 10 m length, leaving the rest of the contraption intact. To make sure there is no air left in the 10-m tube, he pumps on the upper receptacle several times, opening the valve that connects it to the tube. Finally, the level of the water column remains unchanged. Without knowing, he has repeated the Torricelli–Pascal experiment with water, using a completely new technique. It is more direct than Torricelli's, resembling the action of a suction pump. Guericke, by the way, is already aware that water cannot be pumped to a level higher than 10 m; it doesn't take him long to find the connection between this fact and the result of his experiment.

CHANGING AIR PRESSURE

Guericke ascertains, paralleling Pascal's earlier observations (of which he knows nothing), that the ambient air pressure is not always the same. He knows it is responsible for the level of the water column in the evacuated tube, and he notices that it sometimes rises, sometimes sinks, with respect to some initial value. If it only sank, this might be due to an air leak in the tube. But if it rises, this can be explained only by an increase in external air pressure. He concludes: "It is the weight of the air that causes the *horror vacui;* the pressure it exerts on the water causes the latter to penetrate and fill in every bit of space that is not otherwise occupied, and it does so in an amount set by the ambient air pressure. This means the variation in the height of the water column furnishes the best proof for the fact that the water level in all its aspects is governed by an external condition."

The realization that the ambient air pressure varies is, he writes, "an experimental result that I did not at all expect." He notices the systematic connection between air pressure and upcoming weather conditions; this permits him to use his barometer for weather predictions. In particular, a precipitous decrease in air pressure tells him that there must be a rainstorm close by, and that it is likely soon to show up in his city of Magdeburg. Upon hearing about Pascal's experiment on the Puy-de-Dôme, Guericke wants to repeat it on a local mountaintop. But his attempt fails—the handyman who carries the apparatus stumbles and Guericke's equipment is shattered.

Clearly, Torricelli's, Pascal's, and Guericke's contemporaries have a hard time visualizing that they are living on the bottom of an ocean of air that presses down on them with the same force that a mercury column of 76 cm height would exert. Pascal and Guericke explain it—correctly—in this fashion: The air pressure, like all pressure, is not simply directed downward like the weight of a rock; it acts equally in all directions, from the bottom up as well as from the top down. In all hollow spaces, the pressure is equal in all directions—even when it is almost zero, as in Torricelli's vacuum or Guericke's evacuated spheres. Omnidirectional pressure cancels its net effect on us, and we don't notice it. Only when air pressure changes do we become aware of it. Had there been a way to change elevation rapidly in Guericke's time, we can be sure he would have taken the popping sound in his ears as another proof of his assumptions.

Guericke, who is motivated by his vision of the grandeur of the universe to do the vacuum experiments, recognizes the implications of changing pressure:

> We see that air pressure changes frequently at our elevation; it also changes when we scale a mountain. Now, considering the immense distance of the stars, we can conclude that the immeasurable space that surrounds Earth cannot be filled with air—not even as far as the Moon, certainly not all the way to the Sun. It is even less correct to imagine the air makes up this space and that, as a consequence, there would not be any space in the absence of air: In that case, Earth, Moon, and Sun would come in immediate contact with each other and with the stars.

For the same reason, Guericke holds it impossible that there should be fire or ether spheres beyond our atmosphere: No matter how dilute, the pressure of the

Figure 42f Guericke's most spectacular experiment: sixteen horses pulling on his hemispheres are incapable of separating them.

immense amount of matter contained in such spheres would have to dwarf the effect on our relatively thin atmosphere, the height of which he estimates. Guericke correctly assigns the shape of a spherical shell to the atmosphere, with its density slowly vanishing into empty space. Today, our ideas of empty space differ from Guericke's, but with that exception we agree with him.

THE MAGDEBURG HEMISPHERES

We saw that Guericke was fond of spectacular demonstrations of physical phenomena. The experiment that has secured his fame to this day—the demonstration involving the so-called Magdeburg hemispheres (see fig. 42f)—is just one of them. He knows what sets the level of effort that has to be expended by powerfully built men whom he uses to pump the air out of his receptacles: When the vacuum is nearly complete, evacuation depends exclusively on the area of the pumping piston. When he doubles its cross-section, the force with which it has to be pulled also doubles. This is so because external air pressure acts equally on every square centimeter. As a result, he can join two hemispheres with an air seal, pump on

Figure 42g The force exerted by air pressure wins over the force of the men pulling from the left.

the resulting sphere, and produce an evacuated volume that can be pulled apart only by forces that increase in direct proportion to the square of its diameter. He can then easily figure out how big his hemispheres have to be so that as many as ten, sixteen, or twenty-four horses will not be able to separate them. When, with an ultimate effort, the teams finally pull them apart, it produces a spectacular, explosive effect. If, however, he permits air to enter gently into the evacuated sphere, it can be separated into its two hemispheres without any notable effort. In fact, that will happen easily with the smallest pressure level above that of the surrounding atmosphere. It is sufficient for one of his men to blow air into a nonevacuated sphere, and it will separate at the seal.

He demonstrates his experiment during the session of the German Imperial Diet in the city of Regensburg, in the year 1654; using sixteen horses, eight of them pulling on either side, he produces his effect with spectacular success. Had he attached one of the hemispheres to the wall, it would have taken only the eight horses on the other side to pull with the same strength. Whether he knew that, I cannot judge. But clearly the show was better with sixteen horses in the game. In fact, he repeated the experiment later on with larger hemispheres and twenty-four horses, twelve to a side.

Figures 42g and 42h show two more experiments that demonstrate impressively the power exerted by ambient air pressure on large surfaces. The container of figure 42g is closed off at the top by means of an airtight piston. At the bottom of the container, there is a valve that can either let in air or connect the container to an evacuated volume. At first, several men try to pull on the piston while the valve is closed. They succeed in pulling it upward, up to a point. When the valve admits air from the outside, the men have no trouble pulling the piston to the very top of the container—in fact, it's so easy that they'll probably topple over

Figure 42h Guericke's quantitative experiment to measure the force exerted by air pressure.

backward. If, on the other hand, the valve is opened to connect the container with the evacuated volume, the effect is inverted: The suddenly decreased pressure inside the container pulls violently on the piston, the external air pressure pushes the piston into the container, and the men will topple forward instead of backward. Figure 42h shows an apparatus that uses weights instead of the men in figure 42g to obtain similar results.

ROBERT BOYLE

Robert Boyle repeats Guericke's experiments on the vacuum in England; in the process, he improves on them and adds some new features. There have been objections to the claim that ambient air pressure is responsible for the effects ascribed to the *horror vacui,* based on the argument that air alone will not support a mercury column of 76 cm. Boyle makes his point with the experiment shown in figure 43: Using an uneven U-shaped glass tube, the shorter end of which is closed, he pours mercury into the open end. The more mercury he adds, the higher the level in the closed leg of the tube. The air remaining in that leg is being compressed further and further, but it does—and this is the point he wants to prove—support the mercury column.

Obviously, Boyle can use this instrument to measure the ambient air pressure. This pressure, after all, adds to the weight of the mercury column in the longer leg of the U-shaped glass tube, which compresses the air in the shorter leg. As a result, the mercury level in the shorter leg rises as the external pressure increases. A pressure gauge built in this fashion, and called a manometer by Boyle, is used in his version of Pascal's experiment on the *void inside the void.* He also proves that the action of magnets is not interrupted by empty spaces.

Figure 43 Boyle's experiment shows that air is capable of supporting the weight of a mercury column. The glass tube is ensconced in a wooden block to protect the experimenter, should the glass break.

In their tendency to resort to the experimental investigation of theoretical questions, natural scientists starting with Galileo and reaching all the way to Newton make the vital transition from Scholastic approaches to those of modern scientific investigation. A second, completely independent path in the same direction is followed by Descartes's philosophy.

DESCARTES

"Most assuredly, Descartes was a great mathematician; but I would not put him in the same category as a physicist," says Friedrich Hund, one of the founders of quantum mechanics and an expert on the history of physics. It would probably be fairer to say that Descartes as a physicist did not measure up to such luminaries of his time as Galileo, Torricelli, Pascal, Guericke, Boyle, Newton—and a few others whose fame has not held up to our day. Nobody followed Descartes in his approach to physics, but his basic scientific methods were seminal in the thinking about science for a long period after his death. Not that he himself stuck to his own rules. To quote Hund again: "His principles are too narrow even for physics. And Descartes himself is among those who violate them routinely."

Descartes believes that science has to start with statements that are evidently correct. They must be followed by inferences that are irreproachably logical. He agrees with the church that the universe with all its stars, its planets, was created together with all that exists he specifically includes plants—and not just their seeds, as was speculated in antiquity. The resulting world that we observe can, however, also be thought to have originated in large blocks of matter and nothing else. These blocks might have started out with some motion that caused mutual friction, and the fine matter shaved off their surfaces in the friction process pervades the universe. Some of it has clustered into stars. And what remains of the large blocks is still present in the form of the planets, interstellar dust, air molecules—everything that we observe in the space that surrounds us.

These hypotheses, were they true, could explain the world in its present shape, but he calls them *obviously erroneous*—because they contradict his faith. I will not try to judge his faith; scientifically motivated people of his period clearly have a hard time bringing faith and evidence into consonance. As Descartes says, "My faith is the faith of my nanny"—thus demonstrating that he is deter- mined to believe. It is the result of an act of willpower, not of logical thinking. Galileo's fate may have had some influence: We recall that the Inquisition made sure he was aware of the instruments of torture that threatened him; Giordano Bruno had been summarily put to death by fire some three decades earlier.

For the present purpose, let us restrict our interest to those arguments of Descartes's that have a direct physical interpretation; and let's suppose that he actually believed in them. We can infer that from the fact that he includes their consequences in his cosmic physics concept: The stars *are* made up of a concentration of fine matter, and the same matter fills the universe in a dilute state. His theory ascribing the motion of the planets and all gravity to vortices of dilute matter (see fig. 44a) is dependent on this basic tenet. Obviously, Descartes did not see to it that his notions of the stuff that makes up our world really correspond to his tenets of starting only from statements that are evidently correct. His followers softened their criticism of this point by praising the range of his imagination.

His criteria for accepting "evidence" have to be seen in the framework of his time. His starting point was geometry. He accepted Euclid's axioms on the properties of space. In Descarte's interpretation, Euclid's geometry dealt with bodies, not with space—space and bodies were essentially synonymous to him: All space is replete with bodies. For Descartes, there is no space unless it is completely filled with bodies.

The space assigned to bodies is what the Scholastics called *inner space*. That is the only space Descartes knows. Without bodies, there is no space; without space, no body. Thus "empty" space has no meaning: Space occupies a volume— height, width, depth; thus space itself *is* a body, not simply the container of bodies. It exists only if there is matter. Since, in his opinion, there is nothing else, space is synonymous with dilute matter. We might say this matter fills space: "It is obvious that there cannot be a void in a philosophical sense—space that contains no substance; the extent of that space cannot be distinguished from the volume of the body that fills it." In particular, gravity cannot exist in a true void; after all, it results from the multiple interactions of highly dilute matter (see fig. 44a). Descartes considers Galileo's idea of free fall in *empty* space as nonsensical:

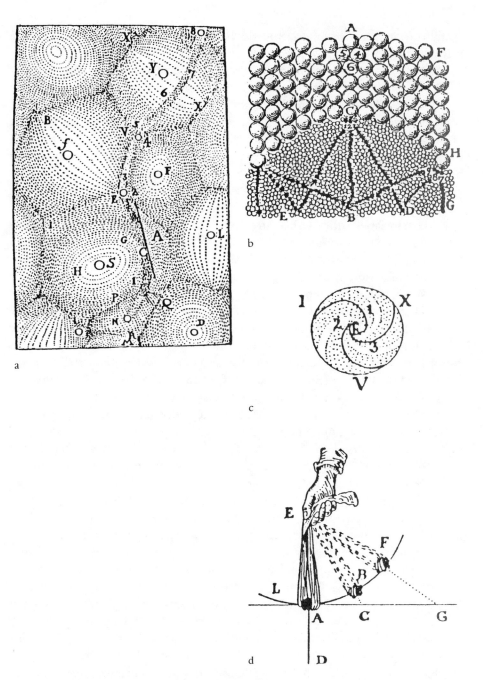

a

b

c

d

Figure 44 The vortices in 44a cause celestial bodies to move as they do, according to Descartes. His idea of the interface between two media, as it diffracts light, is shown in 44b. Figure 44c shows in detail one of the vortices in 44a. Figure 44d redraws this vortex as a sling that illustrates Descartes's ideas of celestial mechanics.

"Galileo should first have determined what gravity is, and if he had known the truth, he would have known that it is nothing in the void."

Today, we would have to approach the world of Descartes's dilute matter in terms of the laws that govern fluid mechanics: Fluids move with varying velocities in what physicists call *laminar flow*—rapid in the center, slower as it approaches the walls, fastest where passages become narrow. This can be expressed as a field theory: We can associate each point in space with a given quantity, and in this case the quantity is the magnitude and direction of the velocity at that point.

We first encountered the concept of fields when we discussed the Stoics. They talked about the propagation of physical effects in terms of fields: Quantities that are associated with neighboring points exert an influence on each other. Think of temblors that propagate as seismic waves; they can be described in terms of fields. The quantities associated with individual locations, such as the amplitude of a vibration, propagate in space. Something that shakes here influences the shaking of neighboring matter.

Similarly, we know that an iron rod that is hot on one end and cold on the other will see its "field of temperature distribution" change with time. The abstract description of the temperature distribution in the rod doesn't get down to the reasons for the effect—that the iron atoms oscillate faster about their average location when the temperature rises. The field-theoretical laws of the changes in temperature distribution concern only this distribution and don't "know" about the atomic effect. We can make similar statements about the field theory of seismic waves, or the fluid motion of, say, the dilute matter with which Descartes imagines the universe to be filled. The remarkable feature in the field description of such observable phenomena is the fact that it does not touch on the fundamental processes on which they are based. It is only the fields themselves that field theory describes; we don't need to know what they are based on. It is possible to define levels of description that deal only with quantities described at exactly these levels—say, macroscopic phenomena that do not betray their microscopic causes.

This approach is not limited to physics. Every discipline of science, from chemistry and biology to medicine and psychology, might be described as a *level of description* using its own concepts, and nothing else, in formulating its laws. If, because of the universal interdependency of all phenomena, it were not possible to define this kind of specific and self-contained level of description, progress in the sciences as we know it would hardly be possible. That nature allows this division of phenomena into levels at all is one of its greatest mysteries.

Fields are not limited to the role of a macroscopic "effective" description of more basic truths—they can also be defined on a basic level beyond which we cannot penetrate. On this level, light can be described in terms of a field. This field is autonomous, in the sense that there is no substance such as the ether which would have properties from which the laws of nature for the field follow. Whoever insists on a fundamental substance has to accept the field of light on that level. To use an image we mentioned before, the field is the cat and not only its grin.

Descartes would have a hard time following us if we left the fluid—his *dilute matter*—out of our considerations after abstracting its basic properties in terms of a field. To Descartes, the fluid itself is synonymous with all the properties that define it, including the space in which we write down the individual field quantities that describe it. In the absence of the fluid, there is no space, and vice versa. In our field-theoretical description, we don't worry about the totality of properties that can be associated with the fluid; we limit ourselves to the quantity *velocity* that we can associate with every point in space. Since Descartes's theory of dilute matter can be interpreted in this fashion, it is, in fact, a field theory. Not only that, it is the first such theory associated with detailed ideas about the properties of this fluid motion.

Descartes believes that Torricelli's vacuum is filled with dilute material. It penetrates the walls imperceptibly; and for this very reason, there is no experimental way of telling the difference between Torricelli's and Descartes's interpretations of space devoid of air. On close inspection, many of the hotly contested scenarios on the vacuum deal with appearances only.

Descartes denies the possible existence of a vaccum. This, on the face of it, clashes with the notion of the atomists, who need a vacuum for the motion of their atoms. But this does not hold true: Descartes's space, while filled with dilute matter, does not resist the motion of heavy blocks of matter, as we said before. And that matter, obviously, consists of atoms.

Another notion of Descartes's is that matter can be subdivided ad infinitum: There is no such thing as an indivisible, elementary atom. This idea arises from his acceptance of geometry as the only true property of bodies. Only geometry is an objective reality; all other properties that we notice are, according to him, nothing but the reactions of our senses to shape and size. Shape and size for us, however, are only what fills space. Take a statue and a plaster cast of that statue: We have no trouble having either of them occupy the same space; they share shape and size, but they fill an identical space with different substances. Descartes does not differentiate between space and the bodies that fill it; for him, the bodies *are* the space they take up. All shapes and sizes are possible, and all can make up objects. Descartes transfers the properties of Euclidean space from geometry to the bodies that are, therefore, for him identical with space.

For all practical purposes, this is another instance where there is no meaningful difference between the notions of Descartes and those of the atomists. While Descartes's particles can be infinitely subdivided—if the creator could not do that, it would curtail his omnipotence. But in reality, this has no meaning. Ever since creation, or since their ablation from primordial blocks of matter, the particles of dilute matter are as small as Descartes's image makes them. Further division, although possible, does not happen. Figure 44b shows light passing from a medium made up of large spheres into one consisting of small ones; the image lends itself to atomistic interpretation. As far as the vacuum and the indivisibility of atoms are concerned, the atomists and Descartes agree for all intents and purposes.

The same goes for their explanation of the cohesiveness of matter. The atomists imagine individual atoms hooked to each other; Descartes wants to explain this cohesiveness exclusively by the need for precisely coordinated motion—or simultaneous rest—of all neighboring parts of a body. Neither of these images has significant consequences; therefore, their differences can be ignored.

We could package the opinions of Descartes and of the atomists, including Gassendi's, in terms of what is alternatively called a *mechanical* or *mechanistic* approach. Without passing through this intermediate state, the Aristotelian views that prevailed in the Middle Ages could never have evolved into the modern natural sciences. This mechanistic philosophy did away with Aristotle's logic, based as it was on the observations possible at his time; it no longer held for the natural phenomena that were now clearly observed. Aristotle, we recall, described condensation and dilution as the realizations of potential states of matter—but that description did not contribute to an understanding of what was actually happening. From our vantage point, we might say that the mechanistic approach overdid the dismissal of earlier thinking. For its version of understanding, a natural explanation of mechanical phenomena had to be formulated in terms of levers and hinges, nuts and bolts.

The only explanation acceptable to mechanistic philosophers sees all mechanical processes in terms of the contact and collision of the participating bodies. This is like mechanics on a pool table: There are no forces acting at a distance, in the absence of individual billiard balls colliding. There is no way, in this world, by which the Sun could act attractively on the distant Earth across empty space. Newton, a disciple of the mechanistic school, finds himself aghast in the face of the implications of his own successful system of mechanics: His system contains gravitation as a force acting over distances between remote celestial bodies, in the absence of visible carriers of that force. He searches in vain for such carriers; he is loath simply to invent hypotheses he cannot test. Today we accept fields as ultimate realities; notions as basic as shape, size, location, and motion that we see in the world of bodies will have to be carefully interpreted in terms of fields.

Newton, whose thinking we will discuss in the next chapter, considers the universe as infinite in extent. Gravitation would have all masses collapse into its center of gravity if it existed inside a finite volume—and that is obviously not the case. We will revert to this point and its physical implications.

In Descartes's view, the world knows no limit, but it is not infinite. To this day, it has not become clear what he meant by this distinction. By coincidence, there are present-day models that see the world in similar terms—unlimited but finite. We have previously given an example of such a world: the surface of a sphere. In chapter 8, we will discuss to what extent this model of a two-dimensional world can be applied to our universe. But Descartes cannot have intended this differentiation between *unlimited* and *infinite,* if only because the surface of a sphere is not unlimited in his interpretation. For him, *unlimited* means that any given space must admit distances larger than any distance previously defined—which does not apply to the surface of a sphere. If we move halfway around a sphere, any further straight motion will bring us back closer to the point of departure. But more important than this detail, the introduction of curved spaces in general was not accepted before the nineteenth century. An unlimited but finite space must necessarily be a curved one.

Once the world is seen to be infinite, its relation to God has to be discussed. This is a requirement of Scholastic tradition: Anything infinite challenges God's perfection. And for Descartes, who does not recognize the reality of any property beyond geometry, an infinite world would have to be identified with God himself.

LIMITED, UNLIMITED, AND NEVER-ENDING UNIVERSE?

The first philosopher to differentiate between what is *infinite* and what is *limitless* was Nicholas of Cusa, cardinal-bishop and expert in canon law. He is best known by his treatise *Learned Ignorance,* which dates to about 1440. By picking up the concept *unlimited,* or *limitless,* from pagan philosophers such as the atomists and the Stoics, who are to wait two more centuries for their vindication by Gassendi, he sneaks the concept of an infinite universe in by the back door. We recall that at the time nothing could be considered infinite except God himself. In open contradiction to common sense, which sees no difference between *finite* and *not infinite,* Nicholas writes: "Even though the world is not infinite, we cannot understand it to be finite either: It has no limits that might contain it."

This argument opens up a tightly constrained discussion. If the universe is infinite, it does not have a center. That means Earth, humanity, is not at the center of the universe—a clear contradiction to the church's teaching. Today, we know that our life happens to be made possible by an accidental occurrence of certain conditions on a planet in some arbitrary planetary system, galaxy, cluster of galaxies. Copernicus was the first to realize that Earth orbits the Sun, and not the other way around. In his system, the Sun replaces Earth as the center of the world. This is acceptable to Copernicus and later on to Kepler; the solar system and, by implication, Earth retain their pivotal location in the universe. Its center has changed, but it has not been lost—and Earth belongs there.

The most important scientific objections to the notion that the stars are far removed from Earth sprang from the belief that the universe orbits Earth once a day. This implies that the farther a star is removed from Earth, the more rapidly it has to move in its daily orbit. The velocities of distant stars would therefore have to be unacceptably large. This objection falls by the wayside if it is Earth that rotates, not the universe. Starting with Copernicus, it became possible to think of an infinite universe replete with stars.

I will refrain from tracing how human thinking about a larger and larger universe has developed since antiquity. Neither the atomists nor the Stoics believed in an observable universe filled with stars. The world of stars that was within their vision remained small and finite—immersed in a more or less empty space without stars; Lucretius went on to surmise the existence of isolated stellar systems beyond our vision and which we may imagine to resemble our own. The stars within our vision are contained inside a spherical shell (see fig. 24a); the space beyond—the nothing—is mere speculation. The first to embed the solar system in an infinite universe that is the home to arbitrarily many stars (see fig. 24b) was Thomas Digges in 1576. Like his contemporary Giordano Bruno, and like Otto von Guericke after him, Digges is positively inebriated with his notion of an infinite universe. He speaks in terms of a "palace of bliss, ablaze with perpetually brilliant lights that surpass our Sun in number and luminous power," about the firmament of stars that "stretches spherically out to infinity." He does not discuss how he sees the compatibility of *spherical* and *infinite*. But there were many notions on this point, some of them approaching mysticism. Blaise Pascal summarizes them as follows: "Our world is an infinite sphere; its center is anywhere, its circumference nowhere."

His fatal end makes Giordano Bruno the best-known advocate of an infinite universe. He loudly proclaims his belief that the universe consists of innumerable worlds inside a space infinite. This heretical belief—or, more precisely, the heresy of its public utterance—are enough to have him burned at the stake in the year 1600. In one of the dialogues we owe to him, his spokesman Filoteo says: "There are many who speak of a finite world; but no one is able to explain what that means—where its boundary is. They say there is no void, no empty space; and yet the way they describe it, their void becomes a 'something.' If there is a void without contents, it must at the very least be capable of accepting something in its volume." In the sixteenth century, it was still imperative for anybody who wrote about matter, even the fiercest critics of Aristotle, to couch their views well inside the accepted standard model. And so even Bruno's infinite space is not really empty outside the material worlds he sets in it here and there; it is filled with a substance that he declines to specify. Is it some kind of matter; is it the ether? He doesn't say.

I borrow much of this presentation from Alexandre Koyré's impressive work on the subject. Let me quote one of the footnotes in *From the Closed World to the Infinite Universe:* "Bruno's space is empty. But this emptiness is not really devoid of content; rather, there is something everywhere. A vacuum that contains absolutely nothing would imply a limit to God's creative powers; it would furthermore mean a violation of the 'principle of sufficient cause' according to which God may not treat some part of space differently from the others."

The atomists saw no reason to prefer one shape or size of an atom to another one; hence, they thought all shapes, all sizes must be possible. Similarly, Bruno feels the creation must contain everything that can exist. To quote Koyré again: "In this case, if something can be, it must be." And Bruno concludes, again in the wake of the atomists, that in order to be able to accommodate this infinite variety, the world must be indeed infinite.

Later commentators will accuse Bruno of not having attained the same conceptual clarity as his Scholastic contemporaries. He does not differentiate sufficiently between space imagined and the space actually created by God. Had he pointed out that his discussions concerned imagined space alone, Bruno, like others among his contemporaries, could have made statements just as he pleased. Thus, there were astronomers who did not incur the displeasure of the church when they stated that in their imaginary space the Sun remains at rest while Earth moves. But Bruno includes God in his theory of space; that means he speaks of real space—and that is clearly dangerous. In Bruno's time, it took an ultimate formal schooling in the concept of Scholastic thinking to discuss so-called real space; this was well beyond the range of an enthusiastic dreamer, as he appeared to be. It was easy to prove him guilty of a whole number of contradictions to the church's teaching—some of which he was certainly not aware of. In Edward Grant's words: "The medieval fear that void space would be interpreted as an eternal, uncreated positive entity independent of God was realized in the metaphysics and cosmology of Giordano Bruno. . . . Did he realize that the infinite space he described and the properties he attributed to it were such as to exclude it from God's creation . . .? Probably not."

William Gilbert arrives at the same conclusion as Bruno—that the universe is infinite in size—for purely scientific reasons. The stars, "those immense and varied lights are situated in the skies at *various* distances from us. They are surrounded either by the thinnest of ethers or by some other dilute essence—maybe even by the void." He infers this because the light of some faint stars becomes observable to us only when the Moon is below the horizon; the reason their light appears fainter can only be that they are farther away than other stars.

Aristotle's objection to the idea of huge empty space between us and the stars is not for Gilbert, who thinks, like Copernicus, that it is Earth that revolves, not the universe. It may well be that Earth orbits the Sun, but he avoids the discussion of that point. In the process, he stays away from one of the most telling arguments in favor of an immeasurable universe: that the angle under which the stars are visible from Earth does not change with the relative position of Earth in its solar orbit. Consider, by comparison, the angle by which we see a tree that is standing close to the railroad tracks as we move by in a speeding train; the angle changes rapidly. When the tree is farther away, the angle changes more slowly. The Sun and the Moon are so far away that the passengers in the train feel they are traveling along with them. Similarly, the stars appear to travel along as Earth orbits the Sun; the naked eye does not resolve a stellar parallax, and from that fact we can calculate a minimal distance to those stars. Stellar parallax, in fact, was not confirmed experimentally until 1837.

Kepler notices the absence of such parallaxes; he concludes that the universe that is home to those unmoving stars must be huge. Not that he believes it to be infinite—he is not convinced by any one of Aristotle's arguments in favor of a finite universe. But he does accept that it has only finite extent. The arguments of the atomists and the Stoics in favor of an infinite universe do not convince him, maybe for psychological reasons. He does not share in the rapturous enthusiasm of Digges and Bruno for infinity. Quite the contrary: The thought of a potentially infinite universe "bears some hidden terror, in whatever form; in this immeasurable space we would be all but lost—it has no center and no boundaries, and no specific location can be fixed in it." This is unacceptable to him, because he believes in the unique mission of humanity in the creation, in the unique arrangement of the Sun and the planets. He takes his arguments in favor of his point of view from the apparent size of the fixed stars. He believes that we see them just as a disk at a large distance—that is, quite correctly, but diminished in size by our relative perspective. This, however, is incorrect because of the limited resolving power of our eyes. It is the latter, together with the light diffraction in Earth's atmosphere, that determines the apparent size of the stars. It took modern optical methods and large telescopes to see some of the stars as actual disks. Galileo's telescope did not improve on the naked eye in that respect.

Starting from the apparent sizes and distances of the fixed stars, Kepler infers that the unoccupied space separating them cannot be greater than the size of our solar system. This inference is mistaken, along with the reasoning it is based on; and he surmises as much: "There is no firm knowledge we have of the fixed stars. Maybe each star is at the center of a world that is like the world that surrounds us. That would mean our world is just one among innumerable locations in this infinite array of stars, and our Sun does not basically differ from any of them." He draws a picture to illustrate this train of thought (see fig. 45).

DISTANCE, EXPANSION, AND THE EDGE OF THE UNIVERSE

Kepler dismisses this concept and continues to believe in a finite world within impenetrable boundaries. His universe is surrounded by a cosmic wall, and the

Figure 45 In a hypothetical world that Kepler considers unacceptable, each of the fixed stars in this figure could be surrounded by a planetary system just like our Sun's. The fixed stars are arranged in an almost symmetrical pattern of triangles.

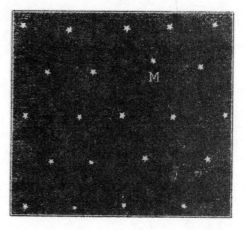

fixed stars are attached to its inner surface. All the way to that wall, a space traveler would not encounter resistance; but the wall is impenetrable.

The Scholastics had surrounded the outer sphere of Aristotle's universe, the one containing the fixed stars, with further space; but that space, in its infinity, is divine by virtue of that very property: Infinity is an attribute of God himself. Just as in Aristotle's original system, they don't believe that a body can penetrate into or beyond the sphere of the fixed stars without losing its identity. Their world of bodies is not surrounded by a wall; we might say it ends in a kind of quagmire. Beyond the sphere of the Moon, they believe, the terrestrial laws of nature give way to what they call celestial laws: Archytas and his staff, Lucretius and his javelin, change as they pass through the spheres from Earth all the way to that of the fixed stars; the matter they consist of changes gradually into celestial matter, subject to celestial law. All motion there is confined to circular motion—which, in particular, excludes any velocity perpendicular to the ultimate sphere; that, after all, would lead beyond its confines.

If there were a second world, as suggested by Lucretius and Bruno, it would have to be separated from ours by more than just empty space. For Descartes, after all, empty space is tantamount to the nothing, devoid of all properties. But volume is clearly a property—more, it is the *only* property Descartes accepts as real. Thus different worlds separated only by empty space would have to touch. If they don't, if there is a volume separating them, then "something," not "nothing," is between them. Similarly, any volume from which all contents are removed would have to collapse; that would have to apply to the hydra when water seeps out through the bottom while the top remains closed, and similarly to any receptacle that Torricelli evacuates. To quote Descartes on the question: "If anyone asks what would occur if God removed the whole body contained in any vessel and did not permit anything else to take the place of the body which had

Figure 46

condition	pressure in mbar	applications/matter density
normal pressure	1000	respiration / at 20 C, 5×10^{19} molecules/cm^3
rough vacuum	1000–100	vacuum cleaner, milking devices, vacuum packaging
intermediate vacuum	$1–10^{-3}$	simple chemical procedures, metal coating, metallurgy
high vacuum	$10^{-3}–10^{-6}$	as above; lightbulbs, picture tubes
ultrahigh vacuum	below 10^{-6}	research chip manufacture, elementary particle acceleration below 10^{10} molecules / cm^3
Earth's atmosphere at 250 km above surface	10^{-7}	/ 10^9 molecules / cm^3 at about 20 degrees C
Vacuum tube of LEP rings Earth's atmosphere at 400 km	10^{-9}	/10^7 molecules / cm^3
Best vacuum reached today	10^{-13}	/ 10^3 molecules / cm^3
matter density in interstellar space		/ 1 hydrogen atom / cm^3
"critical matter density" (see chapter 8)		/ 10^{-7} hydrogen atoms / cm^3
average matter density of visible matter in the universe		/ 10^{-9} hydrogen atoms / cm^3

This table gives the air pressure and, where more appropriate, the number of atoms or molecules per unit volume, for a number of different scenarios relevant to our discussion. The vacuum in interstellar space compares to the best vacuum achieved here on Earth as air under normal pressure compares to a wall of stone.

been removed, the answer will have to be that the sides of the vessel would thereby become contiguous to each other . . . because all distance is a mode of extension, and therefore cannot exist without an extended substance."

PRESSURE AND VACUUM

The discussion of the question about the extent to which the space from which Torricelli pumps all air is truly empty had no influence on the further development of physics. It soon became apparent that the experiments on Torricelli's vacuum did not actually investigate the vacuum but rather the property of gases in Earth's gravitational field. In this context, Torricelli's (and Guericke's) vacuum became a practical problem, nothing else.

But there was a new, philosophical dimension to the question about the vacuum: Is a volume from which all air is removed actually empty? Is it empty space? Probably not—but what then could space contain? Torricelli's experiment opened the curtain that had kept this question from being asked; it separated the problem of the vacuum proper from the problem of air pressure. This is how empty space—the object of philosophers' curiosity ever since pre-Socratic times—was transformed into a subject to be discussed in terms of physics. By contrast, the physics questions raised by Torricelli's evacuated space were devoid of all philosophical implications, notwithstanding various speculations on the subject.

Ever since Boyle's intervention, Torricelli's and Guericke's vacuum has been an object mainly of technological efforts. True, Boyle did try to find ether winds, the existence of which had been postulated; but his experiment produced a null result. He did not engage in philosophical speculations on the vacuum; his vacuum is just space emptied of air, nothing else. A good vacuum contains very little air per cubic centimeter, a better one even less. Today, vacuum technology speaks of four levels of useful vacua: rough vacuum, intermediate vacuum, high vacuum, and ultrahigh vacuum. The chart in figure 46 gives the data on vacuum quality—it mentions the pressure of the residual air in units of millibar (mbar). The normal pressure of ambient air—that is, at sea level and in medium weather conditions—keeps Torricelli's mercury column at the 760-mm level. That corresponds to 1013 mbar, slightly more than 1 bar.

END OF AN ERA, BEGINNING OF AN ERA

I imagine that I am a son of Gilbert, born in 1564, when Gilbert was just twenty years old. I learn in school that the universe is finite; Earth is at rest in its center; the universe itself revolves around this center at the rate of one revolution a day. My father tells me that this is wrong: It is not the world that revolves, it is Earth. At age sixteen or so, I read Thomas Digges's 1576 book with its views of an infinite universe (see fig. 24b). From Kepler's books I learn not only that Earth is turning around its own axis but also that it orbits the Sun. Empty space, my teacher says, cannot exist. Gilbert, my father, believes in empty space separating the stars. I am an old man of eighty when Torricelli manages to produce empty space here on Earth. What changes in my life span! A whole era has ended, tremendous changes have occurred, a new era is well on its way!

CHAPTER 4

MATTER IN THE VOID:

•

ETHER, SPACE, FIELDS

•

ISAAC NEWTON LAID THE GROUNDWORK FOR MODERN NATURAL science in the year 1687 when he published *Philosophiae naturalis principia mathematica*. Prior to this work, nobody had tried to differentiate between the realm of initial conditions, over which we have some control, and that of the laws of nature, which are independent of us. This differentiation is at the basis of the modern sciences.

With overwhelming success, Newton applies these notions, instead of the principles that Aristotle had postulated, to heaven and Earth. The laws remain the same throughout the different spheres; it is the conditions under which they operate that change. As an example of one of these initial conditions, think of throwing a rock into the air; whether it will drop back to Earth or will orbit the Moon depends on its initial velocity. Given the starting conditions, the laws of nature as Newton formulated them in his *Principia* will fix the rock's trajectory as a function of time, just as they do for the Moon or the planets.

The configuration of the world that surrounds the Sun is such that every planet moves almost as though there were nothing in the universe but the Sun and itself. The planets exercise very little influence on each other. This means Newton can successfully study a kind of model world; that which consists of Sun and Earth in an otherwise empty space is a good example. He needs the empty space as choreographers need the stage for their dancers. He can throw in, at will, planets, rocks, moons, suns. Once he has chosen the starting conditions, including location and velocities, the laws of nature, which are the same everywhere and at all times, permit him to calculate what will happen in the future.

NEWTON'S SPACE AND TIME

Newton's *absolute space* is a volume into which he can place worlds. He calls it absolute to distinguish it from a *relative space*, which he also uses. If there are objects in his absolute space, he can define their relative position—as in a solar eclipse, where the Moon is located on the straight line between Earth and Sun. Taken together, the relative positions of objects make up their relative space. Newton's *absolute space*, however, doesn't require objects for its existence. Making up the infinite volume of the world, it exists in complete independence of anything else and is always the same, irrespective of what bodies, if any, it contains. Absolute space acts (and I will have more to say about this later), but is not acted upon.

Albert Einstein considered the assignment of physical reality to space as such, and empty space to boot, a tough imposition. With this objection, he does no more than pick up the argument advanced against Newton's absolute space from the very start. In contradiction to the program of mechanistic philosophy, of which he was a follower, Newton felt compelled to introduce his absolute space as an object that acted in a fashion not caused by physical contact and impact. We will see how ardently he defended his concept without ever managing to make it quite clear. Neither he nor his contemporaries found it plausible—and it has remained implausible to this day.

Still, Newton's mechanics was stunningly successful. Based on a few fundamental assumptions, it was able to explain not just Galileo's laws of free fall but also the motions of the planets. All the great theoretical physicists of the eighteenth and nineteenth centuries, while ignoring philosophical discussions about space, built Newton's mechanics into a comprehensive system of nature; it permitted them to calculate the behavior of complex machinery as well as the orbits of hitherto unknown planets, whose existence could be inferred from the perturbations they exerted on the orbits of known planets.

The great theoreticians of mechanics did not limit the basis of their science to concepts that can be defined singly in a constructive and explicit way; whether or not they knew it, they applied the science of mechanics as a whole—in terms of a system of axioms that defines its objects implicitly. Absolute space does not make its appearance in this system; it cannot even be defined in its terms. Newton was among those who went about their physics in this fashion; but he didn't think he could do without some kind of metaphysics. After all, he didn't merely want to express his axioms of mechanics; he also wanted to justify them and make them understandable.

To do so, he introduced absolute space and absolute time. To absolute space, as far as it can be defined and motivated by mechanics, we can ascribe *acceleration* but not *velocity*. The reader may recall that we touched on this point in the prologue; but we are now in the position to discuss it more clearly. In my opinion, Lawrence Sklar is the only theoretician of science whose point of view is convincing in this respect: He suggests that we may want to understand acceleration as a property of bodies—a property that can be defined in and of itself, irrespective of whether we are able to say to what this acceleration is relative. That makes sense:

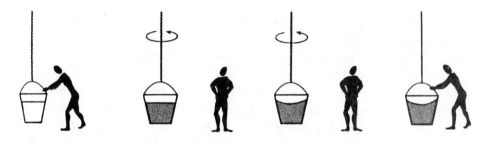

Figure 47 By his experiment with water in a bucket Newton intends to demonstrate that absolute, not relative motion is responsible for centrifugal force. If a vessel hung from a long cord is turned so that the cord is strongly twisted, and then filled with water and held at rest together with the water; thereupon, by the sudden action of another force, it is whirled in the opposite direction, and while the cord is untwisting itself the vessel continues for some time in this motion. The surface of the water will at first be level, as before the vessel began to move, but after that, the vessel, by gradually communicating its motion to the water, will make it begin to revolve and recede little by little from the middle, and ascend the sides of the vessel, forming a concave figure. The swifter the motion becomes, the higher the water will rise, until at last, performing its revolutions in synchrony with the vessel, it becomes relatively at rest in it.

This definition incorporates the observable aspects of acceleration. It acquires a physical meaning only as part of the full machinery of the science of mechanics, as do all the other concepts of mechanics.

SPACE AND MOTION

The continuing discussions about space that have spanned the millennia bear witness to the controversial nature of this concept. If, like Descartes and Newton, we add velocity, the difficulties increase. The Aristotelian definition of location as the *inner rim of a container* was already almost impossible to reconcile with velocity. We recall that, according to Aristotle, the fixed stars have no location, since there is nothing outside the spherical shell they inhabit; it would make no sense to speak of a further container that comprises their sphere. But they do move, they do *change* their location, according to Aristotle. Is that possible if they don't *have* a location?

The analogy to Newton's space, which has an observable acceleration but no observable velocity, is amazing. Relative motion—say, the motion of Earth with respect to the Sun—is obviously observable; absolute motion is not. Galileo had already seen this quite clearly. He said:

> *Shut yourself up with some friend in the main cabin below decks on some large ship, and have with you there some flies, butterflies, and other small flying animals. Have a large bowl of water with some fish in it; hang up*

a bottle that empties drop by drop into a wide vessel beneath it. With the ship standing still, observe carefully how the little animals fly with equal speed to all sides of the cabin. The fish swim indifferently in all directions; the drops fall into the vessel beneath; and, in throwing something to your friend, you need throw it no more strongly in one direction than another, the distances being equal; jumping with your feet together, you pass equal spaces in every direction. When you have observed all these things carefully (though there is no doubt that when the ship is standing still everything must happen in this way), have the ship proceed with any speed you like, so long as the motion is uniform and not fluctuating this way and that. You will discover not the least change in all the effects named, nor could you tell from any of them whether the ship was moving or standing still.

Let's leave aside Galileo's poetic description and instead home in on the pivotal concept of *nonuniform motion*. This is the accelerated motion that Galileo's friend would be able to notice by, say, observing that the falling droplets wound up outside the vessel below.

Newton's absolute space had to remain a chimera, a fictitious object, given that his mechanics contained the fundamental notion that a velocity that does not change in magnitude and direction is unobservable. The concept of absolute space incorporates the notion that rest and motion are defined with respect to it. Newton's mechanics doesn't permit such a definition: In it, the state of absolute rest is nothing but one of infinitely many equivalent states. All systems that do not rotate and move in empty space, in straight-line, uniform motion relative to each other, are subject to the identical laws of nature in Newton's mechanics. If there is no centrifugal force acting on one of them, then there will be no such force noticeable by any of the others that are in constant motion relative to it.

For the same reason, it is impossible to give objective meaning to the notion of a given location at different times. Imagine that a passenger opens his book in an airplane just after take-off from Boston in the direction of New York City: Let him close it just before landing; if the plane moves at constant velocity, the passenger is justified in saying that he opened and closed the book in the same place. Viewed from the surface of Earth, this is obvious nonsense.

Newton's mechanics assigns meaning only to the state of acceleration of physical systems, not to that of their motion. In contrast to this idea, he asks himself whether the relative motions of individual parts of those systems might be the cause of centrifugal force that occurs during accelerated motion. Newton answers his own question in the negative after performing the celebrated experiment with a bucket of water (see fig. 47): When we see the seats attached to a carnival ride slowly rise as the apparatus turns, Newton's mechanics tells us that this is due not to its motion with respect to the fairgrounds below but, rather, to its motion with respect to absolute space.

Rotation is a special kind of accelerated motion. It differs from other kinds of accelerated motion in that no energy has to be spent to maintain it. If there were no friction, a grindstone, once in motion, would keep rotating forever. Centrifugal force is well known in rotational motion, but all accelerated motion generates similar forces. Whenever we attempt to set masses in motion—and let's recall that mass is a fundamental concept in Newton's mechanics—those masses will offer resistance; more generally, they resist any change in their velocity.

All kinds of motions that generate forces are distinguished, in Newton's physics, from those that don't. In order to speak of absolute space, Newton also has to find differences between various motions that are free of forces, in his system of mechanics. Every motion with respect to his hypothetical absolute space is, in his words, a "true motion"; it is distinguished from all "relative motion," where only individual elements of the various systems change their location relative to each other. His attempt to define true motion in terms of observational quantities of his mechanics can be successful only when we deal in *changes* of velocities. And so it is. In his own words:

> The causes by which true and relative motions are distinguished, one from the other, are the forces impressed upon bodies to generate motion. True motion is neither generated nor altered, but by some force impressed upon the body moved; but relative motion may be generated or altered without any force impressed upon the body. For it is sufficient only to impress some force on other bodies with which the former is compared, that by their giving way, that relation may be changed, in which the relative rest or motion of this other body did consist. Again, true motion suffers always some change from any force impressed upon the moving body; but relative motion does not necessarily undergo any change by such forces. For if the same forces are likewise impressed on those other bodies, with which the comparison is made, that the relative position may be preserved, then that condition will be preserved in which the relative motion consists. And therefore any relative motion may be changed when the true motion remains unaltered, and the relative may be preserved when the true suffers some change. Thus, true motion by no means consists in such relations. It is indeed a matter of great difficulty to discover, and effectually to distinguish, the true motions of particular bodies from the apparent: because the parts of that immovable space, in which those motions are performed, do by no means come under the observation of our senses. Yet the thing is not altogether desperate. . . .

After that, Newton gives a new formulation for the argument that the reality of absolute space can be deduced from the fact that force is needed to *change* the velocity of any body with respect to it.

SPACE AND THE FIXED-STAR FIRMAMENT

It must appear mysterious that an experimenter in a closed room can determine whether that room and all its contents are being accelerated linearly or rotationally. This is even more puzzling if no object can be specified with respect to which the acceleration occurs. Two hundred years after Newton, Ernst Mach will postulate that it is the fixed stars that determine the reference frame that is not accelerated in any way. Centrifugal force makes its appearance when objects rotate with respect to the firmament of the fixed stars. Those stars, through their own motion, define—or, in Newton's parlance, show—which motion is without accelaration. In so saying, we have to assume that the fixed stars are sufficiently distant from one another and that, quite unlike Earth, they move in straight lines and at constant velocity. As long as we move the way they do, we will not notice any "centrifugal" forces; but we will if we rotate or are accelerated linearly with respect to them.

Ernst Mach objected to Newton's interpretation of his own experiment with the bucket of water. Its result, he said, merely shows that the motion of the water with respect to the bucket does not generate centrifugal force. But that leaves the question open as to whether we can hold the motion of the water relative to the firmament responsible for the observed centrifugal force.

Mach did not propose any mechanism by which the faraway fixed stars might influence objects here on Earth. It is impossible for an accelerated object to emit a signal that will influence its own motion by a response to it from the fixed stars. Signaling in both directions would take time—too much time, because we know that no signal travels faster than light. Since a mechanism of this kind obviously cannot work, all information about the location and velocities of the fixed stars would have to be accessible locally to the accelerated body. Indeed, the general theory of relativity has space carry this information inherently.

It is astonishing that Newton never tried to define what he means by the term *absolute rest* with reference to the fixed-star firmament. Had he tried, it is fair to say he would not have been able to come up with a mechanism by which the stars exert this influence—no more so than Ernst Mach two hundred years later. And so it already is in the minor case of the gravity of the Sun acting on Earth through empty space. Had Newton tried to hold the fixed stars responsible for the centrifugal forces observed on Earth, the same problem would have arisen much more prominently. There wasn't a mechanism he could have named as an agent for an influence that, unlike gravity, was even unequivocally present. The idea that fixed stars might be the cause of centrifugal forces observed on Earth, devoid of a mechanism that explains it, looks more like a concept of astrology than of physics.

In the prologue, I used the example of two spheres that rotate in an otherwise empty space to explain the physical effects that the motion of one mass has on

Figure 48 The pendulum in the figure is suspended from the dome of the building. If it was located at the North Pole, the motion of the pendulum could be easily predicted: The pendulum retains its plane of oscillation; Earth rotates below it. To the observer on Earth, the plane of oscillation of the pendulum appears to be rotating. Our figure shows Foucault's pendulum in Paris. It also serves to show the rotation of the plane of oscillation with respect to Earth. While the basic reasons are the same as at the North Pole, the details at this latitude are more involved.

another. This gedankenexperiment can be traced back to Einstein's original paper about the general theory of relativity, published in 1916. In that paper he speaks of two fluid bodies, each of which is held together by its own gravity counteracting the centrifugal force.

A realistic example of the possible influence of the motion of one mass on other masses is based on the famous experiment we call *Foucault's pendulum* (see fig. 48): A pendulum swinging back and forth while suspended at the end of a long string will, in Mach's parlance, retain the plane in which it oscillates, relative to the fixed stars; in Newton's words, the plane of oscillation remains unchanged with respect to absolute space. Let me assume, for simplicity's sake, that the pendulum is suspended above the North Pole; then Earth will rotate below it, so that the plane of oscillation will rotate with respect to Earth. An observer on Earth might conclude that she rotates, together with Earth, in an "absolute" sense. She has to resist these forces if she wants to keep the plane of oscillation constant from her vantage point, that is, *with respect to the rotating Earth*. She might do that by making sure the pendulum is suspended not from a string but rather at the end of a metal rod anchored in a hinge—like the pendulum in a grandfather clock. That will assure the plane of oscillation's rotating along *with* Earth. The pendulum will then exert pressure on the mechanism by which it is suspended. It doesn't rotate relative to Earth; but it does rotate in an "absolute" sense, relative to the masses of the universe—or relative to Newton's absolute space.

According to Mach and Einstein, Earth will actually carry the pendulum that is suspended by a string above the North Pole, by a small amount, along with its rotation. This is because our pendulum—like the little sphere we discussed in the prologue—is subject to the gravitational influence of *all* the masses in the universe; it will seek a compromise between following the pull exerted by the large but distant masses of the stars and that of the nearby Earth. All the masses of the universe in their entirety determine the meaning of rotation.

Newton was the first to give a physical meaning to the question about space in terms of the *positional quality* of the physical world (in the words of Theophrastus) or as a *receptacle* (as Strato called it). He does so by including motion in his description of mechanics. He had little interest in the positional quality of the physical world; that was taken care of by the relative location of all bodies, of *relative space.* He introduced in his mechanics the gravitational pull between bodies by brute force, without any attempt at an explanation. The riddle for him and us is: How is it possible that two bodies—say, Earth and the Sun—influence each other across empty space without a carrier of the force? For Newton, this enigma was unrelated to the qualities of space. In his opinion, the presence of bodies has no impact on the space that surrounds them.

Let's recall: To Newton, space is an empty receptacle at absolute rest; bodies may be inserted into it or not. It acquires physical significance by the observability of acceleration with respect to its framework. But this concept of space slipped out of his grip as soon as it was noticed that we cannot speak of any velocity of space: There is no experiment that can determine whether, and how fast, the receptacle we call *space* is moving. Newton's influential predecessor Descartes had emphatically taken the position that there is no space where there are no bodies, and that there are bodies only where there is space. In other words, Descartes considered matter and space to be synonymous.

In Newton's terminology, the only space Descartes knows is *relative space.* The same holds true for all motion. According to Descartes, we cannot discuss the motion of any particular body unless we specify with respect to what other bodies this motion occurs. If we cannot specify this, there is no observable motion, linear or rotational. A traveler at sea in a moving vessel is in motion relative to the coast he sees in the distance, but he is at rest relative to his ship. If the boat moves against the direction of rotation of Earth at a rate that cancels that rotation, the traveler will remain at rest with respect to an appropriately chosen frame of reference in the skies—in other words, the stars. Descartes didn't enter into the discussion that the traveler will, in all other cases, participate in the rotational motion of Earth so that centrifugal force will act on him. Newton was the first to realize the significance of this point. He was the first to differentiate between the two forms of motion—those that are accompanied by forces and those that are not.

DE GRAVITATIONE

In this treatise, Newton shows himself to be a natural philosopher who aspires to a comprehensive view of our world—a view that will fulfill all of Descartes's unrealized ambitions. The metaphysical meaning that Newton sought here, as well as in his equally ambitious alchemy, was to remain beyond his reach. The world such as he meant to comprehend it does not exist. Instead, he discovered a way in which our world can be understood; he introduced this method in his *Principia,* and physics is following this path to the present day. A few points not touched on in the *Principia* were hinted at in the earlier *De Gravitatione;* it showed Newton attempting to make room for the concept of the vacuum, distancing himself from that of an ether. As Galileo had done before him, Newton tried to approach the void by means of extrapolation: "Water resists the motion of a projectile less than mercury would; air resists it less than water, and a space filled with any kind of ether will do so less than an air-filled one; and so, if we manage to remove every resistance against the projectile's motion altogether, we will deny all bodily nature to space. We would have to conclude that there are empty volumes in nature. . . ." In other words, if we approach empty space by having it filled with successively more dilute substances, there is no reason to argue that the concept of motion with respect to these substances should be lost. But so it is. If there is no force that resists the motion of a projectile, a state of motion cannot be told apart from a state of rest. Newton had realized this by the time he wrote the *Principia,* where he tried very hard to give an absolute meaning to the state of motion of empty space.

THE CONTROVERSY BETWEEN LEIBNIZ AND NEWTON

Like Descartes before him, Leibniz did not recognize an inherent quality of an object that might be able to indicate whether or not that body is in motion. He agreed with Descartes that the concept of motion makes sense only when we can specify the reference frame: "If motion is nothing but a change in relative position of neighboring or touching objects, there is no way to tell which of these objects is actually moving." To Leibniz, space means a system of relations between different objects; motion, then, means a change of these relations over time, and nothing else: ". . . and I have said time and again that I consider both space and time to be purely relative, defining an ordering of things that exist next to one another. . . ."

This opinion is at the basis of the controversy between Leibniz and Samuel Clarke, in which Leibniz takes the position that all motion is relative motion and Clarke assumes Newton's point of view; Leibniz does not accept the existence of Newton's void. Both sides bring in theological arguments in addition to their philosophical and physical reasons. To quote Leibniz: "The more matter there is, the more chance we have to see God display his wisdom and his might; and for that reason and others I am of the opinion that there is no void at all." And elsewhere: "If those gentlemen pretend that space possesses an absolute reality, they get in deep trouble. That existence would appear to be eternal and of infinite extent. And this, to some, means it would be God himself—or, at the very least, it has the same attributes as God himself in his immensity. But space is made up of parts; and that is a property we cannot ascribe to God.

Answering Leibniz's theological contention that the existence of more matter affords God more opportunity to exert his wisdom and his might, Clarke picks up the theological stance: "No matter how small a quantity of matter we consider, God will not be lacking of objects where he can demonstrate his wisdom and power." To which, after again waxing theological in his turn, Leibniz answers that "every substance that has been created must, in fact, be material." And in a vitriolic postscript to his fourth letter to Clarke, he writes: "All the reasoning that is being advanced to prove the existence of the void [is] nothing but [an exercise] in hair-splitting." In the fifth letter, Leibniz admits that there is, in fact, a difference between a true, absolute state of motion of an object and a relative change of position with respect to other objects. He makes a point of distancing himself clearly from Descartes: "I'm not saying that matter and space are the same. What I do say is: There is no space where there is no matter, space as such has no absolute reality." With his answer Clarke concludes the exchange of letters: "Too bad that death kept Mr. Leibniz from replying to my most recent letter."

Newton considered all motion to be real with one exception: rest with respect to empty space. To his contemporaries, the idea that motion could be established in ways other than by comparison with neighboring bodies was a novelty. Newton's insistence on coupling the reality of absolute motion with the idea that motion in the absence of centrifugal force may be absolute remains confusing to this day. Maybe his philosophically oriented contemporaries might have stood a better chance of understanding him had he not applied the same description to both types of motion.

He assigns an independent existence to both space and matter. Space does not depend on matter; nor does the extent of a body—that is, its spatial description—give an exhaustive account of all its properties. One irreducible property of an object is, in Newton's mechanics, its mass. But all attempts at a definition of mass in terms of more basic properties were doomed to failure prior to the advent of the general theory of relativity in 1916. Newton's mechanics permits the definition of the mass of a body only when we know how it reacts to the forces applied to it.

MASS AND WEIGHT

Which is heavier, 1 kg of lead or 1 kg of feathers? Among all the questions that Newton's mechanics raised, this is the one that has received the most experimental attention. The answer supplied by Einstein's general theory of relativity is based on the expectation that children have when asked this question: that it is a trick question and that they weight the same. And that is exactly what our experimental knowledge tells us. But it is by no means self-evident; if true, it is a law of nature that should be tested again and again with ever-improving means.

A physicist will say that the measure of any *mass* is a kilogram, and that of its *weight* is called a kilopond (the weight of one kilogram of mass). Logic tells us that the definition of a pond implies that a pond—or a kilopond—of lead and of feathers will weigh the same. The kilogram is the unit of mass; but mass is not the same as weight. Equal masses offer the same resistance to acceleration. If we load 100 kg of lead in our car and accelerate with maximum force, we will reach a speed of 60 mph in exactly the same time as we would had we packed 100 kg of feathers in our trunk. The mass of an object is defined by the acceleration that is enacted by a given force. Its weight is the strength with which the gravitational force of Earth pulls on it, and has no a priori relation to that mass at all.

To weigh an object means to determine the force with which Earth attracts it. The process of weighing measures forces, not masses. Weighing an object by suspending it from a spring illustrates this point well: The elongation of the spring tells us the degree of force with which Earth pulls on the object we are weighing. Similarly, the balance used by the pharmacist compares forces, not masses. It will be in equilibrium when Earth pulls on the calibrated 5-oz reference weight on one side with the same force with which it pulls a carefully measured amount of ascorbic acid on the other. The question of whether both these substances will also offer the same resistance to attempts to accelerate them appears as a different question: Is the velocity that a toy train will reach in one second independent of the type of freight its cars carry—the 5 oz of ascorbic acid or the 5 oz of reference weight?

The answer cannot be furnished by logic alone—that is, by the definition of kilogram and kilopond. But to the best of our knowledge, there is indeed an apparently universal law of nature that equates the two types of mass, *gravitational* and *inertial:* If two bodies weigh the same, their resistance to acceleration will also be the same. Thus the speed reached by the train will be independent of the type of freight it carries. Mind you, this is a law of nature, not a logical necessity. As long as both—law of nature and apparent logical necessity—lead to the same result, we don't differentiate between the two types of mass in our daily life. But in physics, we must.

Ask a stuntman whose job it is to pull a truck by a rope that he holds between his teeth about resistance to acceleration; he knows. There is similar resistance against *deceleration,* as many a motorist has had to experience just before and during the crash he was trying to avoid. In addition, there is resistance to any change in direction of motion; ask the javelin thrower who has a hard time counteracting the centrifugal force just before he releases his missile. According to Newton, in all three cases the force needed to overcome the resistance has the same function: It is needed to change the velocity of the body relative to absolute space.

All these types of resistance are described in terms of what we call *inertial mass.* The gravitational force by which objects attract each other, on the other hand, determines what we call the *gravitational mass.* We know that you, the reader, and Earth attract each other, just as Earth and the Moon do, because of the gravitational masses. You would be lighter on the Moon than you are on Earth, because the Moon is less massive than Earth. Your mass, in the inertial or gravitational sense, however, does not depend on where you happen to be. The greater the distance between objects, the less the gravitational force their masses exert on each other.

The original experiment that tested the equality of inertial and gravitational masses was done by Galileo. Let an object fall to Earth in an airless volume; it will be accelerated because it has *weight,* which we measure in units of kilopond. At the same time, its mass, in units kilogram, offers resistance to this acceleration. The resulting acceleration of the object strikes a balance between the gravitational pull of Earth and the force due to resistance to the acceleration. But since the contributions to the balance are proportional to the gravitational and inertial masses, respectively, and since these masses are equal, it follows that their common value cannot influence the acceleration of any free-falling object. In conclusion, except for friction, all bodies take the same time to fall the same distance from the start, independent of their masses and compositions. In other words, in an evacuated volume, all objects are subject to the same acceleration.

Seventy-one years after Newton's death, Henry Cavendish was the first to determine the strength by which two precisely known masses attract each other. His experiment is often called the first measurement of the mass of Earth—and that is exactly what it was. If we know with what force one pound of mass attracts another pound, Newton's first law will tell us what mass our Earth must have so that, at its surface, it attracts a kilogram mass with the force of a kilopond, as we observe. Once the mass of Earth is known, the masses of other celestial bodies—the Sun, the Moon, the planets—follow from astronomical observation and the application of Newton's law.

Experiments like Cavendish's can also be used to examine the equality of inertial and gravitational masses. The first time this was done to great precision was the year 1886, by the Hungarian physicist and aristocrat Roland Baron Eötvös. His result confirmed the equality of both types of mass. Ever since, and particularly in the past ten years, this result has been checked to ever greater precision in response to various challenges. Experimental flukes have been eliminated. As a result, our best information shows the exact equality of inertial and gravitational masses.

POINT MASSES

In Newton's mechanics, both types of mass constitute primary properties of matter; they certainly cannot be explained in terms of the volume of material objects, although volume is the only property that Descartes granted them. Given that inertial and gravitational masses are the same to the best of our knowledge, we will no longer differentiate between them. We will henceforth speak of mass only, and we will measure it in units of kilogram. Now, let there be spherical objects of one given mass, but let them be made up of different materials—say, feathers, wood, lead. The masses are the same, so the radii of the spheres have to be different. We might ask with what gravitational force these spheres attract mass probes at a given distance. Our experience agrees with Newton's law of mechanics: This force is the same for all three spheres: All of the mass inside their perimeters determines the gravitational field of each sphere. Neither the material that makes up a sphere nor its radius have any influence on the external gravitational effect. We might imagine that we let the radius of a sphere grow smaller and smaller without changing its mass, so that its material has to increase in density.

There is no denser material at our disposal than uranium to test our progress in this direction. But in our imagination we can go to ever higher densities and ever smaller radii, until we arrive at a sphere with radius zero that still contains the same mass. This, of course, is no longer spherical at all; it is a a point, and we call it a *mass point*. With the exception of their behavior in collisions, Newton's mechanics lets such mass points act precisely like rigid spheres of finite radius with the given mass. In his studies of planetary motion, Newton replaced the celestial bodies—which, due to their own gravity, are approximately spherical—with mass points. He did so with astonishingly good results, as we know.

The concept of mass as an independent property is a slap in the face of Descartes's system of thought. On the one hand, the gravitational field of spherical bodies of a given mass is independent of their radii. On the other, different masses with the same radius have different gravitational fields. This means that the mass of an object neither determines nor is determined by its volume. In the final count, what we called a mass point—a massive object that occupies no volume at all—turns Descartes's system upside down: While its gravitational field has considerable extent, the object at the basis of this field has no extent at all.

To the best of my knowledge, none of his many controversies, of which we have discussed only the one concerning Leibniz, forced Newton to address this obvious break with Descartes's concepts. He was not able to get away as easily with his concept of absolute space, or with that of the action of gravitation across empty space. He did not believe in the latter, but he could not do without it, since he was not able to explain gravity in terms of a continuous action from point to point. All these concepts contradicted not only Descartes but the mechanistic philosophy of the time, from which Newton's thought was descended.

THE CARRIERS OF GRAVITY

Newton's laws of mechanics say that the Sun acts gravitationally on Earth across the space that separates them. Although immensely successful in their description of observed phenomena, they fail to make the slightest attempt to explain by what agent gravitational force is carried. Newton's success recognizes the mutual gravitational attraction of masses as an empirical fact, but he declines to give it an interpretation: "I frame no hypotheses."

In fact, he was convinced that the propagation of gravitational forces across space was in dire need of an explanation, but he simply didn't know of one. And it would remain a question until 1916, when Albert Einstein presented his first published version of the general theory of relativity. For Newton, two explanations offered themselves: Either there was the emission and absorption of tiny spheres that race across empty space and act as the carriers of gravity, or some sort of ether transfers the action of gravity by its currents. Both interpretations would make sense in the framework of the inherited mechanistic philosophy, but both had to be abandoned because of the untenable consequences they implied.

If gravity propagates without a carrier, it must do so with infinite velocity. Let's assume that our Sun changes its shape from spherical to oblong within one instant. Newton's mechanics prescribes that its gravitational field will adapt instantaneously from that caused by a sphere to a different one caused by exactly the same mass in a cigar-shaped configuration. That would cause planetary motion to change instantaneously, too: This consequence—in contradiction, of course, to present-day knowledge that signals do not propagate with speeds greater than the velocity of light—was also unacceptable to Newton.

Objectively, Newton's theory might be seen to reintroduce into the sciences *mythical actions* across space, which the natural scientists had only recently liberated themselves from, thanks to Gassendi, Descartes, and many others. The tendency of the time was termed *The Mechanization of the World Picture* in an influential 1950 book by E. J. Dijksterhuis. Among the long-distance actions to be explained in terms of interpolating mechanical influences of oscillations, vortices, and the flow of various material objects is the influence exerted by magnets on iron, as well as the influence of the positions of the Moon on the ocean tides. Despite his disavowal of hypotheses, Newton regretted his inability to come up with a viable mechanism for the propagation and action of the gravitational force. He was convinced that there must be a carrier of that force, as he expressed vigorously in a letter addressed to the Reverend Richard Bentley: "To me, it is an absurd idea that an object would act, across some distance in a vacuum, onto another without having that action transmitted by some carrier." But he declined to specify whether the carrier he thought necessary might be "material or devoid of matter."

With the Reverend Bentley, who obtained a doctorate in theology in 1696, Newton maintained a friendly correspondence. Bentley believed in the spreading of gravitational force across space that is free of "all material objects and fluxes." He developed "some enthusiasm for the idea that our universe consists largely of empty space"; his conclusion was that the force with which matter attracts matter issues from an immaterial and divine source. This latter conviction was also shared by Newton, who saw space and time as attributes of God.

INSTABILITY OF FINITE AND INFINITE WORLDS

In his correspondence with Bentley, Newton was the first to discuss a problem that has occupied cosmologists to this day: If our world consists of massive bodies that attract each other gravitationally, why doesn't it collapse altogether? When Bentley raised this question, Newton answered that if the universe were finite and therefore had a center of gravity, as the Stoics had postulated, then indeed all mass would tend to assemble there. "But if matter is uniformly distributed over an infinite space, it will never congeal into one mass; some fraction will be pulled gravitationally in one direction, others in different directions. This will result in a large number of very massive objects that are distributed all over infinite space at large distances from one another. And this is how the Sun and the fixed stars might have been formed." Newton concludes that the universe is infinite; if it were not, it could not have reached a state of equilibrium.

We might question the truth of Newton's concept of how matter is distributed in infinite space. The conclusions we reach on infinite systems must be based on how we approach them from what we know about finite systems. Newton was aware that his universe was unstable. In this respect, he compared the stars to needles that are standing on end, then fall down in one direction or another. From today's vantage point, we are amazed that Newton did not consider a model in which the so-called fixed stars orbit an overall center of gravity in the same manner in which the planets orbit our Sun. He was aware that, more precisely, the Sun and its planets orbit the center of gravity of our solar system. The centrifugal forces that arise from this orbital motion keep our planetary system from collapsing. Similarly, our galaxy does not collapse by the action of its gravitational attraction, because it, too, is in orbital motion. This was not known in Newton's time but he could have foreseen it, since he was aware of the reason for the stability of our planetary system.

FORCES IN FREE FALL

If a physical system is considered by itself in an otherwise empty universe, Newton's mechanics will say that it moves at constant velocity as long as no external forces act on it. The only forces present will be those between different parts of that system. If the whole system is accelerated linearly in some direction, there must be forces acting on it as a whole. In the absence of such forces, it cannot be accelerated. Forces and accelerations acting in a system that is considered alone in the universe can be defined and calculated from one another. They are essentially two aspects of the same thing.

Now let's assume that there is an external mass acting gravitationally on some system that we observe. Then there are two possibilities: First, the system may be kept from following the gravitational pull as though it were suspended by a rope; second, it may be able to move freely in reacting to the gravitational force. In the first of these cases, the system reacts to the outside gravitational pull with forces that compensate for it—say, its weight pulls on the rope and distorts it. In the second case, the system will freely follow the pull of gravity and will be accelerated accordingly. The centrifugal forces occurring in the process counteract the gravitational force of the external mass such that an observer located within an accelerated system will feel that the system is at rest with respect to absolute space. The implication is that the internal properties of small systems—and we will come back to the notion of "small"—make free fall and absolute rest indistinguishable.

Shuttle astronauts appear to float in their cabin as though they were weightless. As they orbit Earth, the gravitational pull and the centrifugal force from the angular acceleration of the shuttle's circular orbit cancel each other. They act with equal strength in opposite directions, so that there is no net force to which the astronauts are exposed. This, incidentally, would not be so if the ratio of

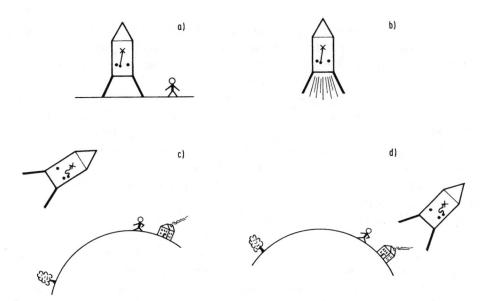

Figure 49 Inside a rocket at rest or in accelerated motion (figs. 49a and 49b), all pendulums oscillate, regardless of the material of which they are made. Inside a free-falling rocket—e.g., one that orbits Earth as a satellite—there are no forces acting on the pendulum; its string appears loose (49c and 49d). A satellite orbiting Earth in the manner shown by the transition from (c) to (d) does not rotate. Tidal forces that have been acting on the Moon through the millennia have forced it to keep the same side pointing toward Earth, so that it takes approximately one complete turn a month.

gravitational and inertial mass, as discussed above, were not the same for different bodies. For another illustration, think of a pendulum that oscillates in the gravitational field of Earth (see fig. 49a). Then imagine we let the box surrounding the pendulum drop in free fall; the pendulum is now no longer subject to any net force and will therefore not oscillate (see fig. 49c). We will later discuss the idea that a pendulum may, in fact, swing inside a spacecraft that is accelerated in an otherwise empty space (see fig. 49b), and also the definition of a lack of rotation of a satellite (or the Moon) when it orbits Earth (see fig. 49d).

Let's note one detail: To be precise, we must acknowledge the fact that an astronaut is not just a point in space; and extended objects that move under the influence of gravity are subject to forces depending on their shape. The gravitational field of Earth decreases with its distance from Earth while continuing to be directed toward Earth's center; it changes in magnitude and direction from one point to the next—from head to toe of the astronaut. Thus the motion due to the centrifugal forces compensating for gravity is not the same at every point. From this it follows that two freely orbiting mass points will not remain equidistant in the process.

This is most easily seen when we compare the motions of the planets as they orbit the Sun in (nearly) circular orbits. They move freely in space, constrained by the gravitational pull of the Sun, but they do not remain at the same distance from one another. If they did, the planetary system could be likened to a rigid rotating disk to which the planets were affixed. In that case, all planetary years—the time in which a given planet orbits the Sun once—would be the same. We know that this is not what actually occurs. The farther removed a planet is from the Sun, the longer its planetary year. Mars, the next planet out from Earth, has a year that lasts twice as long as our terrestrial one.

Gravity and centrifugal force cancel each other only in systems that are small enough so that the gravitational pull acting on them in free fall can be approximately seen as having the same magnitude and direction in all its components. Such systems are important for the considerations that lead up to Einstein's general theory of relativity. The masses of the universe, in their role as the originators of the overall gravitational field of force, determine how such a system moves as a whole; but they have no influence on the physical processes inside these systems. In other words, the laws of nature inside their confines are the same in all of them.

A rotating system undergoes acceleration because the *directions* of the velocities of its parts change. As a result, centrifugal forces act on them. One example, which we discussed, is the merry-go-round. Another is the rocket of figures 49c and 49d. A rocket does not rotate when its direction from tail to head remains the same as a function of time. But how do we define this direction? Surely not with respect to our Earth, which it orbits (and which rotates in its turn); rather, with respect to absolute space, as Newton would say, or to the firmament of fixed stars (in Mach's approach). When the rocket orbits Earth without rotating, that will appear as we indicate in figure 49c and 49d. If, on the other hand, a rocket orbits Earth while keeping the same side turned toward Earth's surface, it will revolve once during each orbit. If a spot on Earth were connected to a fixed star by a thread, Earth would slowly spool up that thread. This is precisely what we mean by "rotation." The Moon always turns the same face toward Earth; that means it revolves once a month about its own axis. If there were a Foucault pendulum mounted on the Moon's surface, it would demonstrate that fact. This illustrates that the laws of nature give an objective meaning to the question of whether or not a celestial body rotates.

LIGHT AND ETHER ACCORDING TO NEWTON AND HUYGENS

When Newton introduced absolute space into physics, he introduced an entity with properties that could be formulated but not observed; Ernst Mach later called it a monster flat out. If we replace this absolute space with a highly dilute substance at rest, and if we accord it properties that make it observable—if not mechanically, then by the way light passes through it—we can define *absolute rest* as rest with respect to this substance. We can identify it with an elastic *pneuma*, as the Stoics would have called it, or with the ether, which has been invoked repeatedly; the ether was seen as the main carrier of light until 1905, when Albert Einstein's theory of relativity made its first appearance.

Newton interpreted light as a current of particles, but he never settled down to a complete, final opinion on the existence of the ether. He doubtless felt the need for the existence of this substance for the transmittal of various effects, but he did not develop a consistent picture of it. As a consequence, his principal written document, the *Principia*, did not introduce it either. But he did mention it in other, private communications, and in his later treatise on optics. There was the problem that his mechanics would not admit that all space is filled with a substance that is subject to its very laws: If that substance—that ether—existed, it would influence the motions of all celestial bodies. To quote from Newton's *Optics:*

> *And against filling the Heavens with fluid Mediums, unless they be exceeding rare, a great Objection arises from the regular and very lasting Motions of the Planets and Comets in all manner of Courses through the Heavens. For all thence it is manifest, that the Heavens are void of all sensible Resistance, and by consequence of all sensible matter. A dense Fluid can be of no use for explaining the Phaenomena of Nature, the Motions of the Planets and Comets being better explain'd without it. It serves only to disturb and retard the Motions of those great Bodies.*

An argument in favor of the existence of ether was, in addition to the fairly complicated part it was supposed to play in the propagation of light, the fact that heat can penetrate an evacuated space. After performing an experiment to this effect which he devised himself, Newton asks: "Is not the Heat of the warm Room convey'd through the Vacuum by the Vibrations of a much subtler Medium than Air, which after the Air was drawn out remained in the Vacuum?" Already Torricelli had discovered that light propagates through an evacuated space. Newton never came up with a definite model for the way in which the ether would support the spreading of light. He felt that light had to be corpuscular, since it spreads along straight lines. But if it is nothing but a current of particles, it cannot produce interference effects, such as the rings that were named after him. Newton's rings are colorful effects produced by light impinging on thin films of matter. We all know how thin films of oil spreading across a water surface can produce such rings; the coloring of some bird feathers has the same origin.

All this means that light either causes or stimulates oscillations. And to have this happen, Newton was obliged to assume that his light corpuscles were accompanied by the oscillation of ether. If light is indeed corpuscular, it is hard to find a principle that would accelerate the light particles immediately after their emission to the enormous velocity of light, and then no further. Newton likened this to what happens when an object falls into water, with the water being analogous to ether, and gravity with the principle that accelerates his light particles: "It is much after the manner that bodies let fall in water are accelerated till the resistance of the water equals the force of gravity."

Newton was not able to incorporate in a unified theory the properties that the ether would need to fulfill its function. Instead, he kept asking questions about it: Needn't there be two different kinds of ether, which influence each other? Or was that impossible? These questions were addressed in very detailed theories developed by Newton's contemporary, the great physicist Christiaan Huygens. By interpreting light in terms of ether waves, he was able to come up with conclusions that appeared to agree with empirical fact. To this day, high school students learn about interference effects—the amplification and damping of light by light—using his method. He explained his concept of light as a wave and not as a stream of particles by stressing the fact that light rays can easily traverse each other, just as people sitting around a table have no trouble talking to their opposites as the sound of their voices crosses above the tabletop.

When one particle current crosses the path of another, the particles collide and the currents perturb each other. This is not the same for sound waves. Huygens knew that sound is in fact not a current of particles; instead, it is an oscillation of air. In the same way, light had to be an oscillation phenomenon; and since it can traverse an airless space, the medium that oscillates cannot be air—thus Huygens settled on the concept of ether for the oscillating medium. Newton had derived his notion of light as a particle stream from the fact that we cannot *see around* a hill, but we can *hear around* it.

Both Newton and Huygens were well aware that a volume that has been evacuated and is therefore free of air will not conduct sound but will transmit light. This observation was at the basis of Huygens's conclusion that this evacuated space had to be filled with something that could oscillate—his ether. It could be argued that Newton's light particles could also traverse that evacuated volume without finding resistance. It was an open question: Their time did not supply them the means for a decision.

A true picture was to emerge only after the year 1900, when Albert Einstein's perception of light as a dual phenomenon appeared; quantum mechanics tells us that light has properties of particles as well as of waves. An experimental phenomenon known as the *photoelectric effect* convinced Einstein that light, notwithstanding its obvious wave nature, also retains particle properties. But up to the year 1900, Huygens's concept of light as nothing but a wave prevailed.

Huygens compared the conduction of sound by air with the apparent conduction of light by ether in order to calculate the latter's property—and concluded that there are essential differences between his ether and air: his ether is much more *elastic,* as can be inferred from the fact that light travels at a velocity that is a million times the velocity of sound.

Another influential mathematician and physicist of the mid-eighteenth century, Leonhard Euler, contributed to the discussion of light, space, and ether. In his attempt to answer Newton's question of how the objects that emit light can accelerate its particles to the enormous velocity of light, he has to assign four different tasks to Newton's ether; and these tasks can hardly be accomplished by one and the same substance. Ether is responsible for all electric and magnetic effects; it also transmits both light and the gravitational force. To do so, it would be easiest to come up with different kinds of ether, characterized by different properties. In particular, magnetic and electric phenomena were seen to be so different that they could hardly be transmitted by the same medium.

MAXWELL'S EQUATIONS UNIFY ELECTRICITY, MAGNETISM, AND LIGHT

James Clerk Maxwell, who devised his phenomenally successful unified theory of electricity, magnetism, and light in 1864, did not solve the problems physicists had had with the ether. Rather, he made them worse. Clearly, the *same* kind of ether had to be responsible for the three phenomena covered by Maxwell's theory. But this demand could not be met by any kind of mechanical model.

We would waste our time were we to try and follow this track any further. But Maxwell thought that it is a matter of identifying an ether with all the requisite qualities that might transmit his electromagnetic effects. Note that his successful equation unified the description of electric and magnetic properties into what was henceforth to be known as electromagnetism. We will illustrate his views in chapter 6 (see, in particular, fig. 68). The crowning glory of Maxwell's concepts was the resulting identification of light as an electromagnetic phenomenon. Whatever ether was needed for the conduction of electrical and magnetic properties therefore had to be responsible also for the spreading of light.

THE VELOCITY OF LIGHT AND OF ETHER

Maxwell's equations provide a quantitative description of all electromagnetic phenomena; they contain only one quantity with the dimensions of velocity. It can be calculated from one purely electric and one purely magnetic quantity, both of which can be experimentally determined in a classroom demonstration. The resulting quantity, the velocity of light, is 300,000 km per second. In terms of Maxwell's equations, the electromagnetic effects that are due to oscillating electrical charges propagate in space with this velocity.

The measured values of the electrical and magnetic quantities on which the calculation of the velocities of light or of electromagnetic waves depend do not change with any possible state of motion of the laboratory in which they are measured. They are one and the same, whether or not the classroom in which they are demonstrated is at rest with respect to the surface of Earth. This implies the same quality for the velocity of electromagnetic waves in Maxwell's equations: It is a universal constant, regardless of the velocity of the system that emits the waves.

That goes without saying when waves propagate in a medium—for instance in ether, if there is such a thing. The velocity with which waves spread depends on the medium, not on the velocity of whatever started the wave. Take a stone you drop into water: The resulting wave will spread at exactly the same velocity—about 1 m per second—irrespective of whether it was dropped from a pier or from a moving ship. A traveler at sea who wants to know the velocity of his ship relative to the water can do so by throwing a rock into the water; the relative motion of the boat and of the wave started by the stone and spreading in all directions will tell him. By adding the known velocity of propagation of the waves in water to the observed difference of the ship's motion and that of the wave that spreads in the same direction as the ship, he can easily determine the relative velocity.

Likewise, when we measure the velocity of electromagnetic waves, this argument tells us that the result will depend on the velocity with which the measurement apparatus moves. The velocity of the waves is always the same with respect to the ether; this means it cannot always be the same with respect to the apparatus that measures it. Consequently, we should be able to find out the velocity of a closed space—a space in which we measure the spreading of electro-magnetic waves—with respect to its surroundings or in relation to the ether. At one, and only one, velocity of the measurement apparatus, the meter will indicate 300,000 km per second; this is the case when it is at rest with respect to the ether. Given that the velocity with which electromagnetic waves move is rooted in Maxwell's equations (i.e, in the fundamental laws of nature), this will be the case when the measuring apparatus is in a state of absolute rest.

Here, finally, we seemed to have the means for an experimental definition of Newton's *absolute space,* or space in a state of absolute rest. I should add that light is a wave emitted by moving electric charges, as Maxwell's equations tell us. The first hint came from the observation that the velocity thus calculated from electric and magnetic properties is "so close to the observed velocity of light that we must conclude that light is an electromagnetic field that propagates according to the laws of electromagnetism." The same, by the way, is true for heat radiation and other radiation, such as X rays. If we take our previous argument seriously, this poses a new problem: Why should our Earth be at rest, or nearly so, in ether?

THE VELOCITY OF LIGHT AND THE THEORY OF RELATIVITY

Out of the flood of investigations in the wake of Maxwell's equations, both experimental and theoretical, let us pick two: the Michelson-Morley experiment

Figure 50 Sketch of the apparatus used by Michelson and Morley for their experiment which proved, in 1887, that the velocity of light is the same regardless of the direction of Earth's motion.

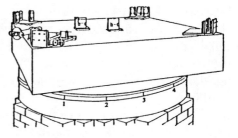

and Einstein's considerations that lead to the special theory of relativity. The first of these provided proof positive that there is no such thing as the ether. Einstein took this result as his starting point. His interpretation of Maxwell's equations didn't need any ether, and revolutionized our perception of space and time in the process.

Let's start with Michelson and Morley's experiment. In the late nineteenth century, experimental technique was sufficiently advanced to make a measurement of Earth's motion around the Sun at a rate of about 30 km per second, observable in its influence on the velocity of light propagation, which we know to be 300,000 km per second. If we start from the notion that light spreads in ether with precisely this fixed velocity, then its velocity as measured on Earth will have to depend on the direction of Earth's motion, which changes with the seasons. This means its velocity in ether, or in absolute space, will be either parallel, antiparallel, or even perpendicular to Earth's motion. The Michelson-Morley experiment was sufficiently precise to notice an adding or subtracting of the two velocities. The result was that no such effect was seen (see fig. 50).

I cannot judge whether Einstein knew of this experiment when he developed his special theory of relativity. I tell my story assuming that he did not. It appeared absurd to Einstein that experiments inside a closed room should give information about the motion of that room; but that is exactly what Maxwell's interpretation in terms of an ether implied: The same experiment in two closed rooms would have yielded, upon comparison of the results, the relative velocity of the rooms. In the case at hand, Einstein applied his deep conviction that this could not be so only to motion with a velocity that is constant in magnitude and direction. Later on, he extended it to arbitrary motion in his general theory of relativity.

Clearly, the measurement of the velocity with which waves spread in the lab where they originate must be independent of the velocity with which that lab moves—if indeed it is impossible to infer the motion of a closed system by experiments that are done inside it. This contradicts Maxwell's theory of an overall ether; but it agrees well with the Michelson-Morley experiment.

This is not the time and place to describe the upheaval of our concept of space and time that was brought about by Einstein's ideas. Space and time are so closely interwoven in his special theory of relativity that one observer's space becomes both space and time to another observer who is at motion with respect to the first. The same holds for time. There is, in fact, a simple formula that tells us how to calculate the relative distance in both space and time between any two events measured by one observer. Both of these distances differ for observers that move with respect to one another.

Einstein's special theory of relativity has survived every experimental and theoretical test to the last significant figure of comparison. As far as our overall topic, empty space, is concerned, there is more to the common elements that join the special theory of relativity and Newton's mechanics than to their differences. There is no experiment in a closed lab that can determine the velocity of its motion, as long as it moves uniformly in a straight line. All measurements in both (closed) labs must lead to one and the same result. This means that if a flight attendant spills a drink on your new suit while you fly at constant velocity and in a straight line, he cannot call on the motion of the airplane to exonerate himself.

Recall Newton's attempt to describe light as a current of particles; its velocity of propagation would clearly depend on the velocity of its source. That turned out to be wrong: Maxwell's equations correctly defined light as a wave. It took Einstein's interpretation to come up with an overall acceptable interpretation of the Michelson-Morley experiment.

THE VELOCITY OF THE UNIVERSE AND THE LAWS OF NATURE

If Maxwell's ether existed, it would determine the meaning of absolute rest. Local disturbances aside, ether would be at rest in and by itself. Rest with respect to ether could therefore also mean rest with respect to the universe as a whole. But observers don't need ether to find out whether they are at rest with respect to the universe—they can get that information from a look at the stars. That means that rest with respect to the ether isn't all that interesting in this context. Rather, it acquires interest from the fact that Maxwell's equations say that an observer's velocity with respect to the ether is a quantity occurring in these fundamental laws of nature. The inference is that the observer should be able to determine velocity inside a closed room, and without a glimpse at the stars, just as the observer is able to determine the acceleration of a closed system with respect to Newton's absolute space.

We discussed this before. There is no way to overestimate the importance of how different a velocity with respect to the whole universe, such as it occurs in the laws of nature, is from some arbitrary velocity. We don't have final knowledge of truly fundamental laws of nature, so we don't know whether the velocity with respect to the universe as a whole, such as we might determine it in a closed room, figures prominently in those laws. The one thing we know is that planet Earth does keep moving in the direction of the constellation Leo at a rate of 390 km per second; the motion of Earth around the Sun, which obviously changes directions, happens to involve a velocity of about 30 km per second, and does not easily lend itself to our best measurement techniques. Rather, we have trouble separating it from the motion with which our solar system orbits the center of the Milky Way. Once we include this, we arrive at the velocity of the Milky Way with respect to the system where the universe as a whole is at rest; this velocity comes in at 600 km per second.

The experiment that gives us this result tells us the velocity with which our Earth moves through the so-called cosmic background radiation, which was discovered in 1964–65. This radiation is, on the average, at rest with respect to what we might call all matter in existence—to the center of mass of the galaxies of the whole universe. The cosmic background radiation fills "empty space" with a termperature of 2.7 degrees Kelvin, or − 270.3 degrees Celsius. When it first came into being, some hundred thousand years after the Big Bang, its temperature was about 4000 degrees Celsius. As the universe expanded, it cooled down.

Just as observers can determine their velocity with respect to Maxwell's ether by experiment, they can do the same with respect to the cosmic background radiation. The difference between the two experiments is that the observers can shield their apparatus from cosmic background radiation. Earth's surface itself is shielded from it, by the surrounding atmosphere, so well that it took precision experiments in 1964 and 1965 to notice it for the first time. But experimenters cannot shield their apparatus from Maxwell's ether. Whatever they do, the ether will penetrate their lab. This is why experimenters can determine their velocity with respect to Maxwell's ether in a closed laboratory, and not with respect to cosmic background radiation. This radiation, in contrast to the ether of Maxwell's mental construct, has no role in the basic laws of nature and cannot penetrate everywhere. Its existence is due not just to the laws of nature but also to the initial conditions and the present age of the universe. We can ask questions about past and present nature—questions that make no sense at all when directed at the laws of nature themselves.

In Maxwell's system, the ether has to be present in Torricelli's vacuum, because the vacuum is translucent to light. Modern notions that pick up the ideas that defined the ether in the first place can be summarized by the generic term *space*, and in this definition ether obviously penetrates everywhere. But if ether is nothing but space, there is no way of determining any velocity that is constant in magnitude and direction with respect to it. We might cite the quantum mechanical vacuum with its fluctuations, or the Higgs field mentioned earlier, as examples of an ether of this generalized description.

MOTION AND GEOMETRY

We have discussed the equality of gravitational and inertial masses, which is valid for all objects; that means we can use probes consisting of any material for the determination of the trajectories of free fall. In other words, these trajectories are the same for all objects irrespective of their makeup. They probe space itself and its properties. This statement is one of the basic tenets of the general theory of relativity.

If we take a large number of trajectories followed by objects that fall freely in a gravitational field, these trajectories form a web that reflects the structure of space in that field. Our knowledge of any one of these paths and of the field itself also tells us the velocity with which the object in question falls. Conversely, the paths and velocities of free-falling bodies let us calculate the field. Take an arrow that is shot vertically into the air from the surface of Earth and turns around at a height of 10 m. These facts are sufficient to let us calculate the velocity with which it was launched (disregarding air friction). In just the same manner, we can take the initial velocity and the height of turnaround to determine the strength of Earth's gravitational field. The relation between these three quantities would clearly give different results on the Moon.

These relations involve velocity and therefore also time. We cannot do without time when we try to infer the gravitational field from the web of trajectories of falling bodies. That is true for Newton's mechanics as well as in Einstein's relativity theories. The difference is mainly that space and time are independent of each other according to Newton's mechanics but not in reality, where relativity reigns. Relativity implies that there is a highest possible velocity in nature, that of light. Thus space and time depend on each other; they cannot be scaled independently. Say that in otherwise empty space a gun shoots bullets at a target 10 m away with a velocity that amounts to 60 percent of the velocity of light. (In reality, bullets do not move quite that fast. But elementary particles in accelerators move even faster.) Now let's imagine we double the distance between gun and target, keeping the time the bullet needs to fly from one to the other the same. That is what we mean by scaling space independent of time. This scenario requires the bullet to fly at a velocity 20 percent above the velocity of light. Keeping the distance the same and asking the time of flight to be halved—that is, scaling time independent of space—leads to the same requirement on the velocity: It must be 120 percent of the speed of light. In Newton's mechanics, this requirement would not at all lead out of the realm of possible events Space and time in Newton's theory are independent of each other in the sense that they can be scaled arbitrarily without ever leading from possible to impossible motions in empty space. Not so in reality and not so according to Einstein: Nothing can ever move faster than light. Thus space and time are no longer independent variables in the theories of relativity; they are fused into a mutually dependent four-dimensional space-time.

Einstein's most important objection to Newton's absolute space was that "it acts, but it cannot be acted upon. This runs counter to scientific reason." Newton's absolute space acts as if it were an *ether of mechanics* since, as we discussed before, all masses resist acceleration; at the same time, masses do not change the nature of the space they enter. They cause acceleration through gravitational fields, but the resistance offered to all acceleration with respect to absolute space remains the same.

This is totally different in the general theory of relativity. Its space is in no way absolute; it is, we might say, adaptable. It adapts to the masses of the universe (and, as we will later discuss, to the cosmological constant). All masses act on space, and space *reacts* by fixing their trajectories. The web of bodies in free fall is identical, in the general theory of relativity, with what we call space-time.

This is not entirely unlike Newton's empty space, which also can be defined by the trajectories of small mass probes: In the absence of other masses, and therefore of gravity, these probes will move from one point to another on the shortest available path—that is, on straight lines as defined by Euclidean geometry. Time runs its course independently of space. Once we introduce real masses other than probes, and thereby gravity, into this space, free-falling probes no longer follow the straight lines of Euclidean geometry. Still, the laws of nature determine their trajectories in space and time.

In this formulation, what the laws of nature say about motion of mass probes in a gravitational field is based on the shortest distances as they were defined in a space free of gravitation. These trajectories once and for all fixed the geometry of space in terms of the shortest line between any two points. Geometry is Euclidean; every triangle consisting of straight lines will see its three angles add up to 180 degrees. The shortest path between two points is always a straight line, possessing precisely one parallel.

This characterization of trajectories in terms of Euclidean geometry appears simple and obvious to us; but it makes for difficult laws of motion. Such trajectories will not be along the lines of shortest distance—in other words, "straight lines"—in all geometries. Einstein's general theory of relativity elevates the *concept of inertia* rather than that of Euclidean geometry to *the* central postulate of its edifice: Mass points, or probes, will move along lines of shortest distances in an appropriately chosen geometry. The shortest distances followed by free-falling mass probes will, of course, no longer be the same as those of gravitationless Euclidean space. Rather, the geometry of this inertial space will have its shortest distances defined by the influences of all masses of the universe. Once its geometry is fixed in this fashion, the trajectories become simple: They again are *straight lines* in a space-time shaped by gravity. What is not simple is this space-time itself, although it is defined in terms of such shortest distances. This space-time is a *four-dimensional* space, where time is the fourth dimension. The fourth dimension was not necessary for the description of Euclidean space, where all shortest distances are straight lines in the three-dimensional space we are accustomed to. It is the inclusion of gravity that changes things. We can illustrate that in terms of a game of tennis (see fig. 51). A tennis player (at least a good tennis player!) can hit the ball toward his opponent's baseline, from his position at the

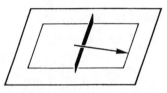

a)   A tennis court with one player at the net, who
      hits the ball directly to the opponent's baseline.

b)   A player at the net, just as in a), who
      serves a lob to the opponent's baseline.

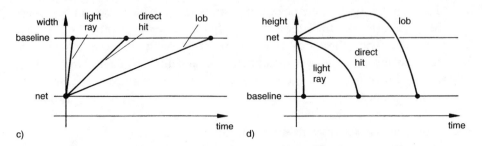

c)

d)

Figure 51 In a tennis game, a straight hit (a) and a lob (b) will result in different times elapsing as the ball moves from the net to the baseline. If we know the point where the racket hits the ball and the point where the ball hits the ground and the time needed for its path, we can determine the ball's trajectory. This is true for all objects: The equality of inertial and gravitational masses sees to that. Once we subtract the effects of air friction, a Ping-Pong ball and a volleyball will take the same time as a tennis ball. This doesn't apply to a ray of light: It will always move with the highest possible velocity, c = 300,000 km per second. Figures 51c and 51d decompose the motions of the ball in terms of their components, length and height, and compare these with the motion of a light ray from the top of the net to the baseline. Gravity doesn't affect motion parallel to the surface of Earth; as a result, the trajectories in figure 51c are straight lines. This doesn't apply to the trajectories in figure 51d, because Earth's gravity acts on the ball and on the light ray. The deviations of the light ray from a vertical straight line are vastly exaggerated in both diagrams, to illustrate our point. The special theory of relativity tells us that all three trajectories are shortest distances in space-time in the curved four-dimensional space close to the surface of Earth.

net, either by a straight hit or by a lob. (Let's leave out fine points, such as the effect of air friction on the ball.) In both cases, the ball's trajectory begins and ends at the same points. But the two trajectories (see figs. 51a and 51b) are entirely different, and so are the times needed by the ball for its traversal. Now let's enter time into the criteria that define the ball's path: If we prescribe that the ball take, for its trajectory, precisely the time that is needed for a lob, then figure 51b shows that the player at the net has no other choice than the lob. The joint demands of space and time fully determine the ball's path.

Once we fix both the location and the time of the beginning and the end of a trajectory, there will be one, and only one, path that a mass probe can take according to Newton's mechanics. There is no analog to the principle of the shortest path in Newton's theory, however, since space in that theory is fixed once and for all by the trajectories of mass probes in empty space without gravity. Newton's theory places gravity into previously defined space as a new and independent reality that can be absent or otherwise without any change incurred by space itself. The four-dimensional space-time of general relativity is altogether ignorant of Euclidean straight lines, because its very definition includes the action of gravity on the probes. Quite irrespective of the distribution of all masses in the universe, the trajectories defined by the four-dimensional space of the general theory of relativity are as simple as they could possibly be: They are the shortest distances in this space. What's complicated is the space—a four-dimensional space with curvatures defined such that the trajectories will, indeed, be the lines of shortest distance.

SPACE, TIME, AND THE METRIC FIELD

Only a book fully dedicated to the general theory of relativity can give all pertinent details. Its four-dimensional space, consisting of space and time, contains a field; to each and every point in it we have to associate numbers that tell us about the curvature of space in that location, at that time. We can characterize the curvature of space by the sum of angles in a triangle, as we explained in chapter 2. The field quantity will be the deviation of that sum from 180 degrees. When that sum is 180 degrees, the space is said to be *flat,* or Euclidean. Another way of tagging the (constant) curvature of a space is by means of the value of its *pi,* the ratio of circumference to diameter of a circle in that space, which is a constant number. (Its numerical value starts with the digits 3.14 in our Euclidean space.) In curved spaces, on the other hand, this ratio will, in general, take on different values—the generalized *pi* of that space characterizing its curvature. We illustrate this in figure 52. If the curvature of a space varies from point to point, so will the value of its *pi.* This means we can use local *pi* values to characterize the field of curvature of any given space. As a two-dimensional space with a curvature that changes from point to point, consider the rubber cloth in a trampoline, as shown in figure 53.

The metric field of the general theory of relativity which implies the curvature of four-dimensional space-time summarizes all relevant properties of it. It is not added to the space, like a color or like a force that might act on electrical charges. Rather, in a somewhat restricted sense, the field is synonymous with the space itself. We qualify that remark because different fields can, in principle, stand for the same space. Take a sheet of paper with a number 6, and nothing else, painted on it (see fig. 54); put that sheet down on the floor between two observers. Both will see the same figure—but to one of them it is a 6; to the other, a 9. They

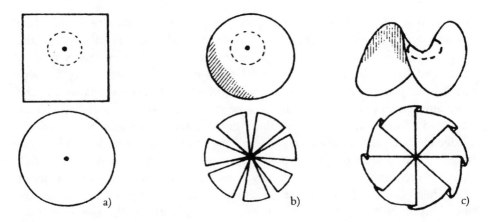

Figure 52 The circumference of a circle of a given radius in a plain (52a) may be smaller (52b) or larger (52c) in a curved space. As a result, the ratio of circumference to diameter will differ in curved spaces from the Euclidean value of *pi* = 3.14. . . .

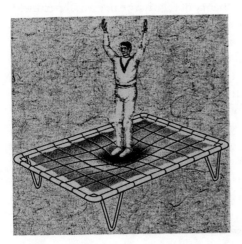

Figure 53 The stretched mat on a trampoline in use is a fine example of a two-dimensional space with a curvature that changes from one location to the next and from one time to another.

see the same figure from different vantage points. Their descriptions of one and the same reality therefore differ. As soon as they change positions, their observations will also change.

The metric field is a *description* of space. But not uniquely so: The same space can be described in terms of different fields. The field, in general, depends on the vantage point of the observer. The space it describes does not. In the general theory of relativity, space is nearly independent of the specific characterization of its points in terms of given coordinates. All fields that describe the same space

Figure 54 Is the figure painted on the ground between the two gentlemen a 6 or a 9?

are physically equivalent. Given the space, its description can be realized in terms of different fields.

As long as two fields that describe one and the same space are identified, the field and the space can be used synonymously. There is no such thing, then, as a quantity called space that can be defined in terms of a set of coordinates and is independent of the field consisting of numbers that tell the distances between any two points with different coordinates. We might say that the four-dimensional field contains all information on the spatial character of our reality. In this sense, there is no empty space: There is no space without a field.

Let's take another look at figures 17 and 20: Anybody who watches the bugs in these figures as they stake out their two-dimensional spaces will get the impression that they are surveying a *field.* Just as geologists measure the magnetic field of Earth with the needles of their compasses, our bugs measure their space with measuring sticks. In another context that we discussed, Carl Friedrich Gauss did a field measurement when he determined the sum of angles of a triangle subtended by three peaks of the Harz Mountains. The quantities measured by Gauss or by our bugs—distances and angles—characterize their spaces. Whether space is or is not a field proper, it can be measured in terms of a locally existing field. Let us include time now; we can measure time in one location in comparison to another location by carrying a clock back and forth. Depending on the structure of space-time and the paths of the clocks within it, they will have advanced differently when brought together again. It should be stressed, however, that their pace at the same point and at the same time is necessarily the same, if each of them is a bona fide clock. Again, when we carry meter sticks and clocks around, we are measuring a field of shortest distances of points in space and time—a field that can also be characterized by the assembly of all possible trajectories of free-falling probes. That space and time are closely interwoven in these paths shows four-dimensional space-time to be the proper arena for the physics of gravity.

A space that is empty in the sense that it doesn't contain masses is nothing but a special case of the above. Looking at it from the four-dimensional vantage point, space and time are interwoven according to the laws of special relativity theory. Space proper is Euclidean in its limit of vanishing gravity—trajectories of free fall are the straight lines already known from Newtonian physics. I will not go into detail considering space and time together in that case. Suffice it to say that this "empty space-time," too, can be described by a field; it is just one specific case among many others, and doesn't stand out.

The gravitational field of Earth will determine the direction of the rock that you drop from your hand. Aristotle concluded, from the simple fact that the rock "knows" in which direction to move, that space cannot be empty where it transmits that knowledge. Both Newton and Einstein would agree—the former because the gravitational field of Earth acts at the location of the rock, the latter because Earth's gravitational field actually curves space in that location.

Newton's physics is all about masses that attract each other gravitationally and resist acceleration. They populate the absolute space of his metaphysics—and this space takes no note of their existence. As soon as there are masses, there will be gravitational fields. These fields are defined everywhere in space and will determine the behavior of probes at every point. But these gravitational fields don't change the space. Rather, like the masses, they simply populate it. In their presence, space is not empty—but it *can* be empty, and will be in their absence.

The same holds for Einstein's *special* theory of relativity. In the *general* theory, the field is all-pervasive; there is no such thing as space without a field. The space and all bodily objects, or masses, are in the final analysis just aspects of the fields. If the space is our principal interest, we can determine its curvature, its *field*, through observation of the free fall of mass probes. I repeat, I'm not just talking about a gravitational field that is simply added to space in a Newtonian context; here we are dealing with the field of curvature, which is a property of space itself. Once we know this field, we know the distribution of masses. Conversely, we might have started with the distribution of masses and used them to calculate the field of curvature. It is the masses themselves that, like pearls on a string, determine the web of trajectories; this web contains all information about masses, space, and its field of curvature.

Newton started from the physical concept of bodily objects. His absolute space, one of the foundations of his metaphysics, was a container of bodily objects and was independent of those objects' presence or absence. In contrast, the general theory of relativity starts with the field as the basic concept. I don't find any metaphysical element at its foundation. Rather, it is a realization of the idea that space is nothing but a properly modified expression for what we earlier called the *positional quality of the world of bodies*. For Aristotle, Theophrastus,

Leibniz, and many others (we have to exclude Descartes), space was determined by the bodily objects present; the inverse relationship did not exist for them. In the general theory of relativity, this is different; the field is the central concept. Space and bodies are equivalent, devoid of a hierarchical order; one determines the other by means of the field.

The inverse relationship—first space, then bodies—makes its initial appearance in Descartes's argument that the physical extent of bodies is their basic property, from which all others follow. Newton's mechanics owes its overwhelming success to the negation of this connection. Although he came closer than anyone else before Einstein to the actual physical content of the general theory of relativity, his metaphysics took an important step in the wrong direction.

FIELD THEORIES: CLASSICAL AND QUANTUM MECHANICAL

For Aristotle, the universe surrounds Earth, which is located in its center. It is finite—beyond its outermost sphere, there is nothing, not even empty space. His space, which, as we would say today, determines the laws that prevail on Earth or in the skies, is relative. Close to Earth, a rock falls back to Earth; farther away, above the atmosphere, the rock will orbit Earth. Earth therefore acts on the totality of Aristotle's universe; there is no such thing as empty space. The position in relation to Earth determines what happens at a given point.

We might also relate the position of that point to the uppermost sphere; for our purposes, this is equivalent. Aristotle's space has no existence of its own, detached from its position relative to Earth and the stars. Mach's principle states that material objects can be influenced only by other material objects. It is only the masses of the universe as a whole that determine the meaning of motion free of acceleration, or of free fall. According to Mach, space cannot be anything but a calculational quantity. If there is a physical cause for resistance to acceleration—which, for Newton, was the manifestation of a separate law of nature—that cause must be the presence of physical bodies. In empty space, different laws might apply. In Mach's context, there is no system that can be shielded from the influence of the physical universe as a whole; there is no such thing as a closed system.

This construction envelops the state of our world in an observational physical framework. It does not make a difference between the laws of nature and its initial conditions for the understanding of the world as a whole. While that is a natural ultimate goal of physics as a science, it remains a program at best, to our day. Newton's absolute space remains an empty container, free of all the complications attending a world of bodies. Such bodies, or masses, may enter into this space or they may not; they determine gravitational forces. But they have no influence on the definition of motion without acceleration; this definition is absolute.

Insofar as the general theory of relativity follows Mach's ideas, it can be seen as expressing the distribution of masses in terms of properties of space itself. It all boils down to geometry—to the web of shortest distance between points in space-time. Here, all that exists can be expressed in terms of geometry. It is the motions occurring under the influence of all the masses of the universe that make up the elements of the geometrical web. The identity and the location of those masses can be read off the curvature of space—that is, off a geometrical quantity. Seen in this way, the general theory of relativity is a model for other, nongeometrical theories where the fields are added to space, just as we might add colors to a blank canvas; this is the way we treat electric or magnetic fields. But let us recall that those latter field theories are more realistic insofar as they quantize the field. It is only through the quantum effects that our hypothetical empty space becomes the physical vacuum.

CHAPTER 5

CROWDED SPACE

•

MOVEMENT ALL AROUND—THE QUANTUM VACUUM

•

THEORY AND EXPERIMENT BOTH SHOW US THAT LIGHT DOES NOT NEED a material medium for propagation. We would have no trouble understanding this if light were nothing but particles zooming along through empty space. But that is not the way it goes: Light is a vibration, an oscillation made up of elementary oscillations we call photons. But there is *no substance* that oscillates when light propagates; thus we cannot impede the propagation of light when we remove substances from some given space.

LIGHT AS AN EXCITATION OF SPACE

Let's try and understand light in terms of an *excitation* of empty space—even if that makes no immediate sense. We might alternatively understand light in terms of a *field*, as we introduced that term in chapter 2. But light differs from the fields of temperature distributions, of sound, or of water in fluid motion described there: Whereas those phenomena are due to the composite action or motion of molecules at a more elementary level, light has its own reality at that level. It cannot be understood in terms of an oscillation of some matter that also exists in the dark—no, light is nothing but just that, *light*. It is an oscillation of an abstract nature, equivalent to a set of numbers that are assigned to each point in space. True, these abstract numbers have implications—most notably, they imply energy. But while a water wave transports energy by the movement of water molecules, the passing of a light wave does not mean that anything material oscillates. The energy of the liquid wave is the energy associated with gravitation and motion of its molecules; the energy of light is energy pure and simple, associated with every illuminated point in space.

I repeat: It takes nothing but space proper to make the oscillation of abstract field quantities possible; and these oscillations can and do transport energy. This may be hard to visualize, but we can read this behavior off the equations that describe light propagation very clearly. We recall from the previous chapter that a sustained search for the carrier substance of light—what was loosely called the ether—led nowhere. There *is* no such substance. Light is an excitation of empty space proper, of the vacuum. It is no more and no less.

OSCILLATIONS OF SPACE, OSCILLATIONS IN SPACE

We do not have a clear idea of the ultimate vacuum—of whatever remains once we have removed from some well-defined space everything that the laws of nature permit us to take away. The reason is that this depends on the laws of nature—all of them, whether we have deciphered them or not. If we were sure we understood *all* of them, we would indeed be able to define the vacuum. As it is, all we can discuss is certain aspects of the vacuum, starting with Newton's *absolute space* and, as of now, winding up with Mach's *relative space* embedded in the space-time world of the general theory of relativity. The implications of quantum mechanics for the vacuum remain to be discussed.

Our question about the possibility of removing all radiation from some spatial volume will have to elicit a negative response. The crux of the matter is the propagation of the oscillatory phenomenon that is *light,* inside a space devoid of any substance that might do the oscillating. Our void is filled to the brim; but just those elements that our perception would traditionally expect are clearly absent. Light is in a category of its own, defined in terms of field quantities: Abstract numbers are assigned to each point in space, and they change with time.

What really transcends our understanding is the mechanism that permits these oscillations of abstract field quantities to carry energy. We know this is true; our relevant equations say so, and the solar energy collectors prove it. A photoelectric cell registers at this instant an energy that left the Sun, in the form of light, some eight minutes ago. Traveling across space, this light did not lose its energy; rather, it is part of the energetic excitations within empty space.

If our main topic were quantum mechanics, I would now stress that the emission and absorption of light by matter occurs in a manner quite unlike what we would expect from a wave; rather, it resembles a statistical process—as though we were dealing with an ensemble of particles. This odd *duality* of waves and particles is a characteristic trait of modern physics. It is with a view to the specific experiment or observation at hand that we have to decide whether to treat light as a particle or as a wave. The one element of quantum mechanics we must deal with is the so-called *uncertainty relation;* we will do so shortly.

Back to light: Let's remember that it is tantamount to an oscillation of abstract field quantities *in* space, not an oscillation *of* space proper. But the latter exists, too.

BLACKBODY RADIATION

In the pursuit of a truly empty volume of space, we have to remove all matter, but also all light, from it—and not just light, which is just one form of electromagnetic radiation. So are X rays, radio waves, thermal radiation, and other parts of the spectrum, distinguished from each other only by their frequency of oscillation.

There is no pump that can suck radiation out of a volume. To keep a volume free of light at room temperature (and we'll soon see why this qualification matters), we just surround it with nontransparent walls. That will work for X rays, too, but not for thermal radiation; the walls of our vacuum container have a given temperature, and will emit thermal radiation according to it. As the temperature rises, so does the frequency of the emitted radiation; it will pass into that of visible light—first as the red glow of smoldering coals. Once the thermal radiation reaches the surface temperature of the Sun—some 6000 degrees Celsius—we perceive it as white light. At still higher temperatures, 8500 degrees Celsius, the light will be blue. Beyond this, thermal radiation passes into the X-ray range. And so it goes, ad infinitum: the higher the temperature, the higher the frequency of thermal radiation.

If, inversely, we lower the temperature, both the frequency and the intensity of our thermal radiation will decrease. A formula derived by Max Planck as long ago as 1899 describes this behavior; it tells us that even close to absolute zero—at 0.0005 degree above absolute zero—the walls of the vacuum container will emit a tiny amount of thermal radiation in the FM radio frequency range. Planck's formula predicts no emission of thermal radiation only at absolute zero—and in that, it errs. There is a remaining "ground state energy" even at absolute zero temperature, which cannot be removed from any volume. This phenomenon, which we will describe in terms of the so-called Casimir effect, was discovered by Einstein in 1909.

The walls of our vacuum volume, traditionally called a *blackbody*, cannot continue emitting thermal radiation indefinitely. To keep in equilibrium, they will also have to absorb thermal radiation from the volume, from the blackbody itself; if that weren't so, they would have to cool down progressively, while the volume they surround would progressively warm up. This is an absurd notion—it would transform the vacuum volume into a power station. No, since equilibrium must be maintained, walls and volume have to exchange equal amounts of thermal radiation per unit time.

As a result, we can assign to every blackbody the temperature of its walls. This may be self-evident inside the volume close to the walls: If the walls were warmer than the volume of the blackbody, they would radiate energy until the volume reached the temperature of the walls; conversely, if the volume were at a higher temperature than the walls, it would generate a heat flow to equalize the two.

We can apply the same notion to different parts of the blackbody volume. Its temperature must be the same throughout. Thus we can ascribe some definite temperature to every blackbody, irrespective of the nature or position of its walls. We are free to imagine every part of a blackbody volume as immersed in a larger such volume instead of surrounded by physical walls; and we can remove the walls arbitrarily far, so that their existence becomes altogether unnecessary. We may even replace them with dust particles that we distribute throughout the volume—and we can shrink their sizes successively. Eventually, just like the mass probes we discussed in previous chapters, we can make them vanishingly small.

THERMAL EQUILIBRIUM AND EMPTY SPACE

In our discussion above, we assumed that there has been an arbitrarily long time for the exchange of heat between different parts of the blackbody volume and between volume and walls. The entire system is presumed to be in *thermal equilibrium*. Any space that is not in thermal equilibrium can surely not be empty; something is going on in there—thermal energy is being transported. By that token, our best candidate for a truly empty space will have to be a volume from which all matter has been removed, and which is in thermal equilibrium at some given temperature. As long as this temperature is not − 273 degrees Celsius, our volume will contain thermal radiation; it cannot, by that token, qualify as empty space. But suppose we were able to reach absolute zero temperature—would that change matters?

First, there is no way to reach absolute zero. The definition of absolute zero is that thermal motion of atoms and molecules comes (*almost*—we will get back to this point later) to a stop. We know that the warmer a substance is, the more its atoms and molecules move about in random motion. We cannot reduce the temperature by simply stopping that motion molecule by molecule. Quite apart from the practical problems associated with the procedure—say, that the molecules are too small to be caught by a tweezer—there is a more fundamental difficulty: The tweezers themselves have a temperature, and when they come in contact with a colder substance, there will be heat exchange. The tweezers will cool down, the substance will heat up. Think of trying to stop the slight swaying of a lamp suspended from a long cable. It is a well-nigh impossible task; our hands tend to move faster than the remaining motion of the lamp, and to excite the swaying instead of stopping it. The same would happen if we tried to stop molecular motion. It is simply impossible to bring a substance down to a temperature of absolute zero; there cannot be a blackbody without radiation because absolute zero cannot be reached. And that means that empty space in its true sense cannot be created.

What we can still ask ourselves is this: If we cannot reach it, can we come arbitrarily close to empty space? We *can* get arbitrarily close to absolute zero, and entire fields of research and technology are fed by this. If we reduce the temperature of a blackbody more and more, we approach our candidate for truly empty space—matter-free space at temperature zero. Thus it makes sense to ask: Does the end point of this process, well defined if not fully attainable by experiment, have the properties we expect from empty space?

GALILEO'S METHOD OF EXTRAPOLATION APPLIED TO TEMPERATURE

Questions of this kind can be answered by means of the extrapolation method. This method, of course, is applicable only when the quantity of interest at zero temperature is well defined at finite temperatures. In that case we might first investigate it at, say, 10 K, then at 5 K, at 1 K, and so on, all the way down as far as experiment will allow. If our observation shows a simple dependence of the quantity measured as a function of temperature, we may safely extrapolate it to 0 K.

This method applies to the properties of blackbody radiation—the radiation inside a volume free of all matter. We will address two questions here: First, how does the radiation intensity depend on temperature? Does it vanish at zero temperature? We find this not to be the case—our extrapolation indicates a remaining radiation intensity at 0 K (or -273 degrees Celsius). Second, what properties distinguish an otherwise empty volume that retains only some blackbody radiation at zero temperature—dubbed *zero-degree radiation*—from truly empty space? We will show that one remarkable feature of zero-degree radiation is shared with empty space: An observer is unable to determine whether he is at rest or moving with constant speed with respect to it.

THE CASIMIR EFFECT

The fact that the so-called blackbody radiation persists even at zero temperature finds its most direct experimental expression in a phenomenon known as the Casimir effect. I briefly alluded to it in the prologue: Two parallel plates arranged inside an otherwise empty volume will attract each other. These plates are configured such as to reflect the blackbody radiation incident on them. That makes them probes for the presence or absence of this radiation. The investigation, of course, cannot be performed at zero temperature proper; instead, we approach it by extrapolation.

Recall that the blackbody radiation the plates reflect is electromagnetic in nature, as are light or radio waves. We all know from high school physics that

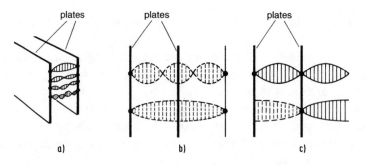

Figure 55 Wave forms permitted between two fixed points.

there are no electromagnetic waves inside a conducting medium. We also know that inside an antenna, which is also a conductor, radio waves will excite the motion of electric charges—particularly electrons. The resulting electrical current is routinely amplified in a radio amplifier; it is then transformed into mechanical motion in a speaker system which ultimately results in acoustic waves that transmit to our ears whatever message the antenna caught—a Beethoven quartet, a commercial, or even a lecture about empty space and its implications. The radio waves give their energies to the charges in the conductor as they set them in motion; that means they will not penetrate any further into the volume of the conductor. If the antenna is electrically insulated, there is no way for it to have the resulting electrical current conducted away. As a result, the charges will simply oscillate and emit radiation to the outside in what we call a reflected radio wave. The analogous phenomenon occurs for all electromagnetic waves, such as light or even blackbody radiation. What is different from case to case is the depth to which the waves penetrate the conducting antenna. If it were *ideally* conducting—that is, if there were no resistance to the motion of electrons inside the conducting metal—this penetration depth for electromagnetic waves would be the same for all materials: It would be exactly zero.

We can regard an electrically conducting plate as a mirror for electromagnetic waves. Because these waves cannot penetrate into a conductor, they will create a condition at its surface. There cannot be an oscillation at the very surface: The oscillating field has to have what we call a *node,* a local zero amplitude at the metallic surface. If that were not so, the field would have to jump from a finite value just outside the conductor to zero just inside, and such discontinuities cannot occur.

Imagine the oscillating electromagnetic field in terms of an oscillating string, like a guitar's. Let it be fixed in location at the surface of the conductor. Figure 55a shows four of the possible modes of oscillation of a string attached to the surfaces of two parallel plates. Every conceivable motion the strings can engage in once they have been set in motion—if no further external force is introduced—can be described in terms of simple harmonic oscillations (known in antiquity to the Pythagoreans). Since the string is held fast at both ends, there are always at least two locations where the amplitude of oscillation is zero—that is, two nodes at the ends. This minimum number of nodes occurs in what we call the fundamental mode of oscillation. As we double the number of oscillations per unit time—acoustically speaking, as we raise the resulting tone by an octave—the distance between two nodes is cut in half. As the figure suggests, we can continue with more nodes in the same way. The important feature is the condition that permits only those oscillations where the distance between two nodes, which is half a wavelength of the oscillation, is a whole fraction of the distance between the reflecting plates. This condition excludes wavelengths such as we show in figure 55b. Figure 55c demonstrates that a doubling of the distance between the plates of the previous figure (55b) will lead to a permissible configuration (see upper wave).

Whenever light impinges on a mirror only to be reflected by it, it imparts a recoil momentum to the mirror; so do all electromagnetic waves and oscillating strings. Let us imagine that of the two plates in figure 55a, the left one is fixed in location and the right one can move freely. This latter plate will be hit from the left, in the blackbody delimited by the two plates, only by those electromagnetic waves that fit into its boundary conditions. It will reflect them back to the left, meanwhile absorbing their recoil momentum. On its other side, the movable plate will be hit by all the waves coming from a blackbody that extends to infinity. In the absence of boundary conditions on the right, there will be a great deal more radiation impinging from the right (see fig 55c). Not only are there those waves that come from the left; there are also those shown in figure 55b that can come from the right but not from the left (fig. 55c, lower wave). The upshot is that the movable plate reflects more waves coming from the right than waves coming from the left; consequently, the recoil momentum imparted from the right and moving the plate to the left is bigger than the one acting in the opposite direction. The right-hand plate will thus be moved toward the left-hand one. In other words, the two plates attract each other.

A computation of this effect from the quantum mechanical theory of electromagnetic waves not only shows that the plates are subject to an attractive force but also yields the numerical value for this force. In particular, it says that this force, even though growing very small, will persist down to zero temperature. This zero-degree force amounts, for plates of 1 sq cm area arranged at a distance of 1/200 mm, to a weight of one-fifth of 1 mg—about the same as that of ten large dust particles. In 1958, ten years after it was predicted by the Dutch physicist H. B. G. Casimir, the existence of this force was experimentally observed by M. J. Sparnaay.

The error margins of Sparnaay's results were so large that his experiment

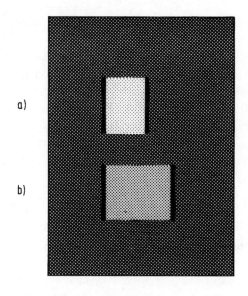

a)

b)

Figure 56 The degree of shading in this figure indicates the energy density of the space we describe. The plates are those used for the verification of the Casimir effect (cf. figs. 15 and 55): The greater the distance between the plates, the greater the energy density in the space separating them. The space in the absence of plates has an even larger energy density; we can imagine reaching that density by removing the plates to infinity.

could hardly count as a quantitative test of Casimir's prediction. Despite the importance of the matter, and its connections with numerous other issues such as the force that molecules exert on each other, it had been put experimentally to rest until 1996. In that year, S. K. Lamoreaux, at the University of Washington at Seattle, performed the first quantitatively successful experimental test of Casimir's prediction.

Let's recapitulate: The space between the plates is not, even at the lowest possible temperature, a mere nothing without properties—it is a blackbody space filled with radiation. Of course, this volume, which Sparnaay investigated, is suspended between two electrically conducting parallel plates. At first sight, the experiment therefore has nothing to say about that volume in the absence of the plates. But things change when we realize that the force acting between the plates is attractive. Given that the lack of balance between the forces from inside and outside the movable plate is responsible for the Casimir effect, there must be *more* oscillations arriving from the outside than from the inside: The smaller the space enclosed within mirrors, the emptier it is. If interpreted in this way, Casimir, Sparnaay, and Lamoreaux really do show that there are electromagnetic oscillations in free space that is as empty as it could be—as cold as possible, and without boundaries.

Looking at the system of figure 55 from the point of view of energy rather than of recoil, we arrive at the same conclusion. In order to increase the distance between the plates, we have to spend energy. As we do so (and we illustrate this in fig. 56), *outer space* becomes *inner space;* the inner space of figure 56b thus contains more energy than the outer space, which was lost in the transition from figure 56a to figure 56b, may have yielded. The implication is that the closer the volume between the plates resembles empty space, the greater its energy density—the amount of energy per unit volume. As we approach the conditions of an empty space in the absence of plates—say, by continuing to increase the distance between the two plates—the resulting space is filled with more energy. The volume within the plates therefore contains less energy per unit volume—less radiation, that is, than it would hold in the absence of the plates. The boundary conditions of the surfaces of the plates exclude oscillation modes that do not have their nodes as we showed above. This means that the blackbody volume is emptier in the presence of the plates than in their absence.

ATOMS IN THE VOID

Progressively more sophisticated experiments have shown, after Sparnaay's initial proof, that there is a physical reality to the changes in empty space due to the reflection on the walls. Atoms in the void between these walls radiate the energy of their excited states in terms of electromagnetic waves. It is possible to prepare them so that they emit waves with certain wavelengths preferentially. In this case, it is possible that one of the appropriately prepared atoms finds itself in a blackbody volume that (recall fig. 55) does not accept the wavelength that the atoms want to emit—and therefore the atom has to remain in a state of excitation. Conversely, if the wavelength fits particularly well into the volume, the atom in question will emit its radiation more readily than in an unlimited empty space. Both these effects have been verified experimentally; as a consequence, it has become clear that the atom and its surrounding space have to be seen jointly to be fully understood.

MOTION WITH RESPECT TO BLACKBODY RADIATION

Blackbody radiation at zero temperature has the special quality that, with respect to it, motion at a constant speed cannot be observed. Therefore the physical vacuum—the best that is allowed by the laws of nature—does not distinguish one constant velocity from another one. But we can observe *acceleration* with respect to the vacuum. In this sense, our vacuum resembles Newton's absolute space.

     To show this, let's take one step back. For one thing, the velocities with respect to blackbody radiation are experimentally observable as long as the temperature is above absolute zero. Add the walls of the blackbody to its radiation above zero in our observational system, and you will expect that we can experimentally determine velocities with respect to it; it is then simply the same as the velocity of the observer with respect to the walls. But even without recourse to the walls, the observer can measure his velocity with respect to blackbody radiation by

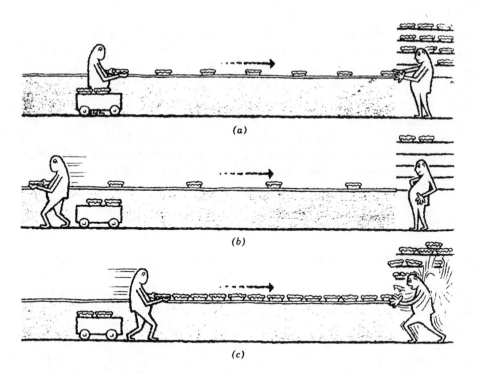

Figure 57 Just as the distance between two cakes each in the production line of the bakery, the distance between the wave maxima of a sound wave depends on the velocity with which the source moves: The shorter this distance, the higher the pitch of the sound in the observer's ear. This explains the dependence of the pitch on the velocity of the source of the sound, commonly called the Doppler effect.

means of the Doppler effect. This effect is known to all of us from everyday life: When a police car with its siren on moves past you, the pitch of the siren gets lower the instant the car speeds by you. The pitch you hear is the same as the frequency of the sound wave from the siren: the higher the pitch, the greater the frequency. The frequency emitted by the siren, just as the pitch the officer in the car hears, remains the same. It is the pitch registered by you, the observer, that changes. It is, as figure 57 illustrates, higher than that heard by the police officer as the police car approaches you, lower as it moves away from you.

Just as with sound waves, light waves emitted by a moving source show the Doppler effect. What we observe instead of the sound pitch is the color of the light: Red light has a lower frequency than blue light. If the light source is moving toward the observer, the light appears bluer; as it moves away, it will appear red. Likewise, an observer moving inside the blackbody volume will register the radiation coming from the direction opposite to his motion as being *blue,* that which comes from behind as *red.* The difference between the two frequencies can tell him his velocity of motion with respect to the blackbody radiation.

This difference, however, will decrease as the temperature is lowered; it will vanish altogether at absolute zero. Regardless of an observer's velocity of motion, the radiation meeting him at zero temperature is the same from every direction. The observer therefore has no way of finding out from the radiation alone in which direction he is moving, or whether he is moving at all.

Once we accept this scenario, we have already fixed the spectrum (that is, the amount of radiation as a function of its frequency) within some constant factor. However, the result obtained in this way does not appear to make sense: It implies that a blackbody at zero temperature has an infinite supply of energy in the form of zero temperature radiation.

The same astonishing result can be derived by means of the quantum theory of electromagnetic radiation, which we call quantum electrodynamics. This theory, the implications of which have been verified in many instances with remarkable precision, tells us that the true vacuum at zero temperature still has an infinite supply of radiation energy. As we proceed, we will see that electromagnetic radiation is in fact only one component, albeit infinite in quantity, of the unfathomable energy supply of the vacuum.

WHY IS THE VACUUM SO COMPLEX? A FIRST SUMMARY

As long as we don't insist on numerical details and on an understanding of the origin of blackbody radiation at zero temperature, we can understand the Casimir effect in purely classical terms. But for a true interpretation of the vacuum in its differences from a nothing without any properties, we have to pull in quantum mechanics and the special theory of relativity, two of the three great scientific theories of this century. The third—general relativity—broadened our thinking about space by interpreting it in the more general terms of a field. We are still waiting for an ultimate unification of this concept with quantum mechanics.

We have seen that the vacuum of physics is distinct from a simple nothing by virtue of the nonremovable activity we called zero-point radiation. Fields and particles originate and disappear; virtual particles appear for the briefest of times—shorter for heavier ones, a bit longer for lighter ones. In this framework, real particles appear—in a fashion reminiscent of what Heraclitus called *panta rhei*—as phenomena accompanied by excitations of the vacuum. We can loosely compare the vacuum to a farmer's field of wheat that is swaying in the wind, and the particles to the wave patterns excited by the wind on its surface. Another telling comparison is the excitation of a string on a musical instrument (see fig. 58).

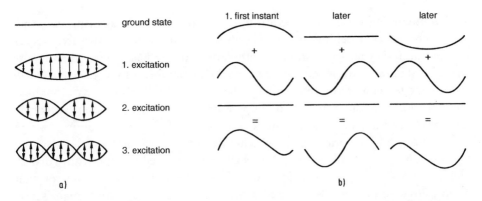

Figure 58 A string fixed at both ends is in its ground state in the absence of oscillation (note that quantum mechanics doesn't allow this). In its excited states, individual points along the length of the string oscillate up and down periodically. These elementary excitations of the string, called the normal mode and the harmonics, are described, in connection with figure 55. There is an infinite number of harmonics. The first three excitations are shown, together with the ground state, in figure 58a. The frequencies of the oscillations increase with the number of nodes—points along the length of the string that do not move up and down in a given mode; the second excited state has twice the frequency of the first one, the third excited state three times the frequency of it.

The sound produced in musical instruments is never "pure." It contains the normal mode as well as harmonics; the string oscillates with several of its frequencies at the same time. Figure 58b details the joint excitation caused by normal mode and first harmonic at different times. The first two lines illustrate the individual oscillation during one period of the first harmonic or one-half period of normal mode. The resulting shape of the string is shown in the third line: The string's deformation moves from left to right; as the wave continues, it will move from right to left, then again from left to right, and so on.

$$E = mc^2$$

There is a saying that mass is, in a way, frozen energy. Take a particle such as the proton—the nucleus of a hydrogen atom—and imagine that it lies at rest on the laboratory table in front of you. Although at rest, it has energy: Einstein's formula $E = mc^2$ says that its mass m is equivalent to an energy E. To calculate that energy E, we have to multiply the mass m by the square of the velocity of light c. We can look at c as simply a conversion factor, similar to the number 100 that translates dollars into cents.

Energy comes in many forms. Einstein's equation simply tells us that mass expressed in the right units is one of these forms. Being nothing but a manifestation of energy, it can be changed into other such manifestations. A simple example is the neutron, another nuclear particle, slightly heavier than the proton. It is unstable and will therefore decay into lighter particles. When unstable particles decay, the decay products are emitted not only with their masses but also with motion energy. We might ask from where that motion energy derives. It is simply a consequence of the law of the conservation of energy: The sum of all energies must remain the same in the decay process as in all other physical processes. In the case at hand, the energies of the masses and the motion energies of the decay products must add up to the mass energy of the decaying particle (which we assumed to have decayed at rest).

This is the same kind of process that occurs in the atomic bomb (or, as we should more accurately say, the nuclear bomb); it is another manifestation of Einstein's formula. Every particle at rest has energy because it has mass. If there were not additional conservation laws that must also be satisfied, energy conservation alone would permit the conversion of all of the mass energy into motion energy, heat, and radiation. One of these additional conservation laws concerns electric charges. Electric charge cannot be made or lost; it remains precisely the same in all imaginable interactions. But this conservation applies only to the total charge—there may well be a redistribution of charges. We know that some particles are positively charged, others negatively. The sum of all charges, or the resulting charge, is the quantity that remains constant in any process. To pick an example, it is perfectly acceptable for a process to start with one negative charge and end up with two negative charges plus one positive charge—their sum is one negative charge, as in the beginning.

Back to energy: The energy corresponding to the rest mass of a particle is, so to speak, frozen and cannot be removed unless the particle is destroyed, but other forms of energy the particle may possess can easily be transferred to other particles. Large machines called *particle accelerators* give motion energy to individual particles. As these particles approach the velocity of light, the accelerators, while still adding energy to the particles, barely add to their velocity—no particle can ever fully reach the velocity of light. Just as we can add motion energy to a particle without destroying its identity, we can have it give up some of its motion energy. That is what happens when one particle hits another one—when a beam of particles from an accelerator hits a target.

In everyday life, we are accustomed to the transformation of one form of energy into another: The motor of an automobile changes the chemical energy of the fuel in its tank into motion energy or, once a car has traveled up a hill, to positional energy. When the driver pushes the brake pedal, the brake pads will heat up; the motion energy of the car is now being transformed into heat (or thermal energy), and so it goes. All forms of energy, with the exception of what is frozen into the mass of a stable particle, can be transformed into other forms without destroying particles. To close the loop, it is also possible to liberate the mass energy of any particle by blowing it up, by destroying it. This process is related to the central interest of this book.

Take an electron at rest. We can liberate its mass energy and transform it into different manifestations of energy by making it collide with a positron. The positron is the antiparticle of the electron: It has the same mass, but opposite charge. The total charge of the electron-positron system is zero; no conservation law forbids the transformation of an electron-positron system into pure energy. As a result, the collision between positron and electron (and, in general, particle and antiparticle) produces *light* of a very high energy, or of very high frequency. We might call it ultrablue light. In the process, the energies of both the electron and the positron, of particle and of antiparticle, have been changed into radiation energy, into light.

ANTIPARTICLES

We are interested in this process for two reasons. The first is the implied meaning of having the electron's antiparticle involved. There is an antiparticle for every particle—that is, for every particle there is another one with the same mass but with the sign of all charges reversed. Nobody has been able to formulate a theory that includes our basic notion of the propagation of effects in a manner consistent with both quantum mechanics and the special theory of relativity, with the exception of the notion of an antiparticle for every particle. Completely neutral particles—that is, particles that have no charges at all, electrical or otherwise—are their own antiparticles. However, all reasonable theories have an antiparticle for every particle. And it stands to reason that the antiparticle of the antiparticle itself will, again, be the particle from which we started. This must be so because reversing the charges twice is equivalent to reestablishing the initial charges.

The second important implication is that we can add to every particle—and indeed to every physical system, whatever it might be—another particle or system where all charges are reversed; the resulting combined system will have all the properties of empty space, save the energy. In particular, the combined system is devoid of all charges, just as the definition of the vacuum demands.

The British physicist P. A. M. Dirac predicted the existence of the positron— the first instance of an antiparticle—in 1928. It was discovered by Carl Anderson at the California Institute of Techlology in 1932. In figure 59, we show Dirac's convincing illustration of the electron's need for an antiparticle. Imagine, he said, that the world is filled with electrons; we might visualize them as a sea full of particles with energies that range from minus infinity to some upper limit. Since they are omnipresent, we don't notice them (see fig. 59a). We will, however, notice when an electron is missing out of this sea; there will be a hole in this infinite distribution of negative charges, and it will appear as a positive charge (see fig. 59b). This so-called hole has all the properties of the positron, the antiparticle of the electron. Figure 59 also suggests that we can revert from the configuration 59b to that of 59a if there is neither the particle nor the hole, but only the filled sea plus the energy that results from the filling of the hole. Seen

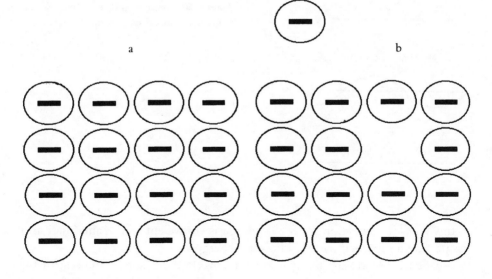

a   b

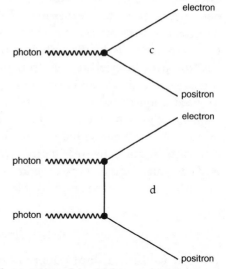

Figure 59 (a) The pervasive Dirac sea of electrons. (b) The addition of energy permits one electron to be taken out and to become an observable particle; it leaves behind a hole, which has all the properties of its antiparticle, the positron. For an observer who is not aware of the Dirac sea, matter has been created from energy. A possible real process of this kind is shown in (d); developments after Dirac's prediction of the positron on the basis of (a) and (b) have clearly shown that there is no need to introduce a sea—the vantage point of the observer who has no idea about its existence is the correct one. Obviously, (d) can be build up from the elementary process (c). The latter cannot occur in nature without the participation of other particles, typically atomic nuclei, which carry away momentum. This process is pictured as (d), with one of the photons attached to the nucleus; it is routinely observed in the lab. Creation and annihilation processes are of many kinds and are also observed routinely in accelerator experiments along with interactions that don't change the number of particles. The opposite of (d)—annihilation of an electron-positron pair into two light particles—is so common that it is applied in diagnostics; the so-called positron emission tomography is based on it. In summary, there is nothing much to the conversion of energy into matter and vice versa. The first actual transformation of light into matter without the participation of other particles, accomplished in 1997, generated a huge response from the media. The technical problems encountered in order to achieve this goal were enormous—but there was nothing particularly new in the result itself. The same holds for the recent assembly of antimatter from antiparticles.

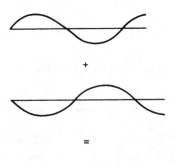

Figure 60 Positive and negative elonga-
tions may cancel each other.

in this way, the transition from 59b to 59a is equivalent to the annihilation of the particle-antiparticle system into pure energy.

In figure 12, we saw the process that inverts the filling of the sea; in this case, the necessary energy was supplied by a photon. An electron was knocked out of the sea in that experimental record—a virtual particle became a real particle, and the hole it left behind emerged as an antiparticle. This elementary process is symbolically illustrated in figure 59c: A photon becomes an electron-positron pair; inversely, an electron and a positron can combine into a photon (or, as we said above, annihilate into a photon). But for the process to take place, we need a fourth particle as a catalyst. In the cloud chamber picture in figure 12, the catalyst was an electron that was knocked out of its atomic configuration in the process. Whereas the process in figure 59c cannot proceed without such a catalyst, the one in figure 59d can: Two photons can materialize into an electron-positron pair; and, of course, the inverse process, electron-positron annihilation into two photons, is equally possible.

THE UNCERTAINTY RELATION

We now consider a third important notion in our discussion of the vacuum: Heisenberg's *uncertainty relation*. Quantum mechanics tells us that one and the same basic object—say, an electron—has both particle properties and wave properties. At a given point in space and time, two waves may well add up to zero—that is cancel each other (see fig. 60), something that cannot happen to particles. The experiment corresponding to our figure 60, the so-called two-slit experiment, proves the wave nature of the phenomenon; our notion that the electron is actually a particle stems from other experiments. But Heisenberg's uncertainty relation states that once we interfere with the wave experiment so as to prove the particle aspect of the electron, we destroy the wave aspect (or, rather, its manifestation). The contradiction between particle properties and wave properties of the electron is thus only an imagined one. The uncertainty relation sees to it that we cannot demonstrate it experimentally. To quote Richard Feynman, the uncertainty relation "protects" quantum mechanics.

Let's see how that happens. We can determine the shape, position, and velocity of macroscopic objects. Now let's move up close and look through some supermicroscope at the edges of, say, a die, so that its molecules and atoms become visible. At this level, we are no longer able to fix shapes, positions, and velocities, as we do with macroscopic configurations. The laws that govern microscopic structure forbid that. The edges of the die seen from this distance look ragged. That in itself is not an argument for a lack of long-range stability; the edge might be built up out of tiny particle configurations that behave like macroscopic bodies, and could even change their relative position with time. But sequences of instantaneous observations with high magnification—that is, observations with precise resolution in both space and time—will show fluctuations that increase with the resolution of our observations. One reason for this effect is more or less obvious: The light we need to obtain these high resolutions will disturb the forces that hold the microworld of atoms and molecules together. But the fluctuations are not due to this effect alone; otherwise we could subtract the influence of light in some way and arrive as a result at the stable properties of matter on that minuscule scale. But in fact we cannot: The smaller an object, the more evidence we find for the quantum mechanical effects that forbid a precise definition of shape, position, and velocity.

Many books have been written on this subject. Even Bertolt Brecht weighs in with a clear (but entirely wrong) interpretation of the uncertainty relation. In his *Discussion among Refugees,* he has the physicist Ziffel argue that there is life, normal life, in the microscopic world. He appears to argue that what we see at the microscopic level is reminiscent of patterns familiar from our macroscopic experience. That, of course, is a misinterpretation. If we define what is *normal* from the behavior of a particle in the macroscopic world, then there is no such normal behavior on the microscopic level. Not only can we not observe such behavior on this level—the very suggestion that whatever we see might be explained in terms of a scaled-down version of macroscopic behavior is in contradiction to experimentally proven tenets of quantum mechanics.

In this context, it doesn't matter that it must be possible to deduce, at least in principle, the laws of nature that apply in the macroscopic world from those of quantum mechanics: it suffices, as a rule, to apply these "effective" classical laws in their domain without ever contemplating their quantum mechanical origin. There are, however, exceptions: Certain macroscopic objects directly display quantum mechanical effects. Superconductors and neutron stars—the latter behaving like atomic nuclei with a six-mile diameter—are macroscopic objects that cannot be described in terms of classical physics.

But let's not stray from our topic. If we manage to force a microscopic particle into a small space, we have to accept the fact that it reacts by moving in an unpredictable fashion. The smaller the space, the larger the random motions it will execute. This is analogous to our observation that when we use our hands to stop the remaining last bit of motion of a lamp suspended from a long cord, we will, instead, contribute to the lamp's motion; similarly, whatever means we use to fix an atom's location will instead increase its motion. No matter what tools we invent, there is no way to fix both location and the velocity of a particle more precisely than quantum mechanics permits as a result of the uncertainty relation.

To illustrate this relation, think of a game of table tennis. As the ball hits the table, it will bounce up and down. When we restrain the upward bounce by holding the paddle parallel to the table at a small height, the ball will bounce up and down at a faster rate between paddle and table. If the paddle, the table, and the ball were perfectly elastic, the ball would not lose any energy in the process. The closer to the table we hold the paddle, the more precisely we fix the location of the ball—but at the expense of having it move faster and faster. If we removed both paddle and table in one instant, we would not be able to predict the direction in which the ball moves—up or down. We simply don't know in which direction the very rapidly bouncing ball was moving at the very instant of removal. Consequently, we might say that this process just prior to the removal of table and paddle fixes the location while failing to give us any good information on the velocity; it makes the velocity *uncertain*. What distinguishes this macroscopic process from the quantum mechanical uncertainty relation is that in principle we can calculate the velocity of the ball at every instant from the initial conditions (how the bounce started) and the boundary conditions (the relative positions of table and paddle). In principle, macroscopic motion is free of uncertainties.

UNCERTAINTY AND PERCEPTION

Our perception of physical structures has been honed on macroscopic objects— say, Ping-Pong balls; that's why we have such a hard time thinking of atomic and subatomic particles as different from Democritus's atoms that move, over the course of time, from one place to another. But we now know that the reality of the quantum mechanical world cannot be described in terms of a classical particle picture, nor in terms of the wave concept. The quantities that describe quantum mechanical reality can only approximately be interpreted in terms of our usual observational quantities.

At the same time, it is both a precondition and a consequence of modern physics that our perception does not extend down to the atomic level. On that scale, it is only one aspect of reality among a number of them. When I say "perception," I mean to convey an image of reality—such as the statement that every atom is built up out of a nucleus and orbiting electrons. This image can be described with arbitrary precision in quantum mechanical as well as classical terms, as long as we limit ourselves to the locations of nucleus and electrons.

But the image is not able to convey the proper information about a *sequence* of events, like a trajectory for the particle, either classically or quantum mechanically. For the classical case, we need added information about the velocities of all parts of the system at a given instant. The implication is that for a full picture we need at least two immediately successive images of the object in question, taken in rapid succession.

Recall our discussion of the pendulum. Its position at a given instant does not fix its subsequent behavior. Suppose a photograph shows the pendulum hanging straight down. This may mean either of two things: The pendulum is at rest, or it is in the process of oscillating through its lowest point. In the first case, it will remain in the observed position; in the second, it is just moving through. But if we simultaneously know both position and velocity, we can predict the behavior of the pendulum for all future times. Knowing only the position or only the velocity will not suffice.

Conclusions based on perceptions need more than one instantaneous image; they need a sequence of images. One photograph will give a proper description of reality; another one, taken a few moments later, describes another such aspect. To draw a conclusion based on perceptions, both photographs are required.

In classical physics, the two observables of a pointlike object—its location and its velocity—are, at least in principle, accessible and measurable. Not so in quantum mechanics. There, a precise determination of one aspect of reality, say, of location, implies that the other one, velocity, must necessarily be undetermined. The uncertainty relation puts this statement into quantitative terms; it nails down the indeterminacy of quantum mechanics. A macroscopic trajectory that is the consequence of a quantum mechanical process cannot be predicted, because no quantum mechanical state permits a simultaneous determination of location and velocity. We might say that an instantaneous picture, like a photograph, freezes a trajectory in a given place. As a result, we gain no knowledge whatever of the velocity with which an object in the picture moves.

Heisenberg's uncertainty relation states that we cannot measure both the location and the velocity of an object with arbitrary precision. The quantum mechanical development of the system must be described in its own terms. The internal variables of quantum mechanics permit us to draw conclusions about the further fate of a system. But these variables are not the classical ones.

It is entirely possible to know the location of a quantum mechanical object— say, an electron. If the measurement that leads to this result—say, a photograph— concerned an electron in classical physics terms, we might note its velocity at the time the picture was taken as additional information. Even when we don't have knowledge of it, every classical particle has some velocity at any given time. If we know the laws of nature that govern its behavior, its location and its velocity in a given moment will together fix its trajectory and the time frame in which the particle moves through this trajectory. As long as our imagery remains classical, the instantaneous photograph might be seen as a single frame in a movie that describes the whole trajectory; the electron will, in fact, like a planet, trace out a certain trajectory.

For the actual electron, we need the laws of quantum mechanics. And these laws do not permit the electron, which is known to occupy a given location at a given time, to also possess any fixed velocity at all at the same time. If a scientist were to mark an instantaneous photograph of an electron with the caption *electron at rest* or *electron moving at 50 mph to the right,* neither of these captions would, in a quantum mechanical context, be consistent with what the picture shows—an electron possessing a certain position. Consequently, there cannot be a film of a freely moving electron with each frame showing the electron at a certain position. If one of the frames shows it at some point in space (as assumed above), the next frame would have to convey the impression that it is nowhere at all. This is because a value of the electron's velocity could otherwise be computed from its two locations in two consecutive frames. Thus there is no such thing as the trajectory of a freely moving electron in space. If we were to try to construct such a trajectory by measuring the electron's consecutive positions, the result would not show the trajectory of it freely moving, as intended, but instead would show the results of measurements, each of which would have destroyed the quantum mechanical state possessed by the electron before the measurement. Physicists can compute the likelihood of finding the electron at different locations as a function of time (as, for example, in fig. 66). But it is not possible to attribute a trajectory to an electron if and when its fate is determined by the internal variables of quantum mechanics alone. It is not only that we cannot observe the trajectory of a freely moving electron, but there is no trajectory that can, even in thought, be attributed to it. If there were, the position and the velocity of the electron could be computed.

What *can* be determined from the state of a system—its description by the wavelike internal variables of quantum mechanics—at a certain time is the *likelihood* for any particular result of a measurement of position, performed at any time. At a certain time, this likelihood might be concentrated around or even at some point. The electron has a certain position at this time, and by virtue of that its quantum mechanical state is fixed. As we know from the last two paragraphs, the future development will then immediately and completely destroy any information about position. If, on the other hand, the likelihood of finding the electron is distributed over a region, this doesn't provide a complete description of its quantum mechanical state.

The implication for our film reel, which is supposed to follow the electron's path, is that the laws of quantum mechanics permit it to catch only the probabilities of finding the electron somewhere in a given frame by a position measurement.

It is not the shape or extent of the electron that can be gleaned from quantum mechanical likelihood distribution—it is sufficient to think of the electron as a pointlike mass. Rather, the question we have to pose is: What kind of quantities develop deterministically in the course of time according to quantum mechanics? These are not position and velocity taken together as in classical mechanics, but rather the admittedly abstract internal wavelike variables of quantum mechanics. Once these are fixed, they develop deterministically—fixing them at some time determines them for all times. Any likelihood for any physical quantity, such as position or velocity, can be computed from them as a secondary property.

PROBABILITIES

The more precisely the location of a system is fixed by the quantum mechanical state it is in, the less can be said about its velocity: The uncertainty relation implies a broad likelihood distribution for velocity in a state having a narrow distribution for position. In short, the observables of quantum mechanics are probabilities, likelihoods—the likelihood of finding a particle in some given interval of space or of finding it moving approximately with some given velocity. Quantum mechanics can tell us the relevant probabilities at any given time from the internal wavelike variables that describe the particle. Although its objects are particles, this probabilistic context is the only one in which they can be described. It is this interpretation of particles in terms of the likelihood of observing their properties that is at the basis of the duality between particles and fields or waves.

Only after this discourse on the probabilistic approach of quantum mechanics can we gain an appreciation of the concrete meaning of the uncertainty, the lack of precision, with which the location and the velocity of particle can be fixed. An *uncertain* location implies not only that we don't know where the particle is but that we *cannot know it* unless we allow its state to be changed. The state fixes the *probability* of finding the particle with a given velocity and, at the same time, at a given location. As a rule, that means the likelihood of a particle being found is different from zero, in a continuum of points: The particle does not have a precisely fixed location; its location is uncertain.

ZERO-POINT ENERGY

In our specific context, the uncertainty relation has important consequences for the energy at zero temperature. As noted, any object described in quantum mechanical terms can have a precisely known velocity only when its location is completely indeterminate—when it can be found with equal probabilities all over space. In all other cases, its velocity must be uncertain. That means it can never really come to rest in any given location—a quantum mechanical pendulum can never just hang there, pointing vertically down. Gravity might constrain it to maximize its probability of location close to this position, slightly oscillating around this known point. For its energy, that dance around the zero point has two consequences: It has motion energy, and any distance from the zero point also implies positional energy (which the physicist calls *potential energy*). Its dance about the lowest point does not permit it to stop there, and at all other points there is potential energy with respect to the zero point.

Positional, or potential, energy is the energy we add to a rock when we lift it from the floor and put it on a table. Except for its mass energy, the total energy of any object is the sum of its motion energy and its positional energy. For a quantum mechanical system that attractive forces keep from moving all over space, this total energy cannot vanish; but there is a minimum permissible energy. This is what we call zero-point energy, the energy of the quantum mechanical *ground state*. This fixed ground state energy means that the smaller the space to which we confine a particle, the higher the velocities it attains and the larger its motion energy. Its positional energy is small, because it cannot move far away from the point with positional energy zero. What we call zero-point energy is the smallest energy to which quantum mechanics permits motion energy and positional energy to add up.

We might find an analogy in the motion of a string, as shown in figure 58: An oscillating string has both positional and motion energy. The positional, or potential, energy is not due, in this case, to gravity; we assume that the oscillation is not influenced by Earth's gravitational field. Rather, this potential energy is due to the *tension* in the string; we might call it tension energy in this case. We will arbitrarily choose to call it zero for the ground state, with no oscillation (see fig. 58a). When the string oscillates, its length between the end points increases, and so does the tension; it now has a positive tension energy. Transferring the arguments above to the present case, we might say that the total energy cannot be less than its minimal, or zero-point, energy; in fact, the string will always oscillate a tiny bit.

We have also discussed the case of the ball bouncing on a tennis table. As we take away energy from the ball, it will bounce less high and its maximum velocity during the bounce will decrease. Its minimum energy is reached when it comes to rest at zero velocity and zero positional energy (with respect to the table); both are precisely known in this case.

In the gravitational field, the reduction of the total energy of the bouncing ball to zero means the reduction of both location and velocity to a given zero value in classical mechanics. Quantum mechanically, the uncertainty relation imposes the conclusion that the ball will be left with a minimal *zero-point energy*.

In microscopic systems—such as in an atom, a molecule, a crystal—this minimal energy means internal motion of its constituents. It is independent even of whether there is any matter at all in a given space volume, and can be found in its absence. The fact that the law of zero-point energy applies to a space that includes material objects, and also to a space that does not, implies that *empty* space is also filled with energy and its carriers; it also implies that all finite volumes are subject to fluctuations in energy. The smaller the volume, the larger the fluctuation. The volumes relevant to our normal experience are so large that the resulting energy fluctuations are below the level of observability. But the scientist dealing with very small volumes has to include energy fluctuations of the vacuum. This means that the differences between the macroscopic and the microscopic world extend not only to the properties of the objects under observation but also to those of the vacuum.

VACUUM FLUCTUATIONS

Small regions of empty space will see large energies appear in the form of these fluctuations. They may be energetic electromagnetic waves. They may even appear as particle-antiparticle pairs—supposing, of course, that the energy of the fluctuation rises above the rest mass of these particles. There is no a priori carrier for the fluctuating energy in empty space, in contrast to, say, the crystal. Rather, the appearance of such a carrier is another consequence of the energy fluctuations implied by the uncertainty relation.

Their short-lived existence keeps us from noticing such fluctuations in our everyday existence. The shorter their lifetime, the larger they get—this is another formulation of the uncertainty principle: It relates energy to time in the same way that it relates location to velocity. Lifetime, range, and magnitude of an energy fluctuation in a vacuum are always related such that the energy uncertainty includes the smallest possible energy value. It is large for short lifetimes and small volumes, smaller when the lifetimes are longer and the volumes larger. Energy fluctuations cannot be larger than what is needed to have them reach the zero level by means of the uncertainty relation; conversely, there *must* be fluctuations within this range. The principle mandates the existence of energetic fluctuations of short lifetimes as well as that of lower-energy ones with longer lifetimes. Electromagnetic excitation of the vacuum, such as light, may have very little energy; that makes these fluctuations carriers of long-lived energy fluctuations. Once the energy is, by dint of Einstein's mass-energy relation, sufficient to create electron- positron pairs, *virtual particles* may, and must, appear for very brief times as part of the energy fluctuation. Since the fluctuations, like every process in nature, don't change the total electric charge, electrons and positrons can be created (and destroyed!) only in pairs.

The larger their mass, the smaller the spatial extent of the volume the pairs can influence. This is due both to the short lifetime the uncertainty relation accords them and to their small velocity; recall that most of the fluctuation energy is taken up by the mass energies of the particle-antiparticle pair being created—little energy is left over for the motion of the two particles.

REALITY OF ZERO-POINT ENERGY

The uncertainty relation implies that there cannot be a region in space totally devoid of electromagnetic fields. They may well be absent *on the average,* but their energy fluctuations have to satisfy the uncertainty relation. We described the Casimir effect; it is nothing but a consequence of the inevitable fluctuations of the electromagnetic field. The fluctuations of zero-point energy are real and measurable, even if the zero-point energy itself is not.

We saw that the thermal motion of atoms and molecules ceases as far as the uncertainty relation permits it, at absolute zero. But it does not vanish altogether; atoms and molecules are not permitted to divest themselves of all motion energy. If they are bound in a crystal structure, they cannot simply remain at rest in a position of minimal energy inside the crystal lattice—they must, to pick an image, perform a dance about that minimal position. They retain both motion energy and potential energy, as we previously described. Their sum makes up the zero-point energy of the atoms and molecules in the crystal. In this case, experimental verification is easy. It may show up as the imprecision of X-ray images of the crystal or as an otherwise inexplicable contribution to the crystal's specific heat close to zero temperature. Similarly, the Casimir effect is closely tied to the specific heat of a blackbody at this temperature. The quantity that physicists call the *specific heat* of a substance is defined as the amount of energy needed to raise the temperature of 1 g of this substance by 1 degree Celsius.

We now know that the uncertainty relation implies not only the absence of empty space but also the nature of the minimal content of every space—fluctuating energy and virtual particle-antiparticle pairs. The fields of the general theory of relativity by themselves are permitted to have zero values, and thereby truly empty space. But a quantum theory of gravity must preclude that; its energy carriers will create fluctuations. These fluctuations may include the value zero but will never violate the uncertainty relation for energy. Since we do not yet have a theory that unifies general relativity and quantum mechanics, anything we might say about fluctuations of the fields of general relativity must remain mere speculation. We can say for sure that quantum theory, when joined to the special theory of relativity, mandates the permanent and ubiquitous existence of fluctuations that include the appearance and disappearance of virtual particle pairs. This mandate includes the emptiest of all imaginable spaces, a void we might surround with an impenetrable wall at temperature zero.

ATOMS, FIELDS, AND ENERGY

As we follow the modifications of our historical notions of empty space and the insights gained with the development of physics in our century, we should not bypass a revision of Democritus's idea of the atom, which he conceived as a particle that cannot be split into pieces. To date, we don't know of any object that cannot, for certain, be subdivided. The chemist's atom consists of nucleus and electrons; the nucleus is made up of protons and neutrons; the protons and neutrons are made up of what the particle physicist calls quarks. That is where we are today, but we don't know whether this is the end of the line; this is not what concerns us in our present discussion. Rather, Democritus's concept of the individual unit of matter was shattered by the discovery that every particle can be made to meet its antiparticle and vanish into pure energy in the process. Particle and antiparticle together blend into structureless energy, and may reemerge as different particles.

There is no such thing as an indestructible material object. Every particle that meets its antiparticle gives up its identity and becomes radiative energy. In the process, the energy is conserved, but the structure of the particle—be it Democritus's sphere or a die, be it charged or neutral—vanishes. It dissolves into pure energy. All identity is absorbed in the sea of pure energy.

What a change in the concept of the atom, the indestructible building block of matter! What is left if there cannot be stable matter? Just fields? No, that would be going too far. Maybe the end of our story will see everything reduced to fields. We will discuss this in the next chapter. But there are obstacles in the way—substantial ones. The proponents of the field as the ultimate reality ignore the fact that we do not really understand the classical limit, where matter arises out of mere fields.

The founding fathers of quantum mechanics—in particular, Niels Bohr—had no trouble accepting the reality of the classical world with its "large" objects, such as planets. They were able to demonstrate that the laws of quantum mechanics, applied to these objects, have the same consequences as the laws of classical physics. If we assume that the planet Jupiter is a particle just like the electron, only much heavier, and we apply the laws of quantum mechanics to it, the uncertainty relation has no observable consequence whatsoever: Due to its large mass, Jupiter *might* possess both a certain location *and* a certain velocity with any desirable precision. But this fact, which works for all practical purposes, does not even touch upon the real problem of the emergence of the classical world: Why *is* Jupiter so precisely situated in space? According to quantum mechanics, it might well be otherwise; indeed, we should *expect* it to be otherwise.

What the fathers of quantum mechanics didn't know is that there are macroscopic systems to which the laws of quantum mechanics do apply fully. Superconductors and neutron stars are such systems—we might hark back to Lucretius and his notion of *large atoms*. The existence of such systems has not removed the dividing line between the classical and quantum mechanical domains; it has simply shifted it. Our observation says that the classical laws of mechanics apply to, say, Jupiter; yet at the basis of classical mechanics, the laws of quantum mechanics operate. What neither level of understanding tells us is the answer to the question: Why is there Jupiter? Why is there such a large classical object, subject to the laws of classical mechanics? The laws of quantum mechanics applied to objects having a large mass are *consistent* with the classical laws, but they do not *imply* them. They do not in any way preclude the existence of objects with the mass of Jupiter which do not have any well-defined position at all; such large objects are quasi-localized, but they don't have to be. Why don't large objects display quantum behavior?

The German physicist Max Born, who received the Nobel Prize in 1954, exemplifies the orthodox interpreters of quantum mechanics. The orthodox interpretation offers no explanation as to why everyday objects never assume quantum mechanical states that do not allow a classical interpretation. Albert Einstein, in a 1953 letter to Born, expresses severe discontent with this attitude. He points out that Born's interpretation *does* imply the possibility of such states and that there *is* a problem: "Seeing things your way, we would have to come to the conclusion that most quantum mechanically permissible processes of macroscopic systems cannot be expected to be described, even approximately, by classical mechanics. This means we would have to be very astonished when a star or a fly appearing in our field of vision for the first time would appear to be quasi-localized."

This question about the reason we do not observe any quantum mechanical uncertainty in the appearance of flies or stars was never tackled by the inventors of quantum mechanics. Given that we are convinced that the basic laws of quantum mechanics are at the bottom of all mechanical laws, and that the beginning of our world can be understood exclusively in quantum mechanical terms, we have to ask ourselves: Why are there any observable systems at all on the classical side of the demarcation line between classical mechanics and quantum mechanics? Why are there these stumbling blocks, these flies, these stars? The demarcation line proper, between quantum physics and classical physics, was clearly stated by the fathers of quantum mechanics: Classical laws govern large and heavy classical systems. But they failed to tell us how and why the world became that way.

WHY IS THE VACUUM SO COMPLEX? A SECOND SUMMARY

If there is, to all appearances, a great void, what does it contain? The infinite sea of electrons in the Dirac vacuum, with their negative energy and their infinite electrical charge, cannot be the answer. They don't even figure in the actual quantum field theory of electrons. They are neither real nor observable; they are an artifact of an evolving theory.

But they help us as we put together our image of the actual vacuum. We glean from figure 59 that every particle has its antiparticle of opposite charge and that particle plus antiparticle are nothing but an excitation of the vacuum, accessible from the Dirac sea once there is enough energy for the transition. Conversely, we saw that every real particle-antiparticle pair can annihilate into a pure energy excitation of the vacuum. These are the results that count; and the uncertainty relation tells us that pair creation and pair annihilation happen in the vacuum at all times, in all places.

If the energy needed for pair creation is not available, the vacuum fluctuations will not create real particles; in this case we speak of virtual particles. The lifetime of such virtual particle-antiparticle pairs in the vacuum depends on the mass of the particle (which, we recall, is the same as that of the antiparticle). The larger the mass, the shorter the duration of the fluctuation at its basis. After all, the fluctuation is the only source for the energy needed to supply the mass of particle plus antiparticle. And we know that the more energy a fluctuation contains, the shorter its lifetime must be. We might interpret this state of affairs as nature's willingness to permit the violation of its basic laws of energy conservation, but only for very brief times. We might even say that nature lends out energy—large loans for short durations, small ones for longer times.

In any event, we can make real particles out of the virtual particle-antiparticle fluctuations of the vacuum if we add enough energy. That happens routinely in the large particle accelerators that cause particles and antiparticles to collide. In these collisions, particles and antiparticles approach so closely that the processes of their interaction are not directly observable. Some of them can be calculated, others cannot. They occur in a thicket, hidden from us by the uncertainty principle. Werner Heisenberg, in fact, suggested that what happens in the minuscule space of the actual particle-antiparticle interaction is beyond our observation on basic principle: an inobservability beyond what is mandated by quantum mechanics. Theoretical interpretation must make do with what our detectors can register far outside the locus of interaction. They tell us what particles exit from the interaction zones; their charges, momenta, and angular momenta; and, which particles were annihilated in the process. Today we no longer accept Heisenberg's so-called *S-matrix theory*. Instead, the creation and annihilation processes, being mediated by fields, are now believed to be accessible to a step-by-step calculation—at least, in principle. But nobody has a precise notion of exactly *what* happens *when* and *where* inside the fundamental space-time uncertainty volume.

LIGHT AND ELECTRICAL CHARGES IN THE VACUUM

If we leave gravity aside for a moment, all the interactions of light involve electric charges. They absorb or emit light in the elementary process symbolically shown in figure 61a. If an electric charge first absorbs, then emits light, as in figure 78a, this is equivalent to a scattering of light by the electric charge; we might say that the light's direction is changed by the charge.

As light is bounced off macroscopic matter, the elementary process is its

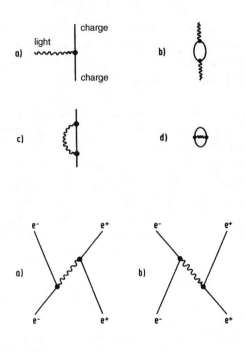

Figure 61 The elementary interaction (a) of light with electrical charges implies the interaction of light with charges fluctuating in the vacuum (b) and of charges with the electromagnetic zero-point radiation (c). Diagram (d) shows a vacuum fluctuation in the absence of real particles.

Figure 62 Feynman diagrams are graphic representations of basic interactions between elementary particles. They can be reformulated as rules for the calculation of the process they stand for. Here they are drawn such that time increases from below to above: The initial state is below; the final state for each process can be read off the top of the diagram.

scattering off the charged electrons that are bound in the atoms and molecules—it can always be reduced to the basic process of figure 61a. If there weren't any charges, there would be neither absorption nor emission nor any scattering. Light would not impact on the matter it passes through nor be impacted by it, and we would have to rely on the action of gravity to notice the existence of light. (If there were no electric charges, we could not exist ourselves, of course.) Light would simply pass through matter (if there were matter, that is). It would not be scattered even by other light; it turns out that light-light scattering happens only because the vacuum fluctuations produce short-lived pairs of charged particles and antiparticles, which can absorb and emit light. We will come back to this in connection with figure 78.

In figure 61, we introduce diagrams that are named after their inventor, Richard Feynman. Using such diagrams, physicists are able to build up interaction processes between particles in a systematic way. While charged particles participate not only in electromagnetic interactions but also in other elementary processes that have nothing to do with their electric charge, the electromagnetic interactions are particularly well understood. Instead of using the word *light*, we might equally well have spoken of electromagnetic waves with arbitrary frequencies. Their propagation in a vacuum is illustrated in figure 61b; they don't move freely, but dissociate virtually into electron-positron pairs. This process is equivalent to the photons being scattered by the electrons and positrons that appear due to the fluctuations of the vacuum. The word *photon* is used for an elementary light oscillation; physicists regard it as the *particle of light.*

There is no space emptier than the vacuum of physics. Consequently, we are unable to measure the resistance the vacuum offers to the propagation of light by comparing it with an even emptier medium. Every particle of light, like every particle of elementary matter, carries with it a cloud of virtual partners in the game of virtual dissociation and recombination illustrated in figures 61b, 61c, and 61d. In the absence of further particles to interact with, this cloud remains unobservable. But as two particles approach each other, so do their clouds. That causes the interaction of the virtual particles, leading to the only thing we can actually see: the interaction of the real particles.

Just as photons are scattered off the charged particle-antiparticle pairs fluctuating in the vacuum, electrical charges will interact with the virtual photons of zero-point radiation (see fig. 61c). Again, the resistance the physical vacuum might be opposing in this way to the propagation of the electron cannot be measured by comparison with any resistance an emptier space might offer; the charge emits, then absorbs a photon in the emptiest imaginable space.

This is what makes vacuum effects observable in the first place—elementary particles don't just interact with each other. Their accompanying clouds scatter off the particles and off one another. The individual processes among the virtual as well as the real particles can all be symbolically described by Feynman diagrams, such as in figures 61 and 62.

DEFINING THE VACUUM

When the atomists of antiquity talked about empty space, their ideas were based on the notion that it actually existed, surrounding sharply delineated atoms of matter. To them, the only difficulty in achieving a truly empty volume consisted in the removal of those atoms from a given space. That kind of empty space, as we have seen, was achieved some two thousand years later by Torricelli. Similarly, the vacuum of classical physics—once the idea of an ether had been discarded, but before the arrival of quantum mechanics and the general theory of relativity—was simply an evacuated volume; without a doubt, it could be achieved in the laboratory given the proper technical means. But those means had to remove not just the atoms but also all radiation.

Today we know that this is impossible. There is no such thing as absolutely empty space. All space contains fluctuating fields and particles. Even in the emptiest space that the laws of nature permit, there are energy levels about which the energies of the fields and particles fluctuate; and these energy levels are never sharply defined. This is where the general theory of relativity comes in. To date, there has been no successful quantum theoretical formulation of Einstein's theory, but we firmly believe that the fluctuating masses of the vacuum cause a curvature of space. We are ignorant of the mass level, if any, at which there is no curvature. We will discuss this point in the context of what is called the *cosmological constant.* What is important here is that for every given curvature there are precisely fixed masses. And since we recall that the virtual mass that fluctuates in the vacuum is not constant in time, we infer that the curvature of space follows this pattern: It must fluctuate. Along with this fluctuating curvature, the distances between two points in space and time must fluctuate. This inference has caused far-reaching speculations. We have no idea around which value the curvature of space in fact fluctuates; but from the observational fact that we can look along straight lines deeply into space, we can deduce that any actual curvature must be very small. It is only when we deal with very small distances in both space and time that the fluctuations of their fields lead to wide uncertainties in the energy and mass of the local vacuum. It is thinkable that at extremely small distances the space-time structure becomes mushy, a kind of foam.

This is a possibility, no more. Let's remind ourselves that there is no quantum theory of gravitation yet. Our practical definition of the vacuum is still that formulated by classical physics. When we theorists ask our experimental friends to remove all matter and radiation from a given volume, we have to add, somewhat peevishly, *as far as possible.* What exactly is it that the experimentalists should get rid of? In tune with the mandates of the special theory of relativity, we might ask them to remove all *energy.* We have generated a vacuum in this definition once we have removed all energy that can be removed from the volume in question. This space is now in its ground state, its state of minimum permissible energy. One of the most interesting questions of present-day physics asks whether there is just one such state or several equivalent states with that same minimum energy. Another remaining question concerns the theoretical possibility of *adding* something and thereby diminishing the energy level.

One interesting facet of this definition of the vacuum in terms of the lowest possible energy is that it can also be applied to numerous other systems. One system we will consider is an electric charge embedded in a void: What is the vacuum state—the ground state, or state of lowest energy—of this system? How are the fluctuations of this vacuum state related to the fluctuations of the vacuum proper without the charge? As a rule, we first define a system by its macroscopic content and then inquire what the laws of nature say about its state—the ground state, or state of lowest energy—of lowest energy. If the system under consideration is the universe as a whole, the inquiry about its ground state should be answerable by these laws without further input.

CHAPTER 6

SPONTANEOUS CREATION

•

PARTICLES AND FIELDS

•

ELEMENTARY PARTICLES OCCUPY A UNIQUE POSITION. THEIR NAME would suggest that they can be neither created nor destroyed, and that they cannot mutate into one another. But that is not really true. My favorite illustration of elementary particles is shown in figure 63. In that image, two strawberries are made to collide. From the resulting mess come one banana, six strawberries, a walnut, seven acorns, and assorted other fruits.

That is about the way it goes with elementary particles—at least, as long as we choose the colliding particles deftly. When we cause two of one kind of particle— say, an electron and a positron—to collide, the final products might be totally different particles; they might be a proton, an antiproton, two photons, plus whatever additional particles the available energy and the conservation laws permit. That is what we mean when we say that particles lose their identity in a collision; as far as the other conservation laws permit, they become pure energy, and the same particles or other particles may result from that energy in the final count.

MEASURING POSITIONS

Elementary particles are emphatically *not* classical ones. Rather, they are best described in terms of fields. The applicable laws of nature are those that govern fields, not particles. True, they are particles as they are being emitted by a material source, and again as they are observed by a material target, or in a detector designed to signal their arrival and their properties; but in the intervening span between emission and absorption, they are best described as fields.

Take a classical particle for comparison. We may not know its location and its velocity, but we know that it does have both a location and a velocity at all times. Maybe our colleague at the next experimental bench knows them and could tell us about them.

Classical mechanics is a theory well rounded in itself; nothing would prevent

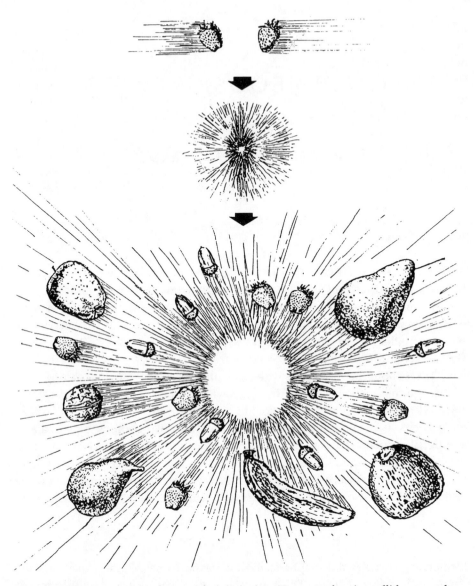

Figure 63 Cartoon picture of a particle interaction: Two strawberries collide to produce a mass of different fruits in addition to more strawberries (this cartoon is owed to a CERN brochure).

it from being applicable down to the smallest and lightest object. We demonstrated that in the case of mass points in Newton's theory. In reality, however, this does not apply to the microscopic world. There, it contradicts quantum mechanics, which, as we know, is universally valid. According to Newton's mechanics, all mass points have trajectories, just like planets in gravitational fields. That is why there is nothing wrong when, in his calculations of planetary motions, Newton treats planets as pointlike. But this is only a model. In reality, pointlike objects such as electrons don't behave at all like planets. To calculate the trajectory of any object, we have to apply the laws of quantum mechanics. According to these laws, pointlike objects, such as elementary particles, will not at all behave like little planets.

The electron is, to a very good approximation, pointlike with respect not only to its mass but also to its electric charge. To date, no experiment has been able to reveal a charge distribution of the electron over an extended volume, no matter how small. The same goes for its mass. As a pointlike particle, the electron has to be treated by quantum mechanics; the laws of classical physics do not apply.

The electron is emitted and absorbed by matter, quite correctly, as a well-defined particle; its detection will always pin it down to one precise location. It never shows up in two different locations at the same time. When we say it shows up, we mean that a detector will be triggered; two spatially separated detectors are never triggered simultaneously by one and the same electron.

Quantum mechanics describes what happens to a physical system between successive measurements. That is where the concept of fields comes in. The laws of quantum mechanics prescribe the behavior of the fields as a function of time. Once a field is known, we know with what probabilities to expect certain measurement results—for example, with what likelihood an electron will be detected in some given interval of space, at (or around) some given point in time.

It is possible that the likelihood of finding an electron at some given time is noticeably different from zero only in a very small space volume around a particular point. But even in this case, it makes no sense to speak of a *particle* in the sense of classical physics. The uncertainty relation sees to it that when an electron's location is very well defined, its velocity is very ill defined and may be very large. As a result, the possibility of locating the electron will rapidly spread over a large volume. It could, of course, be absorbed by matter in the process; and that would trigger a reaction that we would expect a particle to initiate. There might be a flash of light that we could see on an oscilloscope or even on a television screen. But before we get there, we have to change our question fundamentally.

The field of a particle at any given moment always determines the probabilities of the results any measurement we perform on it will give us. We often prefer discussing the *state* of a particle instead of its field. Both the field and its state

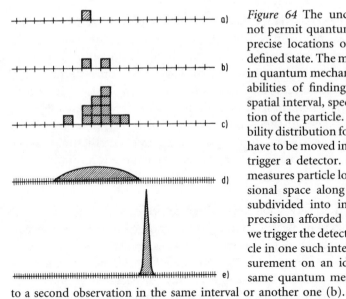

Figure 64 The uncertainty relation does not permit quantum mechanics to specify precise locations of a particle in a well-defined state. The measurement of location in quantum mechanics results in the probabilities of finding a particle in a given spatial interval, specified by the state function of the particle. To measure the probability distribution for a given state, particles have to be moved into this state before they trigger a detector. Our idealized detector measures particle location in a one-dimensional space along the line shown to be subdivided into intervals that mark the precision afforded by the detector. When we trigger the detector, it will find the particle in one such interval (a); a second measurement on an identical particle in the same quantum mechanical state will lead to a second observation in the same interval or another one (b).

After twelve such experiments, the distribution of results might look like that shown in (c). Increasing the number of measurements and their precision will lead to a smooth curve as in (d). This curve gives the probability of finding a particle in a given state somewhere in our one-dimensional space. It is the result of a quantum mechanical measurement of location. The uncertainty principle relates this distribution in space to an analogous distribution when we measure the velocity of particles in the same state: When the spatial distribution is broad, as in (d), the velocity distribution may be narrow, as in (e), where the abscissa now denotes velocity. Conversely, a more precisely located particle has a more broadly distributed probability of being found at a given velocity.

fix the likelihood with which we will find the particle at a given time near a given point. This probability, or likelihood, is a measurable quantity; figure 64 also indicates how the measurement can proceed.

Detectors designed to locate a particle divide space into small intervals, or cells (note that we do that in figure 64 for one dimensional space). Let's imagine that we have a single particle in some given state. A detector will locate the particle in one of those intervals. The interval is not, in general, entirely fixed by the *state* in which we find the particle. Our detector, when locating another particle that has the same *state function* (to physicists, this means "is in the same state"), will find it either in the same or in another interval. The field does not prescribe where exactly the particle will be detected. But when we do many measurements on particles in the same state, we build up a curve from the individual measurement events, and this curve shows the likelihood for finding the particle in a given state at a given location (see figs. 64d and 64e). The state function of the particle implies exactly the shape of this curve.

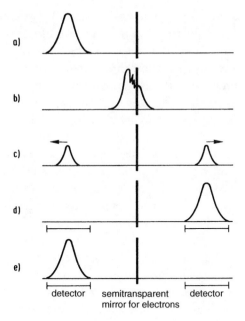

Figure 65 The wave packet of an electron incident from the left (a) is partially reflected by a semitransparent mirror (b).

The statement that quantum mechanics makes about the result of the measurement of a particle's location appears to contradict the sacrosanct principle of cause and effect. Not only is it impossible to predict where our detector will find the particle; even the assumption that this should be fixed prior to the measurement process leads to a contradiction between experimental results and the expectation that every action propagates locally, from one point to a neighboring point. Indeed, we have to conclude that there cannot be a deeper theory at the basis of quantum mechanics—a theory that would be able to predict the precise result of every measurement of a particle's location. But the field can be treated in terms of deterministic and causal equations that determine the future state of a particle from the present one.

Let our particle detector be designed so that it registers the particle's location at a given time without, however, absorbing it. This implies that the probability of detecting it in a second experiment is different from zero only near the space interval within which it was detected in the first experiment. Indeed, this probability equals 1—we can be certain that the particle will be found near the same location if we go through a new detection process immediately after the first one.

In quantum mechanics, the probability of finding a particle extends over large volumes, in general. There is no upper limit to the extent. Figure 65 should explain this. Let's look at an electron, starting when its probability distribution (defined in figure 64a) is that of figure 65a. We let it run—and the expression is literally true—in the direction of a semitransparent mirror for electrons; never mind how that mirror might actually look. Such a contraption actually exists; after some intricate interference effects (see fig. 65b), about 50 percent of the probability that we might find the electron when looking will have moved way to the left and the other 50 percent way to the right. The important thing we can glean from figure 65c about the distribution of probabilities is this: A detector situated on the far right, where the particle has a likelihood of one-half to be seen (the other one-half likelihood is way over to the left), will either find the electron or it will not. If it does, the likelihood of finding it an instant later in the same location is one (see fig. 65d), and if not, zero. So far, so good. Remarkably, there is a definitive implication for the likelihood way to the left, where we did not put a detector. If the electron is seen by the right-hand side detector, the probability way to the left is zero; if it is not seen on the right-hand side, that probability is one (see fig. 65e).

Small wonder that such mysterious long-distance effects across space raised considerable interest; they have been at the basis of many objections to quantum mechanics—unjustly so. The quantity that is being influenced from afar is a probability; and that means we cannot determine *experimentally observable* long-distance effects. Whether we did or didn't perform the experiment on the right-hand side, a detector on the left will have the same one-half likelihood of finding the electron in this location. In the absence of a measurement on the right-hand side, the electron will be seen in the left-hand position by half of all attempts to detect it; figure 65c shows that the quantum mechanical probability of finding it in that location is one-half. And the same goes for the case in which the right-hand side detector is activated: The probability of finding the electron on the left is still either one or zero, with equal probabilities of 50 percent. Even though the result of the experiment on the right-hand side determines what happens on the left for each individual case, the observer on the left cannot infer from his experiment whether his counterpart on the right has activated her detector. There are no observable long-distance effects that could tell him so. The probability of his observing the particle has been changed by the experiment on the right from a quantum mechanical one to the same classical one—that is all. No information can be instantaneously transmitted in this experiment.

It is really curious that we can treat probability distributions as though they were material substances. A few years ago, in collaboration with some colleagues at the University of Karlsruhe, I produced animated computer movies in an attempt to show the behavior of quantum mechanical probability distributions as a function of time. In figure 66, I display a few stills from one of these movies. The probability distribution is compressed by an obstacle to its propagation, just like the pent-up gas flow at a bottleneck. The resulting motion may be slow or

Figure 66 Computer simulation of the quantum mechanical process schematically shown in figure 64.

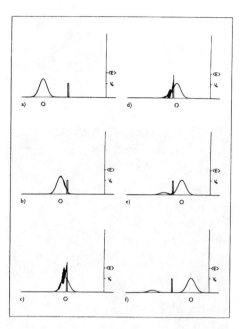

fast, as the case may be. Where the flow is compressed, the likelihood that a detector will see the electron is large.

These movies show quantum mechanical reality. The field of the electron is real because it results from the laws of quantum mechanics and appropriate initial conditions. The laws determine the field's behavior—its propagation in space in particular. Just like the field of light, the field of an electron is nothing but that: a field. With each point in space, we associate a number that signifies it is a probability. The field equations tell us that their solution, the field from which we can calculate the probability distribution for finding the electron, is oscillatory. We know that there is no *something* that oscillates—it's just the field. The field of the electron is nothing but an excitation of empty space. The electron with its charge is just being thrown in—and it is ubiquitous.

The elementary particles are often dubbed the atoms of our day. That may or may not be correct. It might be an acceptable metaphor for a particle that is hidden in its probability distribution; it is less so for the distribution itself. We have no notion whether or not elementary particles such as electrons or quarks are ultimately divisible. We also don't know whether they have any internal structure or we can continue to deal with them as though their entire charge and mass were concentrated in one point. But the probability of finding particles can always be subdivided. Once we include the probability distributions discussed

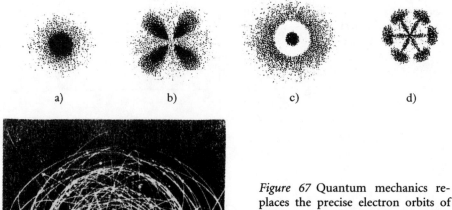

a)   b)   c)   d)

Figure 67 Quantum mechanics replaces the precise electron orbits of Bohr's atomic model by probability distributions for the detection of electrons in the vicinity of the nucleus. (a–d) Probability distribution for the electron states in the hydrogen atom. (e) The Bohr orbits for the ninety-two electrons of a uranium atom. What this representation shows most clearly is the fact that a tiny atomic nucleus is surrounded by a great deal of empty space and a shell-like arrangement of electrons.

e)

above, these atoms of modern times differ from those of the atomists in antiquity by an absence of sharply defined limits that separate things being from things not being, matter from empty space.

ELECTRONS IN ATOMS

In Bohr's model of the atom, the electron is a point or a tiny ball that orbits the atomic nucleus like a planet, separated from it by empty space. The radius of the nucleus, which is also treated like a point or a tiny ball, is at least 100,000 times smaller than that of the innermost electron orbit. A heavy atom has many orbiting electrons. Ninety-two electrons orbit the uranium nucleus (see fig. 67e), suggesting (correctly) that this atom contains a great deal of empty space. In quantum mechanics, of course, there are no sharply defined orbits, but there are the corresponding probability distributions for finding the electrons when placing a detector in a given location of the atomic volume, as shown in figures 67a through 67d.

The probability distribution for the location of the electrons can take any of the forms shown in figures 67a through 67d (as well as others). The density of points in this illustration is a measure of the local probabilities. If an electron is in a quantum mechanical state corresponding to any of these figures, the only correct statement we can make about its location is this: The likelihood that a detector will see it is determined by the density of black dots that our image gives for the location in question.

This figure, in my opinion, provides a good argument for the need to describe an electron in terms of a field in quantum mechanics. Although not as simple and easily accessible as the electrical and magnetic fields in static electromagnetism, the fields of electrons contain everything that can be known about these elusive particles.

FIELDS, ETHER, AND EMPTY SPACE

Magnets have been known ever since antiquity. William Gilbert's image of magnetism (in contrast to electricity) is a *quality* by which substances influence similar substances. This was documented in his *De Magnete,* published in 1600. The notion was replaced by Descartes's theory in which magnetic forces (just like the electrical ones in Gilbert's thinking) are transferred by a stream of substances. Today we know that magnets generate magnetic fields in the space surrounding them; they exist by themselves and don't need a material substrate.

In particular, they don't need any "streaming" substance. The concept of an ether indicates another way of transmitting forces from one point to another. This ether is eternal and ubiquitous; it transmits forces, light, and other phenomena not by moving from here to there but by dint of oscillations. If it is excited in one place, the vibrations will propagate from there, carrying light and forces with it.

It has also been assumed that ether can change its viscosity or elasticity. Details of these notions are bound to be wrong and are of no interest here. But once we disregard all presumed qualities of the ether, then it becomes indistinguishable from empty space; it remains nothing but a bit of ontological hairsplitting to differentiate between the ether and space. At the end of the last century, H. A. Lorentz and Heinrich Hertz were the first to interpret Maxwell's theory this way: The theory tells us the properties of the fields described and also those of their presumed carrier substance, the ether. Seen in this way, the ether in its ground state is nothing but empty space under another name. In the special theory of relativity, Einstein also showed that the ether, seen in this fashion, is not in any given state of motion, just like Newton's empty space. He concludes that "if no particular state of motion belongs to the ether, there does not seem to be any ground for introducing it as an entity of a special sort alongside of space." This means that the ether is certainly not a substance that fills space. There is no reason to introduce it.

Magnets and electric charges are inextricably linked by their fields with the space surrounding them. Once we know the field, we can know the locations of magnets and charges. The closer we approach a pointlike charge, the larger becomes the electric field. At the location of the charge proper, that field becomes infinite.

Charges induce properties in the space surrounding them which can be observed experimentally. If we enter a test charge into an electric field at some given location, a force will act on it. This force points in the same direction as the field. Note that the field *has* a direction and that this force is proportional to that of the field.

Given that forces and fields are directional, they can be represented by arrows. The length of an arrow stands for the strength of the force and the field; its direction is defined to be their direction. The arrow in a given location stands for the field at that point; this means the arrows are affixed to space in given locations.

Graphically, it is more instructive to depict electrical fields by lines that stand for the directions of these arrows at every given point. In figures 68a and 68b, we show an electric and magnetic field, respectively, in this fashion. Maxwell, in his influential treatise on electricity and magnetism, presented drawings of the special fields he had calculated. Figure 68c shows an example. Maxwell wanted to explain the propagation of electromagnetic fields in terms of vortices in ether (see fig. 68d). Heinrich Hertz, the first to observe electromagnetic waves—in 1885, in Karlsruhe—illustrated the transmission from an antenna in terms of the lines of force resulting from the electrical and magnetic fields (see fig. 68e); this meant a notable simplification of Maxwell's representation. All these figures demonstrate the properties of space as a carrier of fields.

PARTICLES AS FIELDS, FIELDS AS PARTICLES

It has become customary to distinguish among three types of particles and fields: They concern matter, forces, and the Higgs phenomenon. This last, involving Higgs particles and Higgs fields, stands out in the process. It adds to the vacuum of matter and forces, which I will describe first. Matter is composed of the electrons and its siblings, and of the quarks that make up protons and neutrons. The forces are known to be transmitted among the particles of matter by the exchange of the force carriers; these carriers are photons, W and Z bosons, and gluons. There is a field associated with each of these particles of matter and force-exchange particles that, together, make up our zoo of fundamental particles. Should we arrive, sometime in the future, at particles that are even more elementary, the above notions will basically remain valid.

There is a field for every particle. If that field is concentrated in some spatial region, the particle of that field is likely to be located in that region once a detector looks for it. We introduced fields as abstract quantities associated with

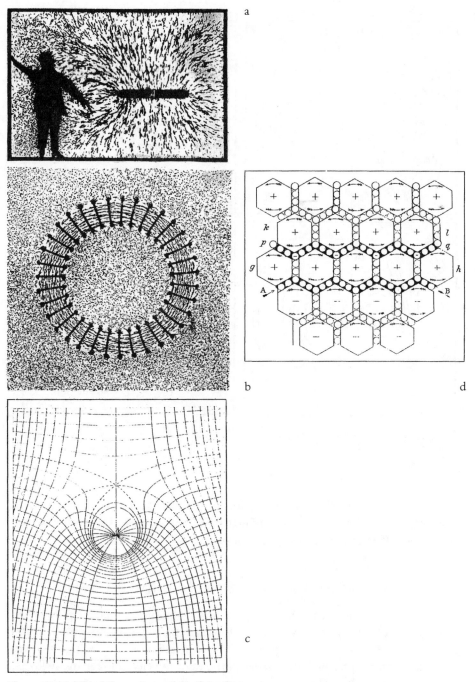

Figure 68 Fields of force, as explained in the text.

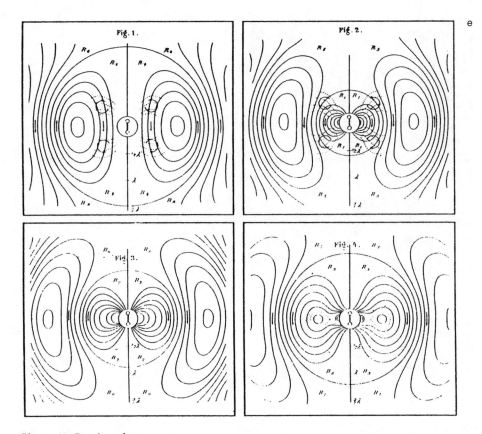

Figure 68 Continued

space; we describe a field as an excitation of space that oscillates like tall grasses in a heavy wind. Recall, however, that there is no material aspect to these field oscillations. Rather, *a field is a field is a field.*

IS THERE SPACE WITHOUT PARTICLES, FIELDS, ENERGY?

Today, this is the way we define the vacuum: It is the remnant left over in a volume out of which we have removed everything that can possibly be taken out. The existence of electromagnetic zero-point radiation implies that this vacuum still contains photons, the carrier particles of the electromagnetic field. The photons fluctuate into electron-positron pairs; that means that their fields also don't vanish in the vacuum. And so it goes for all the fields of all the other particles in our particle zoo—of the quarks and the antiquarks, of the neutrino and the antineutrino, of the muons and the antimuons, of the W and Z bosons, of the gluons. All of them swirl around in this emptiest of all spaces, in all imaginable configurations, and for the briefest of time only. Recall, the larger the mass of any of these excitations, the shorter its lifetime, the smaller the volume in which it is concentrated. The excitations of the vacuum come and go; they oscillate, they fluctuate. It is not easy to conjure up imagery vivid enough to do justice to the fluctuating vacuum.

But we know that all these fields filling the vacuum must, on the average, vanish; the only quantity that remains immutable is the energy, the zero-point energy. The average charge of a vacuum is zero, and so is its average momentum. Any calculation we might attempt to determine the energy of the vacuum must fail; the contributions from some particles and fields will tend to positive infinity, of others to negative infinity. Summing up such indeterminate quantities does not give us a finite answer.

PHYSICS AT SHORT DISTANCES

The uncertainty relation tells us that an infinitely large energy corresponds to infinitely small distances. If some future understanding of physics were to permit the determination of a finite energy of the vacuum—maybe even the energy zero and thus an absence of gravity—then at very short distances it would have to look different from the physics we are familiar with. We must distinguish between two types of experiments in short-distance physics, complementary to each other. On one hand, there are precision experiments: Small deviations of an experimental result from a theoretical prediction may well hint that there are *new* contributions that have to be added to the well-known large-distance effects. In the next chapter, we will discuss such an interplay between theory and experiment in the case of the spectral-lines shift named after Willis Lamb. On the other are the more spectacular experiments that address the physics at small distances head-on by colliding particles at very high energies.

The uncertainty relation tells us that small distances correspond to high energies. Microscopes operating with visible light cannot resolve distances shorter than the wavelength of this light. Electron microscopes operate at higher energies and can therefore resolve smaller distances than ordinary optical microscopes. And so it goes on to higher-energy particle beams: The quantity of importance for the observation of small distances is the wavelength we associate with the accelerated particles, and that decreases with increasing energy. Further acceleration improves the resolution of the resulting experiment.

LEP: A MACHINE TO PROBE THE VACUUM

Our largest available accelerator is just outside Geneva, at the CERN laboratories, and is called the LEP (for Large Electron Positron machine). Figure 69 gives an

Figure 69 Aerial photograph of the CERN Laboratories outside Geneva; the French-Swiss border is shown as a dashed line. The smaller circle is the super proton synchrotron (SPS), the larger one, the large electron-positron collider (LEP). The four LEP detectors are located in underground halls indicated by white dots.

overall photographic view of CERN with added demarcations showing the border between Switzerland and France, and the traces of two ring-shaped accelerators. One of these is called the SPS (for Super Proton Synchrotron); the other is the LEP, which is the largest scientific instrument ever constructed. An evacuated tube of 27 km, or approximately 16 miles, in length runs inside a circular tunnel about 100 m underground. Electrons and positrons, guided by magnetic fields, circulate inside this tube in opposite directions. These charged particles radiate electromagnetic waves as they are being forced by the magnets not to go in straight lines. The resulting so-called synchrotron radiation, signaling their resistance to angular acceleration, leads to an energy loss that has to be compensated by radio frequency cavities that reenergize the circulating beams.

There are four interaction zones along the perimeter of the LEP circle, also marked in figure 69. There, the beams are guided so that particles collide head-on with antiparticles coming in the opposite direction. Each of the interaction

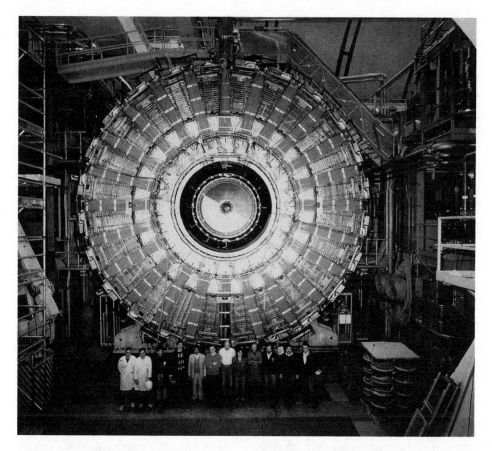

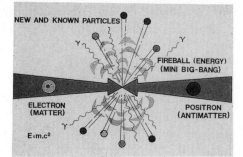

Figure 70 (a) The DELPHI detector. I chose it out of the four LEP detectors because my experimental colleagues from the University of Karlsruhe belong to the international group that built it and operates it. The group consists of 550 physicists from 48 institutes in 21 countries. (b) Schematic view of an electron-positron annihilation event: The detector sees particles, not fields, as we describe in the text.

points is surrounded by an intricately designed detection apparatus that records the details of the elementary-particle interactions resulting from the head-on collisions. Their designers have given poetic names to these high-tech devices— ALEPH, DELPHI, L3, and OPAL; they are huge machines in their own right (see fig. 70a). The particles issuing from the collision reactions—that is, from the annihilation process of electrons and positrons—generate signals in these detectors that on-line computers transform into tracks such as those shown in the inset of figure 70. We show two examples from the DELPHI detector in figures 71 and 72.

INTO AND OUT OF THE VACUUM

To make real particles out of the virtual ones that are part of the vacuum fluctuations, the only thing needed is energy. But the energy inherent in the vacuum is inaccessible; it would have to be extracted from the vacuum, and that is impossible, because the vacuum is already the state of lowest energy. When an electron and a positron collide in the interaction region of the detector, the ensuing final-state volume is overall electrically neutral; in this sense, it is a vacuum. The uncertainty relation keeps us from knowing the precise locations of the particles along with their velocities; our probability of finding them is distributed through a certain spatial volume. If electrons and positrons were classical particles, they could obviously not annihilate each other; no provision at all is made in classical physics for processes of this kind. Quantum mechanics, however, permits us to look at them as though they were both a particle and a hole (as we showed in fig. 59). When the particle drops into the hole—a process that is very likely under the circumstances—the sum of their motion energy and mass energy will be freed.

That makes for a highly energetic interaction volume. Just as in the vacuum, there will be fluctuations of all imaginable fields, of virtual particles, atoms, nuclei, with increasingly small lifetimes, the higher the mass of the particle-antiparticle pairs. Finally, the energetically excited spatial volume collapses into real particles. That might lead, for instance, to a muon-antimuon pair, a quark-antiquark pair, and so on. The muon-antimuon pair has no trouble leaving the interaction region, so detectors can pick up its presence (see fig. 71). For quarks and antiquarks, this is different. They have fractional charges of plus/minus 1/3 and plus/minus 2/3. Before they can become detectable free particles, they have to acquire integral charges, like all the particles that have ever been detected. We will not discuss the mechanism by which they might do that. But we know well that the resulting final state contains all kinds of particles that combine into *jets* before they leave the interaction zone (see fig. 72). These jets help us to visualize the energy balance of the final state.

To sum up: Electron and positron annihilate as a pair, in a particle-antiparticle collision. Their energy heats up the vacuum; when the hot vacuum decays, real particles emerge. The annihilation energy of electron and positron has, in the process, been used to realize the hidden faculties of the vacuum. For individual cases, the results may be muon-antimuon pairs, quark-antiquark pairs, and so

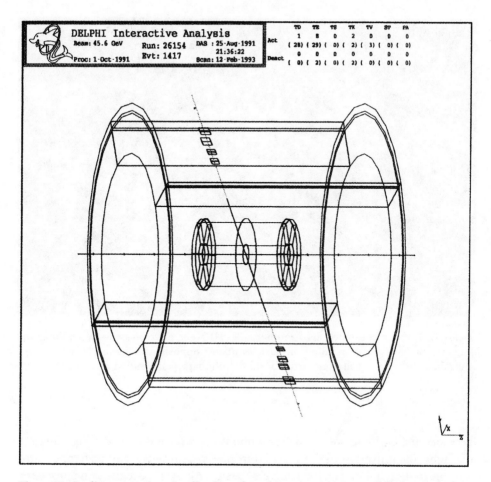

Figure 71 On-line computer visualization of an electron-positron annihilation event into a muon-antimuon pair. The computer image shows which counterelements triggered the electronics, and on-line reconstruction points the muon tracks from the interaction at the very center of the detector in their direction. Electron and positron enter from opposite sides along the axis of the cylindrical detector.

on. In passing through the intermediate state of an excited vacuum, it looks as though the electron and positron actually knew which virtual particles their annihilation might realize. But in fact, electrons and positrons in themselves are pointlike, and can be fully described in terms of a theory that knows neither muons nor quarks. So it is not the electron and the positron that "know" the possible final products of their interaction—it is the intermediate state, our vacuum, that has that knowledge.

The same energy density that we reach today in electron-positron collisions

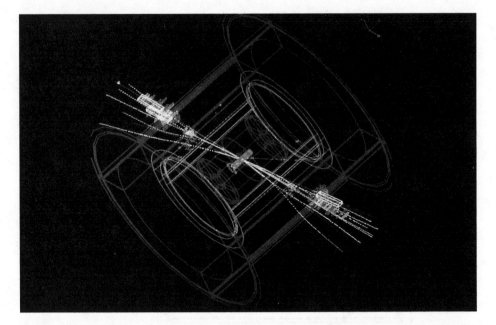

Figure 72 In this reconstruction of another DELPHI event, electron-positron annihilation leads to the emission of bundles of particles in two opposite directions. These bundles are called jets, and usually start from a quark-antiquark pair emitted at the interaction point.

pervaded the entire universe one ten-billionth of a second after the Big Bang. At that time, the universe was the size of our solar system today, and its temperature was one million billion (or 10^{15}) degrees Celsius. All over space the same processes happened that we observe today in the electron-positron collisions we just described. The LEP machine in Geneva has permitted the detection of millions of such mini–Big Bang events—truly, it has been a machine for the exploration of the vacuum!

CHAPTER 7

LET NATURE BE AS SHE MAY

•

SPECIAL SYSTEMS

•

TO PHYSICISTS, THE VACUUM IS JUST ANOTHER SYSTEM TO BE studied and described. They add energy and thereby excite it, like guitarists who pluck the strings on their instruments. Physicists want to know how individual oscillations depend on energy, and they try to investigate the ground state, the vacuum state. These questions about ground state and excited state are commonly applied to all physical systems—to the oscillating string and pendulum no less than to the water-based universe of Thales in antiquity. In the prologue, we said that the ground state of water at 0 degree Celsius is ice, since liquid water at the same temperature contains more energy. We can also inquire about pressure waves and various currents, which are the excitations of water corresponding to fields and particles; we can follow the precise state of water as the temperature rises, as it changes from solid ice to liquid water and on to steam. In abstract terms, the vacuum and a water-filled universe are similar systems; it is only in the details of how the equations of state change with temperature that the states of water and the excitations of a vacuum differ.

SPONTANEOUS MATERIALIZATION

As we showed in figure 5, water molecules oscillate about their rest positions in ice. Quantum mechanics says that this remains true at absolute zero and will not admit a molecule at full rest. Similarly, we saw that what we call empty space is replete with the appearance and disappearance of virtual particles. Add energy, and these particles can become real. Add energy to ice, and it will melt. Moving in the opposite direction, remove energy from water and it will freeze because the water molecules attract each other; ice lets them approach as close as possible. The resulting crystalline structure is destroyed by thermal motion of the molecules when energy is added. Moving at random, the molecules emerge as individual particles, much as an electron-positron pair emerges from the vacuum when we add energy.

Whenever liquid water makes the transition to ice, energy is given off to the surroundings. In the process, the water itself assumes a state that is both more ordered and lower in energy. It is a general rule that any system that can give off heat and thereby assume a state of lower energy will do so. For the purpose of illustration, let's assume that the energy set free by the freezing of water is extremely high—so high that it surpasses the energy that is by virtue of Albert Einstein's $E = mc^2$ connected with the very existence of the water molecules. What would happen? In this fictitious case, it would pay energetically if water in the form of ice were spontaneously created from a space that beforehand contained no water at all. Thus there would be a certain probability for this to occur—never mind that anti-ice would have to be produced too. Let's imagine that it occurs: a crystal of ice is created spontaneously out of the void. Like every crystal, it would have some preferred direction in space and a certain location. Consequently, the perfect symmetry of space would be broken.

These imagined circumstances do not exist in reality as far as ice is concerned, but they apply roughly for one of the most imaginative constructs of physics—the so-called Higgs field. This field appears spontaneously in a void as its walls are cooled down—starting from the absurdly high temperature of 10^{15} degrees. The field will appear in an ordered state; for a poetic simile, think of ice flowers growing on a window. The energy needed for its existence is smaller than the energy liberated by its falling into that ordered pattern. This pattern is not to be understood in terms of spatial geometry; rather, it refers to the abstract space made up of the properties of elementary particles. In geometrical space, it is merely a field resembling a particularly simple distribution; to every point in space, we assign one and the same complex number. This implies that the Higgs field does not break geometrical symmetry—it breaks an abstract symmetry of elementary particles. In fact, it was introduced into modern theoretical physics by the Scottish physicist Peter Higgs for that very reason—to break an abstract symmetry that would not permit elementary particles to have masses.

THE HIGGS BOSON—A VERY SPECIAL PARTICLE

The British minister for scientific affairs, William Waldgrave, promised a bottle of champagne to anybody who might explain to him on one typewritten page the nature of the Higgs particle and the importance of its discovery. In raising this challenge at the annual meeting of the Institute of Physics in April 1993 at Brighton-on-Sea, he said: "The scientist has an obligation to the public to explain his activities if he expects public finances." In fact, the discovery of the Higgs particle (or *Higgs boson*, as it is called more precisely) or, alternatively, a potential lack of evidence for its existence, is among the most important goals of the new generation of particle accelerators, the large hadron collider (LHC) at CERN and a beefed-up version of the Tevatron at the Fermi National Accelerator Laboratory, outside Chicago. "If you can make me understand the Higgs boson," Waldgrave concluded in his speech, "I have a better chance to find the funds you need for its discovery."

All entries had to be sent in by June 1, 1993. John Maddox, the editor of the British journal *Nature,* had immediately suggested that the minister should be told that the Higgs boson can be explained similarly to the way we visualize the W and Z bosons: as carriers of information and effects. They permit us to understand effects that would otherwise look like long-distance action across empty space in the absence of any carrier.

The interpretation of apparent long-distance action by means of continuously propagating effects was vigorously asserted by Isaac Newton. In his wake, the interpretation of all forces or propagation of information over long distances in terms of hidden short-distance effects with finite velocity of propagation is one of the most challenging tasks of fundamental research in physics. We know that oscillations of charges in some given location build up fields propagating in space, causing electric and magnetic action to spread with the velocity of light. The general theory of relativity tells us that masses deform space and thereby cause other masses to move in appropriate directions. The deformation of space, too, spreads from point to point with the velocity of light by means of waves; these *gravitational waves* also have their carrier bosons, called *gravitons.* Their finite, if very large, velocity of propagation implies that it is not the Sun's existence right now that makes Earth follow a curved orbit right now; rather, it is the Sun's emission of gravitons eight minutes ago that keeps Earth from moving in a straight line right now. Photons and gravitons allow us to understand, at least in principle, how electromagnetic and gravitational forces spread from point to point. The same goes for gluons and for the W and Z bosons when it comes to the strong and weak nuclear interactions, respectively.

In the same way, the Higgs boson is used in the standard model of elementary particle theory to mediate certain effects across space. But that is not its principal task. If it were, the minister might rightfully ask why we can't make do with the known particles in that category—the photons, Ws, and Zs. Given that it will cost literally billions of dollars to find the Higgs boson, shouldn't scientists content themselves with the knowledge of the propagation of long-distance effects which they gained from these other carrier particles? Is it really important to investigate this additional corner of the carrier space?

As it is, the Higgs boson is assigned a role that stands apart in the theory of basic particles. The explanations worked out for the minister try to delineate the role it is to play in the overall scheme of the standard model. While this author did not participate in the competition for the minister's bottle of champagne, let me begin my own explanation by noting a contradiction between the theory of elementary particles without a Higgs boson and common observation. The theory needs the Higgs particle to answer the question about the simple existence of objects at rest. It is a relatively easy task to write down otherwise satisfactory theories in which the only velocity that elementary particles can assume is the velocity of light; but we know that this does not represent reality. The standard model starts out from such a theory. We then bring in the Higgs field, which forces the appearance of other velocities, down to zero velocity, upon the particles with which it interacts.

But how can it perform this task? Up to this point, we have not needed to differentiate between field and particle; there has been a one-to-one correlation between them. This may not be true for the Higgs field and the Higgs particle. Most physicists are convinced that the Higgs field pervades the entire universe; it is a notion held both in particle physics and in cosmology. The Higgs field originated as the universe cooled down after the Big Bang—a bit like the crystallization of ice as water cools down. We have already argued that just as ice crystals form an ordered structure, so does the effect of the Higgs field—with the difference that this happens not in geometric space but in the abstract space of elementary particle properties. We know that water molecules drop down into the ordered structure of ice, releasing energy in the process. The most important difference between this notion and that of the ordering by means of the Higgs field seems to be this: The positive energy contribution associated with the field's very existence is more than canceled by the negative energy associated with the structure it causes to emerge. Consequently, a space pervaded by the Higgs field has a lower energy content than a space in its absence.

This means that the Higgs field can spontaneously emerge from empty space, just as ice can spontaneously emerge from water if the temperature is low enough—that is, if the thermal motion that tends to oppose the ordering is weak enough. In the case of ice in water, the critical temperature is 0 degrees Celsius; for the Higgs field, it is 10^{15} degrees. When the decreasing temperature crossed this critical mark some 10^{-10} seconds after the Big Bang, the Higgs field emerged. It has pervaded the universe ever since, and all masses in the universe owe their existence to the ordering it has caused in the process of the cooling down of the universe.

The winning entry for the minister's competition came from David Miller of University College in London; he portrayed the Higgs field in terms of a permanently ongoing party. He meant to show the way in which the field keeps particles from always moving at the highest possible velocity. Take a particle called Margaret Thatcher in the Higgs field—sorry, in the field of partygoers he fancied—and she will immediately be surrounded by a throng of people who want to talk to her. This throng will move with her as she walks across the floor, and she cannot move faster than they do. And while this throng consists of people, it is not actually the people that accompany the former prime minister; rather, the guests remain in place and the excitation that moves along with her is a wave of conversation, hellos and good-byes, applause. This excitation cannot move with the highest possible velocity; it is fed by the interaction of party guests with one another and with the guest of honor. Their exchanges and their interactions tie them to the motion of Margaret Thatcher, who, in turn, must move more slowly in order to acknowledge and respond to all the attention she is arousing. All that interaction makes her move more slowly than she could in the absence of the fellow revelers. She might even be stopped in place—an occurrence we assume should be impossible for her without the party surroundings. And in fact particles in empty space in the absence of a Higgs field cannot be stationary.

The parameter we hold responsible for the ability or inability of a particle to move with a velocity other than that of light is its mass. When the mass is zero, such as that of a real photon, the particle has to move with the velocity of light at all times; for finite masses, its velocity can take any value below the velocity of light, and can also be zero. We saw that quantum mechanics treats light as a current of mass-zero particles called photons, which have to move with the greatest possible velocity across space. That space, real space, contains the pervasive Higgs field. The photon is one of the particles—it might even be the only one!—that does not interact with the Higgs field and can therefore move in this field without being slowed down. We might redefine the influence of the Higgs field on the velocities of the particles it interacts with in the following way: The Higgs field imparts effective masses to the particles—which, above the critical temperature of 10^{15} degrees, are massless like the photon.

When we look at a streetlight, at the Sun, the Moon, the stars, we *look* straight through the Higgs field: Mass-zero particles—such as the photons and, possibly, particles we call neutrinos—do not interact with the Higgs field and therefore travel with the velocity of light. If they leave some distant origin at the same time, they will arrive at our detector, our eyes, simultaneously. That's the way it was with the famous 1987 supernova. Particles that acquire mass through their interaction with the Higgs field—electrons, protons, atomic nuclei— propagate across space with velocities below that of light. But this propagation is not further hindered in any way by the Higgs field. Our theory knows no such concept as friction in the Higgs field; even massive particles move with a constant velocity across an otherwise empty space that is pervaded by the Higgs field. It is the energy of a particle that determines its velocity. If it were slowed down by the Higgs field, it would be transferring energy to it.

The mechanism invented by Peter Higgs and his followers is not limited to telling us that most elementary particles have masses; it also assigns specific values to different masses. The W and Z bosons have masses about 200,000 times heavier than that of the electron. We might say that at David Miller's fictitious party the prime minister is the Z boson and some guest in the ranks is the electron. To gain a detailed understanding of the masses, we will have to discover the Higgs boson experimentally. And that is why we need huge machines like the LHC and the Tevatron. The Higgs field is always in the wings in the space that we inhabit; it undergoes a local perturbation as a particle interacts with it upon its passage. This perturbation—which Miller's imagery likened to the throng about the former prime minister—is the mechanism that gives the particle its effective mass. If we try to accelerate the particle, we will have to do the same to the perturbation caused by this interaction. It is the resistance offered by this perturbation that our acceleration has to overcome, not that of the originally massless particle itself. The Higgs mechanism therefore permits us to understand all particle masses in the same terms: In the absence of a Higgs field, we would have to assign the observed masses to individual particle groups without having the slightest notion as to why they take on the values they do. It is the Higgs field that may enable us to understand the values of the masses. Its presence hides an inherent elegant simplicity.

The as yet unproven existence of the carrier particle of the Higgs field, the Higgs boson, should then be seen as an itinerant perturbation of the Higgs field without an extraneous particle to interact with. In David Miller's allegory, we might say that the party is livened up by the spreading of rumor among the party guests:

> Those near the door hear of it first and cluster together to get the details, then they turn and move closer to their next neighbours who want to know about it too. A wave of clustering passes through the room. It may spread out to all the corners, or it may form a compact bunch which carries the news along a line of workers from the door to some dignitary at the other side of the room.

These local perturbations without an actual center, which pass on the news in our story, should be seen as the Higgs particle. And since density fluctuations of the Higgs field impart mass, they have to be massive themselves. It makes sense that the Higgs boson itself would also turn out to be massive. How massive, we can only speculate; we can even build models without an actual Higgs boson—models in which the Higgs field will not evince local density fluctuations such as we described above.

It will take enormously complicated experiments at the particle accelerators of the next generation to re-create local energy concentrations that will approximate, in tiny volumes, the conditions prevailing right after the Big Bang. We can then hope to observe the melting of the all-pervasive Higgs field at the critical temperature, and that will permit us to detect it. Just like the cosmic background radiation, the Higgs field is witness to the very early development of the universe; its discovery is liable to tell us in scientific terms what the creation myths have tried to couch in imagery throughout all of human history. At the very least, the successful entries in the competition launched by Minister Waldgrave convinced him that finding the Higgs boson was a worthwhile endeavor; as a result, the British government lined up with the other European partners to finance the LHC.

To the physicist, one of the preeminent missions of this machine will be to answer the fundamental question "Whence the masses of elementary particles?" To do so, particle collisions will have to result in energy densities above that at which masses must disappear—densities that were present in our universe only once, a tiny fraction of a second after the Big Bang.

Essentially all models of the initial moments of our world share the feature that it started out in an extremely hot, dense state. This heat and this density, well beyond what we can imagine today, are not imaginable in terms of phenomena we know. But all models agree that the universe as a whole subsequently expanded and cooled down to an average temperature of some -270 degrees Celsius, only a few degrees above absolute zero, -273 degrees Celsius, which we know to be unreachable. Just as water, whether liquid or crystalline, has a state of lowest energy that depends on the ambient temperature, there must also be a ground state of the universe. The standard model of elementary particle physics assumes that the laws governing the Higgs field also are temperature dependent. It does not give a reason for this assumption; it may be due to some fundamental property, or else it may be a collective effect, such as the transition of water to ice.

We know that at very high temperatures, the ground state of the universe does not include the Higgs field; its state of lowest energy is that of an empty universe with nothing but thermal radiation. The universe at those temperatures does contain particles and fields, including the Higgs field; but they do not pervade all space, and the Higgs field does not lead to structure. That happens only as the universe cools below a given critical temperature. As the universe crosses this temperature, the laws of nature assume a new form: They now dictate that the Higgs field be all-pervasive and lead to structures that break symmetries hitherto unchallenged. This symmetry breaking means that in an abstract space of particle properties, one direction will be preferred to others. The energy freed up in this ordering of the universe is then used to produce massive particles; the transition generates matter as we know it. In this transition from a kind of false vacuum to a vacuum in the way we understand it today, the universe passes into a state in which it has lost the symmetries of nature originally present. If we were able, with the help of the giant machines that will explore the properties of the Higgs field, to find those symmetries, we would glean a look at truly empty space. This true vacuum, which no longer exists, has all the symmetries of natural laws—symmetries now hidden by the state of lowest energy, the ground state, of the universe as we know it.

We think there have been at least two such phase transitions—such crossings of critical temperatures—in the history of the universe. Seen qualitatively, it might seem astonishing that we don't witness the transformation of virtual particles into real particles all around us. The experiments at the LEP collider have shown us many times that all virtual particles need in order to become real is energy. Now, we might argue that the energy needed for the passage from virtual particles to real ones could become available by that very passage.

We know that masses attract each other. The closer they get, the more negative their positional energy (what physicists call *potential energy*). For mass points, it can become arbitrarily negative. In the prologue, I explained this fact in terms of an almost pointlike rock that would drive a generator while being lowered down to a pointlike Earth. The Moon does not dash off to infinity because its negative potential energy is, in absolute value, larger than its motion energy (which physicists call *kinetic energy*). If the distance between Earth and Moon were infinitely large, the potential energy would be zero; the total energy would then be positive (or, at least, zero)—motion energy is always positive, tending to zero as the motion slows down to rest. For the Earth-Moon system, the total energy remains constant; and that means the Moon has to remain close to Earth.

We can now infer that at this initial cooldown of the universe across the critical temperature of 10^{15} degrees, a mini-planetary system might spontaneously appear in the hot empty space, just as water can form in ice and the Higgs field can form in the vacuum, always liberating energy in the process. I repeat, this is quite compatible with the energy balance. If the laws of nature permitted the existence of mass points, it would be energetically favorable to have two or more of them materializing simultaneously. They would pass from virtual to real existence.

The particles that appear are, of course, expected to keep the overall energy balance constant. Other conserved quantities, such as electric charge, might require our including the surroundings in order to maintain their balance. If two mass points have opposite but equal charges, their appearance does not present a problem; there will be some added attraction due to the Coulomb electrostatic force. That implies an added energy advantage for the creation of a pair of oppositely charged masses when compared with neutral ones.

The idea of having virtual particles pass spontaneously into reality is a controversial one. Once we get to the level where quantum mechanical laws have to be heeded, it probably falls apart. As two mass points move closer and closer to each other, they give off energy, while the combined system gets smaller and smaller. At some point it will pass the level where the application of the laws of quantum mechanics becomes mandatory. The uncertainty principle then tells us that the smaller the system, the larger its motion energy. The total energy of the system cannot fall below the ground state energy. The best-known example is, of course, the hydrogen atom: Only quantum mechanics can guarantee its stability, and thereby the stability of matter altogether.

The laws of quantum mechanics don't permit the removal of arbitrary amounts of energy from a system consisting of two mutually attractive point particles. The laws of classical mechanics do predict that we can keep removing energy from a pair of point masses or point charges when bringing them closer and closer together, but this is a fallacy quantum mechanically. The laws of quantum mechanics see to it that the construction of a *perpetuum mobile* based on this process must fail, and not only for technical reasons.

The particles whose existence the laws of nature, such as we understand them, appear to permit have masses so large or charges so small that the conservation of energy is not compatible with the creation of real particle-antiparticle pairs from the vacuum. This process does become possible in the vicinity of other very large masses, like black holes, and maybe also in the neighborhood of very large charges. In the next chapter, we will discuss a model credited to the American physicist Edward Tryon, which takes these ideas to very large scales. This model sees our universe as one gigantic fluctuation of the vacuum, with total energy approximately equal to zero. Tryon does not tell us how this fluctuation might actually have looked, and our present-day imagery would no doubt be hard put to describe it.

VACUUM EXCITATIONS

The uncertainty relation mandates that the vacuum *loan* energy to local fluctuations; these energy loans may be large for very short terms, but must be small over long duration. Virtual particles have their origins in this borrowed energy: As soon as the principal of the loan is due, they will vanish. Their fields remain, and their task is to remind us that the probability of finding the particles they describe never goes to zero.

There are many different ways the vacuum can be excited. We mentioned particle accelerators: They give virtual particles enough energy to become real. In all the LEP experiments, particle-antiparticle pairs appear routinely: electrons and positrons, protons and antiprotons, and so on. To make similar pairs of heavy particles, like the W and Z bosons, the LEP energy had to be raised beyond the level of two W and/or Z masses, which happened in 1996–97 for the first time.

In principle, very high temperatures can do the same. They imply sufficiently large velocities of random motion for whatever real particles exist in a system, so their collision energies can take the place of what the accelerators accomplish with their particle beams. In other words, high temperatures are also able to provide the energy needed for the transition of virtual particles into the real world.

THERMAL RADIATION AT DIFFERENT TEMPERATURES

When we heat up a cavity that has been pumped empty, it becomes an oven for fields and particles. We discussed the Casimir effect, and we know that even at zero temperature this "empty" space still has radiation. To that, we add fields of probabilities and virtual particles. The existence of thermal radiation at temperatures above absolute zero can be verified by a thermometer that we suspend in the cavity without having it touch the walls. This is the experiment that Newton used—having it performed by Desaguliers, the experimentalist of the Royal Society, to demonstrate that empty space conducts heat. All the way up to a temperature of 10^{10} (10 billion) degrees—a temperature a thousandfold hotter than the temperature of the center of the Sun—thermal radiation remains a form of electromagnetic radiation. Its basic nature does not differ from light; it's just radiation with different wavelengths, starting with extremely long ones at very low temperatures. As the temperature rises, there are the different ranges of radio waves, the thermal radiation we use for room heaters, and on to visible light at a few thousand degrees, to X rays, to the exceedingly short wavelengths of gamma rays. We use the term *gamma rays* somewhat loosely for all electromagnetic radiation with wavelengths shorter than those of the X rays.

As far as we can tell, electromagnetic waves may have arbitrarily short wavelengths. The shorter its wavelength, the more important the particle aspect of an electromagnetic wave. As the wavelength decreases, the energy of the particles—the photons—we associate with the wave will increase. At a temperature of 10^{10} degrees, the energy of the photons of the electromagnetic/thermal radiation becomes that needed for the creation of an electron-positron pair, according to Einstein's formula $E = mc^2$. We know that virtual electron-positron pairs, like all other particle-antiparticle pairs, always exist in the vacuum. It is just that at temperatures below 10^{10} degrees, they lack the energy to pass from virtual to real existence. It is at this temperature that the transition becomes possible; the energy density is now sufficient to produce electron-positron pairs from the vacuum. This means that our thermal oven is now changing into the playground for real particles. The previously virtual electrons and positrons show up as real particles. They move about, collide with others; charged-particle detectors can give electronic signals upon their passage.

As the temperature grows in our oven, heavier particles appear pairwise: Proton-antiproton pairs show up at about 25,000 billion degrees. Neutrinos are lighter particles, and may even be massless; we have not discussed them, although they have appeared already in pairs at much lower temperatures in our oven. Nor have we said much about quarks and antiquarks, which make up the protons and antiprotons. The expectation is that as the temperature increases, the quarks will be freed from their bound states inside other particles to form a kind of soup consisting of quarks, antiquarks, and gluons—a mysterious brew called a quark-gluon plasma.

When we raise the temperature further—to 10^{15} degrees—nothing that is qualitatively new will happen. At that temperature, the Higgs field vanishes; it melts like ice. The ordered state in which it filled all space is no longer able to

Figure 73 This apparatus, product of a theorist's concept of how to obtain an experimental proof of pressure due to radiation, has to be much refined to be capable of discovering that physical effect.

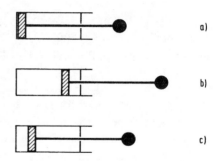

resist the chaotic thermal motion. The Higgs field disappears as soon as the energy of the thermal motion is sufficient to melt it. The quarks and other particles that interact with it will lose their masses and move about exclusively with the speed of light. That, at least, is what the standard model of elementary particle physics would have us believe.

At this point, I will stop discussing my gedankenexperiment with the oven. If I raise the temperature even further, any discussion of what will happen turns to speculation. The Big Bang theory tells us that the universe passed through all these stages in the opposite direction—starting at infinite temperature and cooling down to the ultimate cosmic background radiation temperature of 2.7 degrees above absolute zero. But now, let's turn to the properties of blackbody radiation, since they are experimentally observable.

EXPERIMENTS ON BLACKBODY RADIATION

We know from numerous experiments that blackbody radiation exists, and that it has the properties we discussed above in terms of baking ovens and their scientific variants. It can be likened to a gas that seeps slowly into an empty volume and is exceedingly difficult to get rid of.

Let there be an ideal piston pump, as shown in figure 73. In a gedankenexperiment, let us pull on a piston against the resistance of ambient air pressure and temperature, and keep it there for a while (see fig. 73b). During this time, the evacuated space behind the piston will be filled with blackbody radiation, or thermal radiation of the ambient temperature. This radiation cannot be rapidly compressed without offering resistance; when we release the piston, a tiny remainder of the evacuated volume will be left over for a brief time (see fig. 73c). Before the piston moves all the way back, the thermal radiation has to adapt to the reduced volume. The piston is being pushed from the outside not only by the surrounding air but also by the ambient thermal radiation. The outside and inside pressures due to thermal radiation at equal temperature will then cancel each other.

Experiments with the COBE (Cosmic Background Explorer) satellite, which was launched in 1989, showed in 1990 with overwhelming precision what was already known from previous experiments—that the cosmic background radiation filling the universe has all the properties of blackbody radiation at absolute temperature 2.735 degrees, save some tiny deviations. It would be very surprising if Earth were at rest with respect to this radiation. The velocity of Earth relative to it was first measured in 1977 from an airplane by investigating the influence of the Doppler effect (fig. 57). The blackbody radiation as received by an observer who moves relative to it displays what is called a "dipolar asymmetry": The radiation coming from the direction in which the observer moves is shifted to higher frequencies, the radiation from the opposite direction to lower frequencies. This shift has the remarkable property that the radiation arriving from *any* direction has all the properties of a blackbody radiation; only the temperature is shifted—to higher values in front, to lower in the rear. By measuring this temperature difference of about 0.0035 Celsius, scientists have established that the solar system is moving toward the constellation Leo with a velocity of approximately 250 miles per second relative to the background radiation. By properly adding velocities, it follows that the Milky Way itself moves at a speed of about 500 miles per second relative to the background radiation.

COBE was the first to detect a tiny deviation of the cosmic background radiation from the blackbody form—a barely noticeable dependence of the temperature of this radiation on the direction from which it comes. Other than the dipole asymmetry, this dependence is not an artifact of the observer but a property of the radiation itself. Since the background radiation originated in the early phases of the universe, it has to carry the imprints of the density fluctuations at the time. As a consequence, the radiation cannot be entirely isotropic. It came as a relief to cosmologists when, after some initial confusion, density fluctuations were finally detected by COBE in 1992; these fluctuations in the early universe are needed to explain the origin of the important inhomogeneities that are at the basis of the formation of galaxies and galaxy clusters.

Under the conditions that normally prevail on the surface of Earth—an air pressure of 76 cm mercury and a temperature of 20 degrees Celsius—the molecular motion of the ambient air molecules has much more impact on physical objects and their motions than the thermal radiation. Every cubic centimeter of air contains about 10^{19} molecules. The pressure in the 27-km-long beam tube that contains the electron and positron beams inside the LEP tunnel at CERN is lower by a factor of 10^{12} than the ambient air pressure. This implies a reduction of the number of particles per cubic centimeter by about the same factor. These are the same pressure conditions that prevail in the outer atmosphere of Earth at an elevation of about 400 km above its surface. When an electron or positron orbiting inside the LEP tube collides with a molecule of the gas remnant, it will most probably be ejected from its orbit. But its orbit is also subject to influences from the ubiquitous thermal radiation. Under the conditions prevalent in the beam tube, the influences of remnant gas and thermal radiation are of comparable importance. Together, their action results in a slow disappearance of the beams. The beamed particles move at a velocity close to the speed of light; that means that in their five-hour lifetime as beam particles, they traverse very large distances: 5 billion km, or 5×10^{14} cm; this is equivalent to seven hundred round trips between Earth and the Moon.

EXPERIMENTS ON THE QUARK-GLUON PLASMA

In our gedankenexperiment on thermal radiation, matter is produced from energy. This is in accordance with Einstein's equation $E = mc^2$, and with the countless experiments that have proven its validity. All the predictions concerning reactions at the LEP accelerator—based on the standard model of elementary particle physics—agree with the experimental results on the final state products issuing from the annihilation of electrons and positrons. These tests of the standard model, together with many other pieces of evidence, speak for the correctness of the basic theory behind this model—including, in particular, the existence of the Higgs field and the origination of the final states we observe from the annihilation reactions by way of the intermediate state of an excited vacuum.

Encouraged by this success, many physicists expect that we will soon be able to create, in our particle laboratories, a "soup" consisting of quarks, antiquarks, and gluons at very high temperatures. Recall that the quarks are the constituent particles that make up protons and neutrons; antiquarks are simply the antiparticles of quarks; and gluons are the carrier particles of the strong field of nuclear interactions. To make this "soup" appear, protons and neutrons have to be compressed sufficiently to make them overlap.

This quark-gluon plasma can develop when heavy atomic nuclei are directed at each other at high speed. Figure 74 shows the tracks of charged particles liberated or generated by the collision of a sulfur nucleus and a gold nucleus in a CERN experiment; this collision happened at energies at which the formation of a quark-gluon plasma is not yet allowed.

Two nuclei that collide like two rapidly moving cars in a head-on collision will burst. In the instant before the burst, they occupy one and the same tiny region of space. According to the theory, the quark-gluon plasma will form in this space and at this instant. This may also lead to a phase transition of space; a speculative theory says that the vacuum might take on a new highly excited electric state, and remain in it ever so briefly even after quarks and gluons have diffused away.

Around the year 2000, we will have results from CERN, from the German High-Energy Heavy-Ion Research Laboratory (GSI), and from the Relatively

Figure 74 All the numerous tracks in this picture originated in the collision of a sulfur nucleus with a gold nucleus at high energies.

Heavy Ion Collider (RHIC) at Brookhaven National Laboratory on Long Island, all of which will collide ions at high energies. Those results should be able to tell us about the quark-gluon plasma. These vacuum effects depend not only on the vacuum itself but also on the nuclei and on their quarks. If they did not exist, there would be no quark-gluon plasma. The quarks, unlike the electron-positron pairs at LEP, are not annihilated in the collisions of the nuclei. They just pass into a different state. Therefore, the information on empty space that they give us is only indirect. Nevertheless, they are convincing to physicists. Figure 75 illustrates the transition of quarks inside protons and neutrons into a quark-gluon plasma. The quarks that were generated in the early universe must have formed such a quark-gluon soup about 10^{-5} seconds after the Big Bang. This soup was so dense that the energy present within the volume of one proton would have been sufficient to generate several protons. The pressure generated by this energy density did not permit the quarks to congregate into the atomic nuclei we currently see. The experiments planned today will go the opposite way: They will create, starting with heavy nuclei, the energy density corresponding to three to five proton masses per proton volume. The difference between the Big Bang fifteen billion years ago and the miniature bang in our laboratories will be mainly this: Our laboratory bang can expand into the already existing empty space that surrounds it, while the expansion after the original Big Bang had to generate that space in the first place.

Figure 75 A nucleon—a proton or a neutron—consists of three quarks (a). Their specific interaction properties don't permit them to appear singly: The poetic term for this fact is *infrared slavery*. Inside atomic nuclei, the protons and neutrons are densely packed, but they probably do not overlap (b). When nucleons do approach sufficiently to overlap, their quarks might form one unit consisting of more than the usual three quarks. In some extremely dense stars, the gravitational pull compresses matter to the point that their nucleons overlap at least partially (c). This happens in neutron stars, which contain an entire solar mass within a sphere of radius 10 km. When the quarks overlap, they may form a quark-gluon plasma (d). Before the particles that we observe today materialized, all matter in the early universe must have gone through this stage of a quark-gluon

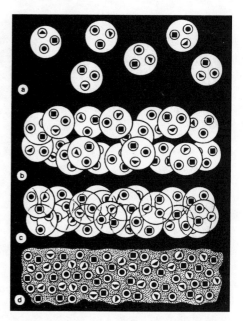

soup, or plasma. Our figure shows only the quarks. If our theoretical estimates hold, we will be able to produce this kind of plasma in the laboratory within the next few years.

As our universe expands, galaxies don't fly off into space that has been empty so far; instead, they tell us that space is growing. They are test probes, like Archytas's staff, which we discussed in chapter 2; their motion indicates to us the development of the universe. We might say that their motion *creates space*. Both of these descriptions are really synonymous, indistinguishable. The initial momentum imparted to matter in the Big Bang provides for the expansion of the universe to this day. And that is the way it was when the universe was nothing but a broth consisting of quarks and gluons; expansion at that time also meant that the universe grew in size, that space was being created. The quark-gluon plasma as it forms in our accelerators will not create new space; rather, it will expand into existing space, and in the process it will fizzle and ultimately vanish.

VACUUM POLARIZATION

We know that the vacuum contains charged particles and antiparticles; they don't make their appearance in reality because their energy doesn't permit it. This is the basis for the vacuum effects that I will now discuss. There are two such effects that are indirect: the shift of spectral lines and the scattering of light from light. There is a more spectacular effect that, upon experimental verification, would give direct evidence of the various states *virtually* contained in the vacuum; this

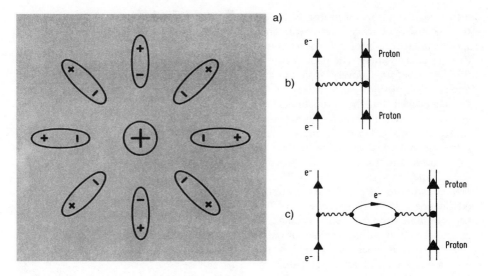

Figure 76 A single electric charge that we introduce into a vacuum will attract, out of the vacuum fluctuations of particle-antiparticle pairs, the particles with a charge opposite to its own; it will repel the opposite charges (a). In this fashion it *polarizes* the vacuum. The cloud of charges surrounding it will now weaken the effect of the original charge on outside probes. A recent experiment at the TRISTAN accelerator in Japan has shown that the measured charge of the electron increases with decreasing distance from it. For figures 76b and 76c, see the text.

is the spontaneous production of real electron-positron pairs in strong electric fields.

Every object is embedded in the vacuum with which physics surrounds it. This means we don't observe the *naked* object—rather, we see it as the surrounding vacuum will *dress* it. Take, as an example, a positive electric charge, such as the proton's. Its presence modifies the surrounding vacuum. The effect appears because the fluctuating electrically charged particles and antiparticles in the vacuum move apart during their lifetime. The proton attracts negative charges and repels positive ones. On the average, the charge distribution will then look much as we illustrate in figure 76a. The field of what is now a *polarized vacuum* weakens the proton's electric field. Its true naked charge is larger than what it appears to be from some distance; the closer we get to the proton, the larger the charge we observe.

This affects the scattering of electrons off protons. What we mean by *scattering* is this: An electron that is aimed at a proton will change its direction of flight. It will be diverted from its initial path by the proton, just like a ball that hits a bigger ball. The deflection and energy loss of the electron in this process are calculated in terms of the Feynman diagrams shown in figure 76b. The proton emits a photon, which the electron absorbs (or vice versa). (Note that in the diagrams, time runs upward.) But this is not the whole story, neither theoretically nor experimentally. We have to add numerous corrections. If the diagram of figure 76b were all that was relevant, electrons and protons would attract each other with a strength that is a product of their charges multiplied by a universal function of their distance. This attraction fixes the angle under which the electron will be scattered from its original path.

In addition to the process in figure 76b, there are others. One of them is depicted in figure 76c: The photon emitted by the proton is not absorbed directly by the electron but by an electron-positron pair that appears in a vacuum fluctuation. This pair exchanges another photon with the passing electron. Qualitatively, the consequences of figure 76c can be seen in terms of the sketch of charges in figure 76a: The closer the electron gets to the proton, the bigger the electrical charge it actually sees. As a result, the product of the charges in the formula for the attraction of electrons and protons shown in figure 76b is modified and becomes a function of the distance of the two particles. The ultimately observed experimental findings depend, in addition, on the inner structure of the proton. I won't go into the details of this further complication at this time. As it is, we can associate the virtual electron-positron pair of figure 76c with either the electron or the proton—it makes no difference to the calculation. During the scattering of an electron off a proton, the charge clouds of both particles interact with each other. The cloud of the proton alone would have to be probed by a hypothetical particle with an electric charge so small that it has on its own no charge cloud to speak of.

COLOR CHARGES IN THE VACUUM

In figure 61 in chapter 5, we illustrated how the electromagnetic interaction depends on the electric charges of a particle involved. In this interaction, a photon is either emitted or absorbed by an electrically charged particle. The photon itself, the carrier of the electromagnetic interaction, is electrically neutral. That means all interactions among photons proceed through the mediation of other, electrically charged particles (see fig. 78).

Today we think that all fundamental interactions occur essentially in analogy to the electromagnetic one—that is, by means of an exchange particle or carrier that is emitted and absorbed by the interacting charged particles. These charges need not be electrical—they differ from interaction to interaction. For one thing, there is the strong interaction among the protons and neutrons that make up atomic nuclei; this interaction holds the nuclei together, and the charges it depends on are poetically called *color charges*. Needless to say, they have nothing to do with actual colors. Color charges are carried by the building blocks that make up protons and neutrons, and which we call quarks. Each proton and neutron is composed of three quarks. We will have more to say about the zoo of elementary particles in the next chapter.

Protons and neutrons interact strongly with each other because their building blocks do so. The exchange particles of the strong color interaction are called gluons. Instead of the one photon that is the carrier of the electromagnetic interaction, there are eight gluons for the strong interaction—for reasons beyond our present discussion. But let's remember that there is more than one carrier particle for this interaction. We already know that the photon is electrically neutral—it does not carry the charge to which it is coupled. It has to be neutral, since the electromagnetic interaction has only this one carrier particle. If it were not neutral, there would have to be another carrier particle with opposite charge, the antiphoton. Every elementary particle, after all, has its antiparticle. As noted, however, the antiparticle of the photon is identical with the photon itself; the photon is its own antiparticle. For the gluons, this is not the case. The gluons themselves carry the color charges to which they couple. Thus gluons, in contrast to photons, can interact directly with one another.

The color charge carried by the gluons has a remarkable consequence in the context of our main topic. In order to transfer figure 76 from the electromagnetic to the color interaction, we first have to substitute a quark-antiquark pair for the electron-positron pair. In figure 76a, we didn't need to show the ubiquitous photons themselves; they interact with the charges but don't carry any. The charged gluons, on the other hand, have to be shown: They transmit color and thereby *smear* the color charges of the quarks. In contrast to the polarization of the vacuum by means of virtual quark-antiquark pairs that shield the charges just as in the electromagnetic case, the smearing leads to a weakening of the shield. The closer a color probe approaches a central quark, the more of the color barrier it has left behind. At the very point where the center of the quark resides, there is none of its color charge left for the probe to interact with.

Consequently, the strength of the color interaction depends on distances quite differently when compared with the electromagnetic interaction. The same holds, by virtue of the uncertainty relation, for its energy dependence: While the effective electrical charge of a proton decreases with distance, the strength of the effective color charge of a quark *increases* with distance. (See fig. 89 in chapter 8.)

We associate two different terms with this phenomenon: *asymptotic freedom* and *infrared slavery*. At one end of the scale, the short distances, the quarks barely affect one another; rather, they act like free particles. At the other end of the scale, the strength of the color interaction sees to it that the quarks remain enclosed inside their particles—the neutrons or protons for which they act as building blocks. This is what we call *infrared slavery*. When we try to separate one quark from another, the force with which they hold together increases. The energy expended for their separation will then be transformed into the creation of added particles according to Einstein's formula $E = mc^2$. What this means, in short, is that it is impossible to set a quark free.

An alternate way of expressing the phenomenon is this: The matter-free space surrounding our nuclear matter—that is, the physical vacuum—exerts sufficient pressure on the modified vacuum containing the color charge inside the protons and neutrons to make the quarks stick together. We know that the separation of one quark from its accompanying quark will necessarily lead to two separate regions with a net color charge—the single quark on one hand, the remainder of the original system on the other. Space with net color is more energetic than color-neutral space. Therefore, energy is needed to separate one quark from its partner quarks; the color-neutral external space exerts pressure on the quarks inside their overall color-neutral vacuum bubble.

ZITTERBEWEGUNG AND THE LAMB SHIFT

Just like any quantum mechanical system confined to a finite region, the hydrogen atom cannot assume every imaginable energy level—only certain levels are possible. This is the most famous quantum mechanical effect—the *quantization* of energy. There are gaps between the individual energy levels of the hydrogen atom—or rather of its electron; they are separated by inaccessible energy gaps, so that the allowed energy levels are arranged like the steps of a somewhat irregular ladder.

This fact was well known to Niels Bohr, whose descriptive model of the atom, formulated in 1913, is taught in all schools to this day. According to this model, the hydrogen atom resembles a mini–planetary system, with a proton as its sun and an electron as its only planet. The electrical attraction takes the role of gravity in this atomic planetary system. But while any orbit is accessible to the planets in gravitational systems, the electron in the hydrogen atom has only a limited set of orbits to choose from. These orbits define the energies of Bohr's ladder. His model was refined in 1926 by Erwin Schrödinger. Instead of Bohr's orbits, Schrödinger's quantum mechanical model assigns certain shells surrounding the proton to the allowed orbital motion of the electron. He didn't consider the effects of relativity; these were included in 1928 by P. A. M. Dirac in the famous Dirac equation, for which he received the 1933 Nobel Prize in physics together with Schrödinger.

Dirac's theory says that the electron can never ignore the fact of the existence of its antiparticle, the positron. In figure 59, we depicted space filled with electrons, the so-called Dirac sea. Recall that this sea, if unperturbed, filled, and at rest, would be unnoticeable, all-pervasive. But the sea in itself is not at rest. It fluctuates: Electrons appear spontaneously outside the sea, each leaving behind a hole in the sea; this hole is equivalent to a positron. Any additional real electron will polarize the virtual pairs, as we showed in figure 76.

We will now discuss a more subtle effect. This effect would exist even if the electrons were not charged. It is due to the fact that the electrons vie for each other's physical space: Where there is one electron, there cannot be another one occupying the same quantum mechanical state. The real electron has to circumnavigate the electron that has left the sea. The simple fact that there are other electrons in its vicinity, limits its possible motion. Consequently, an electron can never fly across space like a small sphere at constant velocity or simply stay at rest; instead, it has to remain incessantly in a rapid, periodic motion that physicists call *Zitterbewegung,* literally, "shaking movement."

This oscillatory motion is observable when we look at the electron in the hydrogen atom. To use Bohr's language, the electron does not simply move around the proton in one of the allowed orbits; rather, it oscillates about one of the orbits. Its energy, therefore, is not exactly that of an undisturbed orbit: Its *Zitterbewegung* displaces the steps of the energy ladder ever so slightly.

There was a hiatus in basic research from 1938 to 1947, largely due to the Second World War. Theory and experiment appeared to agree, until the US physicist Willis Eugene Lamb provided experimental proof that one of the steps of the energy ladder is actually composed of two steps a very small distance apart, a discovery that earned him the 1955 Nobel Prize. Lamb's proof showed that Dirac's prediction of one and the same energy for two different electron orbits of the hydrogen atom doesn't hold true.

The interpretation of this effect followed its announcement immediately; in 1947, Hans Albrecht Bethe (who received the Nobel in 1967) showed that the energy fluctuations of light in the vacuum act on the hydrogen electron in the manner observed by Lamb. Space is filled not just with electron-positron pairs that appear and vanish but also with fluctuations of the electromagnetic fields, the photons. They don't compete for space with electrons but interact with them. These interactions, based on the elementary interactions of the electron with the photon, as we discussed above, are illustrated in figure 77. Figure 77a shows the interaction of an electron in the vacuum with an energy fluctuation of the electromagnetic field, which is shown in terms of the emission and reabsorption of a photon; in figure 77b, the same interaction occurs with an electron bound inside the hydrogen atom.

Bethe's calculation reduced the behavior of the electron, both in the vacuum and inside the hydrogen atom, to the properties of the bare electron—that is, the electron free of all external influence. The bare electron, the object of Bohr's theory, scatters off virtual photons in both cases. The result is the same as shown above: The electron in the atom appears to dance about its Bohr orbit; individual steps of the energy ladder prove, upon closer scrutiny, to be two closely neighboring steps.

Figure 77 Interactions of an electron with the vacuum fluctuations of an electromagnetic field (a) for a free electron, (b) for an electron bound inside a hydrogen atom. Obviously, the graph shown in 77b must also be included in a computation of the *scattering* of electrons and protons we have described in connection with figures 76b and c.

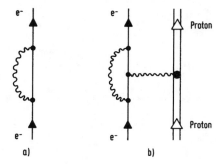

Bethe's calculations contain *divergencies* (a mathematical term for infinities that cannot be physically interpreted) in both cases, that of the free and that of the bound electron. Bethe showed, using a procedure that is now called *renormalization*, that these divergencies that appear in the energy shifts in fact cancel each other almost completely. There is only a very small resulting difference that escapes cancellation, and this difference causes Lamb's splitting of the step in the energy ladder of the hydrogen atom.

The agreement of Bethe's theory with experiment was good, but not perfect. There was an additional need to consider vacuum polarization, the effect shown in figures 76a and 76c: The effective charge of the proton, as seen by the electron, depends on the closeness of approach—clearly more so in the inner orbits of the hydrogen atom than in the outer ones. There is a virtual electron-positron pair that shields the proton from the electron (see fig. 76c). Its effect is smaller than that of the photons, because electrons have mass, whereas photons do not, and we know that the larger the energy of a vacuum fluctuation, the more rarely it occurs. An electron-positron pair can be produced only by a fluctuation that provides the energy as needed according to Einstein's $E = mc^2$ equation; massless photons, of course, can appear in frequent fluctuations with arbitrarily small energy. That is why there are more photons than electron-positron pairs in our not-so-empty space. Also present—but, again, much rarer—are virtual pairs of heavier particles and antiparticles—quarks/antiquarks, protons/antiprotons, and so on. All in all, today we have a very good understanding of why the steps of the energy ladder of our hydrogen atom are exactly in the locations we observe. The vacuum effects that are relevant here are well understood, and experimentally uncontroversial.

LIGHT SCATTERS LIGHT

The fluctuating charges in the physical vacuum are also the reason that light rays cannot simply penetrate each other without perturbing each other. And so it is for all wavelengths of electromagnetic radiation.

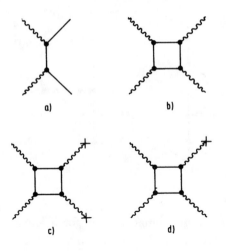

Figure 78 Charged particles can scatter light by absorbing and reemitting it (a). Light can scatter off light only by the mediation of charged particles—the virtual particle-anti-particle pairs fluctuating in the vacuum (b). This effect is very small; there is no experimental proof yet. The scattering of light in the strong electric field of an atomic nucleus was experimentally verified with gamma rays of very short wavelengths. This is shown in figure 78c, where the crosses indicate the coupling of the photons, virtual in this case, to the nucleus. Particle-antiparticle pairs fluctuating in the vacuum also permit the *fusion* of two real photons into a single one, in the presence of an external field (see fig. 78d).

We know that photons (and recall, light is nothing but an assemblage of photons) couple only to electric charges: Charges emit and absorb light (see fig. 61). All light effects can be reduced to this elementary process. One example is the scattering of light off an electric charge (see fig. 78a). Light penetrates easily wherever there is no charge.

Recall that photons are electrically neutral; that means light can interact with light only by means of intermediate carriers of charge. The charges appearing in vacuum fluctuations will serve in this capacity. The most important process for light scattering off light by means of virtual particle-antiparticle pairs is shown in figure 78b.

SPONTANEOUS PAIR CREATION IN STRONG FIELDS

The scattering of light by light can show the reality of the vacuum effects only indirectly. It takes a slew of theoretical arguments to interpret it. But as far as the spontaneous formation of electron-positron pairs in a strong electrical field is concerned, this is quite different. Let's assume there is a large electrical charge surrounded by the physical vacuum; we might choose the charge number 184 of two uranium nuclei. This charge, like all others, will be surrounded by virtual pairs of electrons and positrons. The electrons move toward the positive charge, the positrons move away from it; all in all, the virtual pairs conspire to shield the large central charge.

As has been noted, the virtual particles lack nothing but energy to make their appearance as real particles. This energy is available to them in the field of the large central charge. To illustrate this, let us assume the virtual electron-

Figure 79 Spontaneous formation of a real electron-positron pair from a virtual one (a) can happen in the strong electric field of a large positive charge. Subsequently, the electron can join the positive charge to form an atom, while the positron is being emitted (c). The energy freed in the process is larger than the mass energy $E = mc^2$ of the electron-positron pair that appeared spontaneously from the vacuum. Once the electron arrives in its orbit close to the large positive charge, it can interact with it in a process called *inverse beta decay*. This conventional nuclear interaction is without interest in our present context.

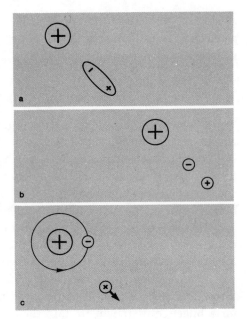

positron pair of figure 79a passes into the real world (see fig. 79b). To make this happen, we need an energy that corresponds to twice the electron mass, following Einstein's equation $E = mc^2$. Once the pair is real, the electron may join the positive charge in the center to form an atom, while the positron will be repelled to infinity (see fig. 79c). The total energy of the final state may well be *less* than that of a positive charge by itself. This is because the energy of an electron in the field of a positive charge is negative. More precisely, it is lower than that of an electron at infinite distance. If it were otherwise, the electron of a hydrogen atom would escape to infinity—which it doesn't—just as the planets would, were they not confined by the Sun's gravitational field.

The larger the positive charge, the bigger the energy gain when an electron is close by. If it is large enough, the energy gain is above the energy needed for electron-positron pair creation in the vacuum. The implication is that there should be spontaneous electron-positron pair formation in the vacuum if there are sufficiently strong electrical fields.

Once we observe pair formation in the strong fields, we have direct proof for the vacuum effects at hand. We can calculate such pair formation in the fields of individual nuclei that we are familiar with. It will take the collision of heavy atomic nuclei (see fig. 74) to generate, if very briefly, charge concentrations that will permit pair creation. The experiments that deal with the quark-gluon plasma check on pair formation as a freebie. Notwithstanding a concentrated search conducted at the GSI accelerator in Darmstadt, there has not been any pair creation of this kind to the present day. The appearance of a positron without an electron outside the reaction volume would provide the smoking gun for the pair creation effect we are seeking.

HAWKING RADIATION OF BLACK HOLES

An analogy to presumed pair production in strong electric fields is the so-called Hawking radiation of black holes. A black hole in classical mechanical physics is a concentration of masses so large that nothing, not even light, can escape from it against the resistance of its gravitational pull: It can gobble up matter and radiation, but it cannot spit them out. The gravitational force of a black hole is so large that it curves time and space to such an extent that black holes and the space-time surrounding them can be understood only with the help of the general theory of relativity. But their very existence can be established by Newton's mechanics and the special theory of relativity alone.

Let's consider a concentrated mass in a given volume, such as Earth. A rock thrown into the air from its surface will fall back to the ground. The same goes for a bullet shot straight up into the air from a rifle. It rises farther than the rock, since its initial velocity is larger, but as both these objects reverse direction, they will remain at rest for an instant. All of their energy, which started as nothing but motion (or *kinetic*) energy, will be positional (or *potential*) energy at that instant. This positional energy is determined by the mass of the object and the gravitational pull of Earth. It follows that the height reached by an object is independent of its mass; this is so because both its motion energy at ground level and its potential energy at its zenith are proportional to its mass. This means that the height to which it rises depends exclusively on the other relevant parameter of its motion, which is its initial velocity. If this is larger than the so-called escape velocity of 11.2 km per second, the object can leave the gravitational field of Earth and move out into space. Were we to increase the mass of Earth while keeping its radius at its present value—or, alternately, if we kept its mass the same but diminished its radius—the escape velocity would have to increase. If we continued the process, this increase would ultimately lead to values beyond the velocity of light. That is where the special theory of relativity comes into the picture. No object can exceed the velocity of light; as a result, nothing can escape from the gravitational field of a mass so concentrated that the escape velocity at its surface is larger than the velocity of light. This mass will then form a black hole.

In order to show that not even light can escape from the gravitational pull of a black hole, we have to hone the argument further. Light, as we know, always moves at the same velocity. When it rises against gravity, light will therefore not slow down. Unlike a bullet, it cannot come to rest and turn around. But by rising it will lose energy and thereby its wavelength will increase. Thus the light becomes dimmer and dimmer—with the result that outside the sphere at which bullets turn around the light cannot be seen anymore. We may therefore conclude that light, just as bullets, cannot leave the gravitational field of a black hole.

Every mass, even a very small one, can form a black hole. All we have to do is concentrate it inside a sphere of sufficiently small radius. Let's not worry about the practical possibility of achieving this; in our context, we are concerned with the theoretical concept and its consequences. The black hole we are discussing need not be part of the cosmic bestiary.

As we let the radius of the sphere in which we have enclosed our mass increase beyond the value that corresponds to the critical escape velocity, it ceases to act as a black hole. This limiting radius is called the Schwarzschild radius of that mass. Conversely, as long as our mass remains concentrated inside a sphere with this radius, it must form a black hole, and no object, not even light, can leave it, according to the rules of classical mechanics. The larger the mass, the larger its Schwarzschild radius. The Schwarzschild radius of an atomic nucleus and that of its atom must be nearly the same, on the order of 10^{-51} cm. Compare this with the actual radii of nuclei and atoms—about 10^{-13} and 10^{-8} cm, respectively. The Schwarzschild radius of Earth is 1 cm; that of the Sun is 3 km. The radius of densest stars we know of are neutron stars; their actual radius exceeds their Schwarzschild radius of 10 km by a factor of only 3.

To make a black hole out of Earth, we would have to reduce its radius by a factor of one billion in order to reach its Schwarzschild radius of 1 cm, without reducing its mass. In the process, its density would increase by a factor of 10^{27}. In principle, this could happen if Earth were to collide with a small black hole and be gobbled up by it. Here is our science-fiction scenario: A small black hole penetrates Earth and starts absorbing its mass. What happens next will depend on the velocity of the black hole. If the black hole enters Earth at low speed, it will get stuck in Earth's volume and will start hollowing it out. If its velocity is large enough to traverse Earth, it will exit on the other side after having dug a tunnel and absorbed the relevant mass in the process. If its velocity is still sufficient (that is, if it hasn't taken on too much mass), it will take off into space. It would be more interesting and, indeed, catastrophic if the black hole tunneled right through Earth only to become too heavy in the process and thus too slow (remember, momentum is conserved!) to escape into space again. In that case, it would reach a certain height and then would fall back to Earth. But since Earth would have continued in its rotation, the falling black hole would miss its old tunnel and would eat its way back across Earth, creating a new tunnel. This game can be repeated, continuously braking and adding mass to the black hole, until it eventually gets stuck inside Earth, while continuing to hollow it out. Ultimately, a black hole of 1 cm radius and with a mass roughly that of Earth would orbit the Sun in almost precisely the orbit previously followed by Earth; and this black hole would in turn be orbited by our Moon with its 1700-km radius. If this isn't a fantastic fate for our Earth, I don't know what would be: It has been gobbled up by a black hole, and the only thing left to observe would be its gravitational pull, essentially unchanged.

Stars with masses in excess of ten solar masses will, according to the ideas of astrophysicists, eventually collapse into black holes, once their nuclear fuel has been used up. Their Schwarzschild radii may be 30 km or even more. In addition, there might be mini–black holes left over from the Big Bang. They may have been created by the high pressure prevalent in the incipient universe, which would have compressed matter sufficiently to make black holes appear. Their masses might have been as low as that of an atom, in which case they would have exploded and vanished long ago. But some may have had the mass of a mountain on our Earth; these could have survived for explosion in our day. Why and how black holes explode will be described later in this section.

If they were as we described them, the existence of black holes would contradict the principles of thermodynamics. One of these principles says that heat always flows from hotter to cooler bodies. The classical non–quantum mechanical black hole will eat up energy in any shape, including thermal energy. Therefore, it is colder than—or at least as cold as—any other matter existing. Thermodynamics will admit temperatures that are barely above absolute zero, but not this lower limiting temperature itself. The problem that a classical black hole poses is therefore this: It cannot be warmer than absolute zero; otherwise there might be colder objects to which it would pass thermal energy. But like all other objects, it cannot reach absolute zero itself. Thus it is impossible for us to assign any temperature at all to a black hole, and that clashes with the laws of thermodynamics.

It takes quantum mechanics to resolve this contradiction. Stephen Hawking devised an answer in 1974: While the black hole itself does not radiate, its immediate surroundings can give off thermal energy by radiation, like any other mass. To an observer at some distance, the radiation appears to issue from the black hole itself.

We know the physical reason for the radiation. If a virtual particle-antiparticle pair forms in close proximity to a black hole due to vacuum fluctuations, it can pay off energetically if one of the partners crashes into the black hole while the other escapes to infinity. Figure 80 gives a schematic picture of this process, in analogy to figure 79. The electric charge is irrelevant here; mass and gravitational force are at the basis of the process. Electrically neutral particles of light—photons—are generated in the Hawking process, just like electrons and positrons. In our figure, we assume that the electron crashes and the positron is emitted. The reverse is just as likely, since the black hole is electrically neutral.

Hawking was able to demonstrate that radiation generated in this way has the characteristic spectrum and intensity of thermal radiation. The temperature is determined by the mass of the black hole. The reason the black hole of figure

Figure 80 Because of the quantum mechanical tunnel effect, particles that are
transformed inside the Schwarzschild radius of the black hole from virtual to real
can be emitted into the outside world.

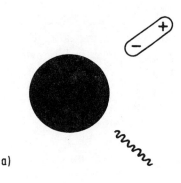

a)

b)

80b is drawn smaller than that of figure 80a is to illustrate the fact that the black
hole will decrease in size through the emission of radiation. This is a simple
consequence of the law of energy conservation: The energy of the emitted radiation
can come only from the black hole; therefore the black hole is losing mass and
must shrink in size.

At first glance, it may appear odd that the black hole loses mass upon
absorbing another mass—that of the electron. Actually, this seeming paradox is
at the very foundation of how our process functions: The potential energy of the
electron close to the black hole is so large, and negative to boot, that its total
energy is negative. By absorbing the electron, the black hole actually *loses* energy,
and thereby mass.

As it loses mass, the black hole diminishes in size and will finally vanish
altogether. What remains behind is the physical vacuum. This disappearance act
of the black hole is truly curious: The smaller it is, the hotter its heat radiation,
and thereby the warmer it becomes. Indeed, its temperature increases as it radiates
energy! This looks like an upside-down world; but for systems that hold together
by virtue of the gravitational force, it is perfectly normal.

As a first example of this effect, let's take the planetary system. The closer a planet is to the Sun, the faster it orbits. And this is true in absolute terms—measured, for instance, in kilometers per hour. The planetary year of an inner planet would obviously already be shorter than that of an outer one if both moved at equal velocity; in reality, the inner one moves faster, which makes its year even shorter.

The same goes for a satellite orbiting Earth. When a satellite loses energy because of friction in the atmosphere, it will descend to a lower orbit—and in that lower orbit, it will move faster—thus an attempt to slow a satellite down by means of friction will actually accelerate it. For its motion, the descent to a lower orbit is more important than the fact that it will have to give off energy in the process.

In general, the removal of energy from gravitationally bound systems accelerates the movement of their components. This also goes for systems whose components move about randomly: the faster their random motion, the higher their temperature. This means that the temperature of gravitationally bound systems increases as they give off energy.

This is true for black holes, too. Once they have diminished in size while radiating energy, the remaining energy is concentrated in a smaller volume. If the black hole had shrunk without giving off energy, its temperature would, of course, have increased because of the resulting higher concentration of energy. But, in fact, energy radiated off; the net effect is an increase of the temperature of the black hole notwithstanding the thermal radiation it emitted.

Considering the energy balance, the reduction in size and the rise in temperature of the black hole form a feedback mechanism that ultimately must lead to its explosion: The more energy it radiates, the hotter it gets. The hotter it is, the more energy it radiates. And as its energy rises, the ultimate outcome is inescapable: The black hole must explode.

This result follows if the assumptions made by Stephen Hawking for his calculation hold all the way down to zero diameter as a black hole shrinks in size. But do they hold? Let's examine these assumptions. A consistent quantum theory of gravitation has not been formulated to date; but Hawking radiation of black holes can be described only in such terms. Thus it is particularly astonishing to see a successful theoretical treatment emerge. The theory makes sense only for black holes with diameters well above the distance at which quantum theory and gravitation become mutually dependent. The critical distance is presumably the so-called *Planck length* of 10^{-33} cm. We have good reason to believe that space and time begin to merge into quantum mechanical uncertainty at distances this small; they must be treated quantum mechanically from this scale downward. If the diameter of a black hole is well above the Planck length, we are justified in treating the particles that fluctuate about its Schwarzschild radius in terms of quantum mechanics while treating space, time, and gravitational field in the classical terms of general relativity.

This is exactly what Hawking did, and his theory makes sense for black holes with diameters well above the Planck length. We will see what that means for the mass, temperature, lifetime, and thermal radiation of a black hole. But we don't know whether or not (and, if at all, how) it shrinks in size below this Planck length. In particular, we don't have any idea whether it will actually disappear in an explosion, leaving behind a flat, undistorted space, or will simply propagate through space as a mini-mini–black hole with a diameter of 10^{-33} centimeters.

The wavelength of the Hawking radiation emitted by a black hole roughly agrees with its diameter. That means large black holes are heavy and cold, and small ones are light and hot. It also means that large black holes emit little radiation energy per second, and small ones emit a lot. The latter will therefore decrease in size rapidly; the former, slowly or barely at all.

Let's examine a few examples. We already know the Schwarzschild radii of neutron stars (10 km), the Sun (3 km), Earth (1 cm), and atoms (10^{-51} cm). Let's look at a black hole with a presumed Schwarzschild radius of the Planck length, 10^{-33} cm. Hawking's calculations, we recall, apply only in the case of much larger radii. Our imagined black hole would have about the mass of a dust particle, on the order of 10^{-5} g; its temperature would hover around a fantastic 10^{31} degrees, and it would vanish from existence within about 10^{-44} seconds. From here, let's turn to the experimentally more interesting case of larger black holes. In addition to those that contain many solar masses, there may be a category that lives on the order of 10^{10} years, about the age of our universe. If such a black hole originated at the time of the Big Bang, it might be ready to explode any day now. Its initial mass might have been 10^{15} g—about that of a mountain on our Earth. Its temperature would have started at 10^{11} degrees and been increasing ever since, faster and faster as time elapsed. It would be the source of an intense, highly energetic radiation today. And in the last one-tenth of a second prior to its explosion, it would give off the energy of about one million hydrogen bombs, each with the explosive power of one megaton of TNT.

A black hole of one solar mass would emit radio waves with wavelengths in the kilometer range. It would be at a temperature only imperceptibly above absolute zero—at about 10^{-7} degrees K. We would have to wait 10^{66} years for its explosion. Indeed, we would probably have to wait longer, because it would act as a kind of cosmic vacuum cleaner, absorbing more matter through gravitational attraction than it could reemit via thermal radiation; thus it would grow instead of vanishing. A black hole with the mass of our Moon would be in thermal equilibrium with the 2.7-degree cosmic background radiation. A much smaller black hole—say, with the mass of an asteroid—would radiate light in the visible region. If all of the presently observable universe pooled its matter to create one giant black hole, its radius measured in light-years would amount to the same number as its age in actual years: approximately 10^{10}.

Black holes with upward of ten solar masses can be imagined as final products in the fate of stars. Our arguments above showed that they might be so cold that they can be detected only through the gravitational force they exert. There are several candidates for such black holes in our cosmic bestiary. We don't really know whether there are black holes that emit observable amounts of Hawking radiation. Nevertheless, there is no way to overestimate the importance of Hawking's discovery; it not only showed that the physical principles of the general theory of relativity, of thermodynamics, and of quantum mechanics are mutually consistent; it also proved them to be interdependent. If we leave quantum mechanics out of the game, thermodynamics and the general theory of relativity are mutually inconsistent. Hawking restored our confidence that we can apply the principles of the two signal theories of our century—the general theory of relativity and quantum mechanics—simultaneously and without creating contradictions. We have no overall theory that unifies them—or, put differently, all the above remains somewhat conjectural. But let's remember that we cannot develop a true understanding of the early universe without a quantum theory of gravity. It was small enough so that quantum mechanics had to apply, and it was massive enough so that its ultimate matter density mandated the action of gravitational forces between its constituents.

CHAPTER 8

NOTHING IS REAL

•

THE UNIVERSE AS A WHOLE

•

THE IDEAS OF THE ANCIENT NATURAL PHILOSOPHERS WERE THE precursors of the exact sciences that have evolved since the seventeenth century. It was only then that the need for observational verification of scientific ideas became an accepted element of the sciences. Ever since, fruitless ideas have quickly vanished and successful ones have been built up into appropriate formalisms.

The ideas of modern theoretical physics are formulated in terms of equations as a matter of course. This novel manner of speech often tends to obscure the ideas as such. In the present chapter, we will try to follow the train of thought of the mathematically formulated concepts that mark the development of theoretical physics in the past thirty years.

Many of these developments stand on proven ground. We will limit ourselves to the topic of this book, the physical vacuum. Two ideas stand out—both deal with the ways in which the symmetry of the vacuum is broken, and both are named after the people who first formulated them, the physicists Jeffrey Goldstone and Peter Higgs. Other, related ideas that concern our topic insofar as they deal with the origin of the universe—the models formulated by Hartle and Hawking as well as by Tryon—have not advanced beyond the status of suppositions, models of thought. But even at this stage, they capture our imagination.

We have made significant progress in our attempts to answer the questions asked ever since antiquity; we have couched our answers in equations that give precise formulations to processes beyond our imagination, and that thus render them approachable. But if we want to speak in everyday language, we have to be able to *imagine* the object of our discussion. In order to bridge the gap between equations that capture what the objects of our thinking really are and our imagination as formulated in colloquial language, we tend to increase our vocabulary. But at the bottom of everything we might achieve this way still linger experiences and imagination. Many of the strange objects of today's physics— particles without any definite position, mini–black holes, confined quarks, the

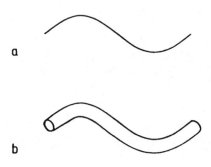

a

b

Figure 81 In addition to the three dimensions—length, height, and width—that we observe in our daily lives, our space may contain additional curled-up dimensions. Whatever appears to be a line from far away (81a) would actually be a tube (81b).

void—don't admit an exact description based on our experiences and imagination. To describe what we know about such matters without complicated mathematics, we must devise analogies that capture their essence in everyday terms. Nevertheless, colloquial language, however advanced, is not really able to delineate the progress of physics from the level where the ancient scientists pose the question about the void to our present understanding.

Wherever possible, we will rely on models accessible to our imagination. Everyday experience and imagination are part of the basis of the most abstract concepts in mathematics and physics. Imagination is not sufficient for scientific progress, but it is not irrelevant either: It constitutes the heuristic foundation for attempts to reach well beyond what is directly imaginable.

GEOMETRIC THEORIES AND THE DIMENSIONS OF SPACE AND TIME

The general theory of relativity, with its geometric formulation of the gravitational force, has served as an example for theories of other interactions. Einstein tried to formulate a unified geometric theory that comprises gravitation and electro-magnetism. But he failed in this attempt. In this context, the word *geometric* implies that the trajectories of particles are shortest paths in appropriately defined spaces. One idea of unifying electromagnetic and gravitational interactions is named after the mathematicians Theodor Kaluza and Oskar Klein. Their theory is formulated in a space with more than the four dimensions of our accustomed space-time. The reader may have encountered mention of speculative theories dealing with *strings* or *superstrings*. These theories are based on the notion that our space started out with more than one temporal and three spatial dimensions as we know them. The additional dimensions never made it into the world of our experience. Right after the Big Bang, they curled up into minuscule tubes with radii of the Planck length of 10^{-33} cm, or maybe much less, never to be noticed again. This mathematical disappearance of a higher-dimensional reality is called *compactification*. As figure 81 illustrates, the curved line of figure 81a might turn out, upon closer inspection, to be a curved tube. Every attempt to move an object in the direction of one of the compactified dimensions will see it revert to its original position after about 10^{-40} cm; and since this path is unimaginably short, we do not know whether our object has moved at all. It would take a microscope with the fantastic magnification factor of, perhaps, 10^{30} to resolve the curled-up dimensions. Since there is no such microscope, they remain unobservable to us, just as we cannot tell a sphere with radius 10^{-33} cm from a point.

The string and superstring theories of the last few decades describe elementary particles not as individual points but rather in terms of pairs of points. The distance between those pairs is again the minuscule and unobservable 10^{-33} cm; therefore, the pair of points looks like a single point for all intents and purposes that are observable or energetically distinguishable. These theories do away with a number of problems encountered by traditional field theories. Most important, they permit the inclusion of the general theory of relativity in a consistent quantum theory. We have previously noted that all theories that give a successful unified quantum formulation of the electromagnetic, weak, and strong nuclear interactions cannot include a conventional but consistently quantized treatment of the general theory of relativity; it is only this new set of string and superstring theories that manages to include gravity in an overall quantized treatment.

Once we accept this new set of theories, the actual dimensionality of space and time becomes a question posed to the physicist: Why would the process of compactification have led to a space of one temporal plus three spatial dimensions, and none other? The philosophical question about the three dimensions of our space has been around ever since Aristotle: Why do all objects have length, width, and height? Why not just length and width? Why not length, width, height a, and height b? To this day, this question has not been answered. Neither do we know why the temporal dimension is not simply another spatial dimension, and why there is only one temporal dimension.

THE ANTHROPIC PRINCIPLE

The anthropic principle attempts an answer to these questions. It includes the fact that humankind exists; the question about three spatial dimensions will be asked only in a world shaped such that it permits the development of beings that are able to ask it. A two-dimensional world would be fundamentally different from our three-dimensional world. It would pose many problems that don't exist in three dimensions. To give an example, it would quite obviously be impossible to tack things together with a nail; in two dimensions, nails separate things instead of joining them together (see fig. 82). A digestive tract would have to run through a living being in a complicated pattern in order not to split it in two (see fig. 83).

There are more serious physical problems concerning the development and

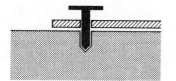

Figure 82 In a two-dimensional world, a nail does not join elements together; instead, it divides or separates them.

Figure 83 We don't know whether life is possible in two-dimensional space. This figure shows that the digestive tract of a dog will be complicated if we don't want the dog to separate into an upper and a lower part.

maintenance of life in any non-three-dimensional world. Immanuel Kant argued in the eighteenth century that the number of spatial dimensions fixes the dependence of gravity on distance; if there were four or more dimensions, the attractive force between two stars or planets would increase faster when they approach each other than we actually observe. That would mean there is no stable orbit for a planet around a sun: The smallest perturbation—say, a meteorite—would cause the planet to fall into its sun. The opposite would be true for two dimensions: Even the smallest perturbation would cause the planet to take off into two-dimensional space.

It is quite obvious that life could not exist in a world that has but one space and one time dimension. The three spatial dimensions have the specific property that they permit life on planets that can move in stable orbits around fixed stars that provide radiative heat to them. But it takes more than those three dimensions to create the wherewithal for the development of life. A number of detailed relations among forces, particle masses, and other properties of the universe must be fulfilled before life can develop. We might ask why, indeed, those conditions prevail. Was the universe tailor-made for our appearance? That may or may not be so. It might well be that there are a great many different worlds, bubbles in some gigantic superuniverse, with different numbers of dimensions that lack the detailed relations that permit our life to appear and persist.

The anthropic principle, which has become popular in the past couple of decades, is centered on the connections between the makeup of our world and human life. Predictably, different authors color this principle according to their own philosophical or religious leanings; but they all agree that our world has a number of properties that are a priori very unlikely and without which life as we know it could not exist. Should it be true that there are many different worlds with different dimensionalities and other properties, then the fact that ours fits us is no more astonishing than that we may walk into the ready-to-wear section of a department store and emerge after twenty minutes with a perfectly fitting new suit. The reason is not that the store provides tailors to attend to all our needs; rather, it is the huge choice of offerings that permits the customer to find his or her fit.

WHENCE THE UNIVERSE?

"The universe may be the ultimate free lunch" is a famous expression coined by the American physicists Alan Guth and Paul Steinhardt. What they mean is this: Our universe may well be but one particular form of the vacuum; it can be distinguished from other vacua only by its local properties, not by global features. Just as, in our bank, debits on one account may well (almost) compensate credits on another one, the global energy of the universe may balance out to zero while locally there may be positive and negative energy regions.

There is no dearth of ideas that illustrate these possibilities. Edward Tryon, in 1973, formulated a scenario in which the universe is one gigantic vacuum fluctuation with total energy equal to, or close to, zero. We are already familiar with the basic notions of this model. Recall that quantum mechanics and the special theory of relativity predict virtual particle fluctuations in the vacuum, which lack nothing but energy before they can make their appearance as real particles. Recall also that there may be *something* that has less energy than *nothing*—charges or masses at close proximities can realize these conditions. Last, let's remember that quantum mechanical fluctuations have a lifetime that increases with diminishing energy: The vacuum *loans* energy in large amounts for short times, in small amounts for longer times. Taking these points together, we understand the possibilities of Hawking radiation in black holes, and of the possible appearance of electron-positron pairs in strong electric fields, as we discussed in chapter 7. We can translate this idea, blown up into gigantic dimensions, into what might have happened to the universe as a whole: If its total energy equals or approximates zero, it may have originated as a spontaneous vacuum fluctuation. We might imagine that there is an approximate cancellation between the negative potential energies of all the masses that attract each other in the universe and the motion (or *kinetic*) and mass energies of these configurations, keeping Einstein's $E = mc^2$ in mind. In addition, there is the possibility that the vacuum does have a net energy; we will discuss this point further. If the universe as we know it is nothing but a vacuum fluctuation with a purloined total energy that is not much different from zero, it may exist for quite a while before the vacuum decides to call in the loan. If the total energy is zero or negative, the universe may persist ad infinitum.

Most physicists believe that the universe was packed with an unimaginable matter density some fifteen billion years ago. But beyond that, there is no agreement as to how that had come about. There is the so-called standard model of the hot Big Bang: If we reverse the motion picture called *The Universe,* we see the soup of matter get hotter and hotter; it ends at the instant when the temperature reaches infinity. This is the moment of the Big Bang, of the beginning of the world. If there was time before the Big Bang, time in which the world originated, we will never know within our model; the first frame of our motion picture is independent of anything that might have preceded it. Increase of temperature eliminates information, as we discussed in chapter 2; no information whatever can be passed on at infinite temperature. To repeat: We have no way of telling whether anything preceded the Big Bang, whether time had its origin together with our universe.

Correct or not, the model of the hot Big Bang serves as a point of reference even to those who don't believe in it. There is no agreement about the state of the world some 10^{-33} seconds after the Big Bang. That is when a process called *inflation* ended—we will discuss it below. We have no way to develop credible models of an arbitrarily dense universe; we wouldn't even know which laws of nature might apply in such a state of matter. In the realm of our experience, tightly packed matter is subject to the laws of gravity, embodied in the general theory of relativity, and to those of quantum mechanics—and at this point, there is no consistent theory that combines these two. Whatever theory we have is still incapable of describing the hot Big Bang; any theory assigns different roles to time and space right after the Big Bang, and that must be fallacious. At extremely short distances in time and space, where the quantum theory of gravity takes over, the overall smearing of quantum mechanical uncertainty ties time and space inextricably together. Virtual particles at small distances will be ubiquitous; a further shortening of distances subjects four-dimensional space to interrelated fluctuations of both space and time in a manner to be treated by a quantum mechanical general theory of relativity. Quantum mechanics and the general theory of relativity describe space and time jointly, just as quantum mechanics and the special theory of relativity deal jointly with fields and particles.

It is the uncertainty relation that is responsible for the breakdown of the traditional notions of space and time at very short distances. We know that, at minuscule time and space intervals, energy and momentum lose their precise quantitative definitions. It is the general theory of relativity that ties the concentration of energy to the curvature of space and time, and so the smearing of energy at very small distances entails an uncertainty in the curvature of space and time.

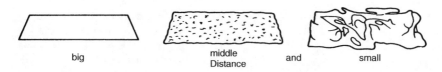

Figure 84 This figure shows the waves and foam that are barely visible on the water surface except at small distances. Waves and foam represent the curvature of space that may become noticeable at very small distances.

Figure 85 The expanding universe resembles a balloon with galaxies affixed to its skin. As the balloon expands, the distance between the galaxies grows. This is because the space—the balloon's surface—gets larger when inflated.

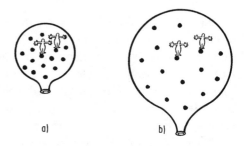

What we see is a fluctuation of what the mathematician calls the *metric,* the basic formulation of distances.

We try to illustrate this concept in figure 84, taking a water surface as an example. Because distances in space and time fluctuate at extremely small values, these different dimensions can no longer be told apart. Space and time commingle; space becomes time, time turns into space. They become equivalent partners in a general space-time, just as north and west are equivalent directions in geographic space.

The expanding universe is often pictured in terms of a balloon; as we inflate it, pieces of matter like pennies we have affixed to its surface might stand for the galaxies, which move apart without expanding themselves as the balloon's circumference increases (see fig. 85). The two-dimensional surface of the balloon is, in this picture, all that we need to describe three-dimensional space—it is a closed space without a limiting margin. If there are inhabitants on the balloon's surface, they will never find out about the third dimension. Recall the flat-bodied bugs we described in chapter 2; space, to them, is only the surface of the balloon. In the same fashion, we experience only the three-dimensional space in which we live. To the bug, the balloon surface is both unlimited and finite. The third dimension, which is perpendicular to the two dimensions it experiences, a bug called Archytas cannot experience: His space does not contain the direction in which he would have to extend his staff to expand his space.

If we run the video called *The Universe* backward, the size of the balloon

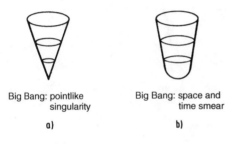

Big Bang: pointlike
singularity

a)

Big Bang: space and
time smear

b)

Figure 86 In the model of the hot Big Bang, in which quantum mechanics is not considered, the world started at a well-defined instant, in one point (a). Taking into account that the uncertainty relation and the general theory of relativity intertwine space and time inextricably, we might, however, argue that the universe did not start in a singularity (b).

decreases. The galaxies move closer together; matter becomes denser and moves toward the soup we have described. Whether there is an end to the video film running in reverse, we actually don't know. In the model of the hot Big Bang, there is no stopping of this reverse motion: The balloon shrinks, becomes a point, vanishes. The point was first defined when the Big Bang *created* space and time. It is a singularity, which we define as the beginning of our world. Prior to it, there was nothing; ever since, there have been space, time, matter.

There is another possibility that might be easier to understand: The world was infinitely extended from the very beginning. How then could it expand? Any two points in it could, just like the galaxies, simply move apart. Think of an infinitely large cake with raisins distributed in the dough. If all the raisins keep moving away from their neighbors, the cake becomes bigger and bigger. Since the size of an infinitely large cake or space is not defined, it has no trouble increasing further. It is only modern mathematics that permits us to put this into a framework. Mathematics is needed to ask meaningful questions about infinity, to tell the difference between questions that make sense and questions that don't: It is perfectly reasonable to inquire about increasing distances in an infinite universe, but asking where to take the additional space from makes no sense at all. If the world is infinitely extended, the Big Bang happened simultaneously all over that infinite space.

In the biblical image, the world was created out of nothing; the authors of other creation myths in antiquity had similar notions. The Big Bang model differs from all of these by having an empirical basis to build on: It implies elementary particle interactions in the early cosmos that are open to experimental verification or repudiation.

Every circle in the Big Bang model illustration of figure 86 signifies, at a given time, a reduced image of three-dimensional space. Each of these circles is a one-dimensional world, in which the flat bugs of figures 17 and 20 would now have to become one-dimensional worms. The farther away from its origin a circle is, the larger its size—the figures are meant to illustrate the expansion of the universe after the Big Bang. Thus there is no need for an independent parameter of time in the model of the Big Bang; the size of the universe can fill in for it.

The time since the Big Bang on any slice through a cone of figure 86 can simply be defined by its circumference.

According to figure 86a, time ends and space ceases to exist in a single point of space-time. It is difficult to see why this property of the classical Big Bang model should persist in quantum mechanics. Both space and time fluctuate, according to quantum mechanics, and their fluctuations are coupled to each other. At small space-time distances, the well-defined space and time of classical physics are transformed into the foam of figure 84. This implies that there cannot be a final point of space and time, such as the tip of the cone in figure 86a. We have to deal with the situation in figure 86b, where space and time have been smeared out in such a way that they cannot be distinguished anymore.

The reversed video depicting the origins of our universe sees it shrink but does not really stop, because space and time lose their separate definition. In the curve that replaces the pointed tip of that cone, time and space shrink together and become indistinguishable: The *before* and *after* of time degenerate into a more general *elsewhere* in space-time.

THE HARTLE-HAWKING UNIVERSE

The hypothesis of the unification of time and space is presumably inescapable in any quantum theory of gravitation. Hartle and Hawking added the assumption that the universe has no defined limit in either space or time. Their model, first published in 1983, basically states that the universe is not developing in space and time; it simply *is*. Space and time in this image—well beyond the powers of our imagination—form the four-dimensional surface of a five-dimensional sphere.

Hawking says that he has trouble imagining spaces with more than *two* dimensions. Taking a cue from him, we try to gain insight into the five-dimensional sphere by replacing it with a three-dimensional one, where time and space make up the two-dimensional surface (see fig. 87). The third dimension of our image, which stands for the fifth Hartle-Hawking dimension, serves as a crutch to our imagination; the surface of the sphere is embedded in it. But just as the bug living on the sphere knows nothing about the third, the Hartle-Hawking universe knows nothing about the fifth.

Here again, we have only one spatial dimension. The image makes sense when we order the points on the spherical surface in figure 87 in terms of the sequence of circles of latitude. In our space, they stand for a given time, so that the three spatial dimensions are represented by the one single dimension of one of these circles.

If we look just at the south pole of our sphere and its immediate vicinity, it does not appear distinct from any other point on the spherical surface. But once we have singled out this point, all concentric circles about different points

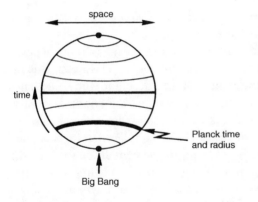

Figure 87 The drawing depicts the universe according to Hartle and Hawking. Between Big Bang and Planck time, quantum mechanical uncertainty envelops both space and time. This ceases when inflation sets in.

differ from the circles around this given point by our ability to tell space at a given time along their perimeter, and different times by the circles' distances from each other. These concentric circles about the south pole, identified as *Big Bang* in figure 87, stand for the expanding universe. Every circle of latitude, with circumference that is no longer small enough to be subject to the smearing of space and time, now contains, we might say, a clock that tells us which particular circle we are dealing with—that is, how much time has passed since the Big Bang.

Let us note that this clock does not tell us time in its ordinary definition; rather, this is an imaginary time—a mathematical construction that goes beyond the five-dimensional space, reduced in our image to two, as far as the lack of a definite interpretation is concerned. It differs from normal time more and more, as the distances under discussion diminish.

The third dimension, which acts as a crutch to help us make sense of the Hartle-Hawking universe, doesn't really exist. This model universe does not possess any kind of limit or surface in any one of its two (in our image) or four (in the full model) dimensions. It has neither beginning nor end, no limit in space or time. There is no creation. To pick one of its coordinates as *time* means an interpretation of what happens inside this infinite, closed universe. There is nothing outside it. More precisely, there *is* no outside.

THE COSMOLOGICAL PRINCIPLE

What we call the *cosmological principle* is the tenet that an observer placed at a random point in our universe will not be able to distinguish, by a look in any direction he chooses, this point and this direction from any other. There may be local fluctuations, but on the average the observer will see the same density of galaxies moving away at the same velocity, surrounded by the same background radiation, in any direction, no matter what his vantage point.

If the universe moved rotationally about an axis, this motion would violate the cosmological principle. The direction of the axis of rotation could clearly be distinguished from other directions, and an observer would be able to tell how far away she is from this axis by, for example, measuring the centrifugal force she is exposed to. But the cosmological principle does permit the universe to rotate like some viscous soup, which is deformed as it moves.

We will discuss rotational motion of the universe further below. It should be noted here that the cosmological principle agrees with the Copernican principle in that neither Earth nor our Sun nor our galaxy can claim a unique position in the universe. The same goes for all other galaxies, near or far, according to the principle—if observed at one and the same cosmological time.

We have established what we believe to be a plausible model for the origin and the temperature of the cosmic background radiation in an expanding universe. The carriers of the information we receive from distant objects in our universe are electromagnetic waves, be they in the wavelength range of visible light or of radio waves. All these waves have been moving with the velocity of light ever since their emission, and this implies that the most distant objects we can see are also those of the greatest age. The sunlight we see right now has been on its way for eight minutes; what we observe of the Andromeda Galaxy today is the way that galaxy looked some two million years ago. The quasar light that our telescopes make out today left those objects up to ten billion years ago.

The most ancient of these objects is the universe itself at the time of the Big Bang. The earliest direct messenger that reaches us is the cosmic background radiation; it originated about a hundred thousand years after the Big Bang. At the time, the universe had cooled down to 4000 degrees Celsius and had just begun to be transparent to radiation. After this so-called decoupling, matter and radiation developed independently of each other in the expanding universe, at ever lower densities and temperatures. Whereas matter conspired to form comparatively unimportant hot spots, such as the Sun or exploding galaxies, radiation continued to pervade the universe unimpaired—unimpaired, that is, by the matter of the universe but not by its expansion. When the radiation decoupled from the rest of the universe, both had the same temperature of 4000 degrees Celsius. Expansion since then has lowered the temperature of the radiation down to the presently observed 2.7 degrees Kelvin.

The effect the expansion of the universe has on temperature is most easily understood when we consider radiation. When the universe expands by a certain factor, all distances grow by that factor except for distances between bodies that are kept together by attractive forces such as gravity. Thus, galaxies themselves will not expand, but the distances between them will. As far as radiation is concerned, there is nothing that can keep its wavelength from growing by the same factor as the universe. The reader might recall from high school physics that the energy of radiation is inversely proportional to its wavelength: The energy of the cosmic background, therefore, diminishes as the universe expands. Lower energy implies lower temperature. That's why the cosmic background radiation has cooled during its history, from 4000 degrees Celsius to the presently observed 2.735 degrees above absolute zero.

To see how expansion acts on matter, we might introduce the concept of matter waves, and then proceed as we did in the case of radiation. But it is easier to note that expansion against the gravitational pull that masses exert on each other means that energy of motion is transformed into energy of position so that, in the expanding universe, motions are slowed down and temperatures decrease. As a consequence, the impact of collisions between particles diminishes when the temperature is lowered; below approximately 4000 degrees Celsius, atoms can form without being destroyed by the impact of collisions with their surroundings. We know (fig. 61a) that radiation interacts with electrical charges. Above 4000 degrees Celsius, in the dense soup of negatively charged electrons and positively charged nuclei we call a plasma, radiation cannot move freely and independently of matter. This changes once atoms form out of the isolated charges: Atoms also scatter radiation, but not nearly as effectively as isolated charges do. Radiation and matter decouple, and the universe becomes transparent to radiation.

The background radiation that hits Earth today originated in the surface of a sphere with a radius defined by the velocity of light multiplied by (almost) the age of the universe; this radius now amounts to some fifteen billion light-years and provides some basic measure for the observable universe. We have no observational means to determine whether or not the universe extends farther than those fifteen billion light-years. There remains the possibility that our universe, albeit immensely extended on scales humanly imaginable, is nothing but a tiny bubble in a much larger superuniverse.

And then again, all of this reasoning may be fallacious. The universe may have remained the same, at all times, on the average. There are theories that have matter formation and disintegration balance each other at all times. I, together with the majority of scientists, don't believe this to be a viable model. But beware of majority rule among scientists; the reader would do well to be open to alternatives that might replace the accepted model of the hot Big Bang.

A universe that conforms to the cosmological principle will consist of a space with—on average—equal curvature in all places. We know of three types of space the cosmological principle can accommodate: There is flat space, with zero curvature, such as we experience in our direct observation; then there are spaces of constant curvature, which may be either positive or negative. Two-dimensional spaces with a constant positive curvature are spherical surfaces; two-dimensional flat space is a plane; a two-dimensional space with negative curvature cannot be embedded in a three-dimensional space.

This is why a space with negative curvature cannot be imagined in its entirety, but only piece by piece. To realize it, Feynman's flat bugs of figure 20 use measuring rods with lengths that vary (to outsiders) from place to place. I won't go into it again. But let me emphasize a relation between the curvature of the universe and its final fate: If the curvature is either negative or zero, the universe will keep expanding forever. A positive curvature, on the other hand, implies that the universe goes through an expansion phase but will subsequently contract.

Recall that the presence of matter means the presence of gravitational forces, and that gravity curves space—the more matter, the stronger the curvature. A very strong gravitational pull may curve space such that it becomes finite, like the surface of a sphere. As a result, the matter density of the universe may brake the momentum initially imparted by the Big Bang to the extent that, ultimately, the universe will have to collapse. But then there remains the possibility that there is less matter, leaving space either flat or with negative curvature. In either case, the gravitational pull is not sufficient to stop the expansion of the universe, and it will keep expanding forever.

These three types of spaces are illustrated in figure 52; all of them conform to the cosmological principle. The ultimate fate of the universe, however, also depends on the value of something we call the *cosmological constant,* which we will discuss at the end of this chapter.

EPOCHS

The physics that describes what happened in the early universe is largely the physics of elementary particles, given the enormous temperatures then prevalent and the highly energetic particle collisions that occurred in its framework. That's why we have no trouble calculating the reactions that governed the development of the universe as early as some 10^{-10} seconds after the Big Bang, when its temperature was some 10^{15} degrees.

At still earlier times, close to time zero, and at still higher temperatures and thus energies, we can only speculate about what happened. As has been noted, the closer we approach the moment of the Big Bang, the less we have to go on. From time zero to time 10^{-44} seconds we know basically nothing whatever—so little, indeed, that we may well ask whether this epoch ever existed at all; and, mind you, the question includes the Big Bang with temperature infinity itself. This possibility cannot be excluded; the models of Tryon and of Hartle and Hawking don't, in fact, have any need for a hot Big Bang. The table in figure 88 summarizes our knowledge of the very early universe and its subsequent fate. This table should be read qualitatively, and its time scales are indicative rather

Time in seconds	Temperature in degrees Kelvin	Energy of collisions in giga-electron-volts	Distance of collisions in cm	Size of today's observed universe in cm	Characteristics
0	infinite	infinite	0	0	Origin of the universe in Big Bang model. Start of TOE epoch.
10^{-44}	10^{32}	10^{19}	10^{-33}	0	Planck time: TOE breakdown. Start of the GUT epoch.
10^{-36}	10^{28}	10^{15}	10^{-29}	0	GUT breakdown. Start of the inflation epoch.
10^{-33}	10^{27}	10^{14}	10^{-28}	10 cm	End of inflation. Start of the epoch of the standard model of the electroweak and strong interactions.
10^{-10}	10^{15}	100	10^{-16}	10^{15} (Solar system)	Breakdown of electroweak symmetry. Epoch of the four separate interactions (continuing to this day). LEP physics.
10^{13} (one million years)	4000	10^{-10}	10^{-5} (1/1000 of the diameter of an atom)	10^{25}	Creation of cosmic background radiation.
10^{18} (15 billion years)	2.7° Kelvin	10^{-13}	0.1	10^{28} cm (15 billion light years)	now

Figure 88 Development of the "effective" laws of nature in the history of the universe.

than quantitatively compelling. We all know that there is no accepted value for the age of the universe; this makes no difference to our overall concept. There is no need for a precise definition of what we call the *size of the universe*—radius, diameter, the third root of the volume. And we don't care whether the first light was emitted from a single point at time zero or from a volume the size of a soccer ball at time 10^{-33} seconds.

INTERACTIONS

The table mentions times, temperatures, energies, distances, and forces. We know that all phenomena we observe in today's universe can be explained in terms of four forces: gravity, the weak nuclear force, electromagnetism, and the strong nuclear force. It may be preferable to denote the strong nuclear force as the *color force,* since we observe it exclusively among colored quarks (see chapter 7).

Forces *couple* to the appropriate charges of elementary particles. As for the gravitational force, the charge it couples to is their energy or, for particles at rest, their mass. For the strong nuclear force, it is the color charge of the quarks that counts; for the electromagnetic force, it is the electric charge. The *strength* of a force is defined by the amount of influence a particle exerts on another by that force, at a given distance. Each force has a characteristic range that determines its action as a function of distance.

Both the strength and the range of a force may well depend on the energy of a particle-particle collision; and this dependence is due to vacuum effects. The higher the energy, the less of a vacuum effect will be present. As the energy increases—or, equivalently, as the temperature rises—the four different kinds of force that we know become more similar. And the hot Big Bang model tells us that prior to the critical time 10^{-44} seconds, the temperatures were such that all four forces were but one single ur-force.

This ur-force had a unique strength and a unique range, and its range was that of gravity. It had masses attract each other across arbitrarily large distances, so we attribute an infinite range to it. We know hardly anything at all about this unified ur-force; it has to satisfy both quantum mechanics and the general theory of relativity, and these two demands have not been successfully blended up to our day. The one thing we do have for this theory of the ur-force is an impressive name: the theory of everything, or TOE.

In the realm of one, and only one, unified force, the laws of nature have the highest symmetry imaginable. In the abstract space spanned by the properties of elementary particles, all directions are equivalent. When we replace an elementary particle participating in a process that is permitted by the unified law of nature with another elementary particle, we end up with a process that is equally permitted by the law.

To gain any intuitive understanding of the breakdown of this ultimate symmetry through vacuum effects at the critical 10^{-44} seconds after the Big Bang, we will have to resort to examples in the space of our experience. Starting at that critical instant, gravity assumes a part of its own, distinct from the other three forces; these remain unified up to another branching point at 10^{-36} seconds after the Big Bang. Up to it, they are jointly described by what the physicist calls a GUT, a grand unified theory. This theory joins the conjoined *electroweak* force and the strong force by means of an interaction we know very little of, and which we will call the GUT force. At 10^{-36} seconds after the Big Bang, the strong force split off from the unified weak and electromagnetic forces. The range of the GUT force then is minuscule, close to zero, while that of the other forces remains infinite. The theory that describes the development of our universe from the second branch point at 10^{-36} seconds to a third one at about 10^{-10} seconds after the Big Bang, we know quite well, and it has acquired the familiar name of the *standard model* of particle interactions. To be more precise, we should specify "of the strong and electroweak interactions."

The breaking of GUT symmetry is accompanied by an effect of enormous importance for the development of our universe. This effect, called *inflation*, describes the unimaginably rapid growth of the universe by a factor of 10^{50} in the minuscule time span of 10^{-33} seconds. We will discuss this inflation together with the breaking of GUT symmetry. The overall symmetry breaking across the three branch points we mentioned was accomplished by the time our universe had reached the mature age of 10^{-10} seconds; by this time, the forces were much as we know them today, with their diverging strengths and ranges. Of the present forces, only gravity, electromagnetism, and the color force retain infinite range, just like the unified force prior to the first branch point. It is the Higgs field that must be held responsible for the fact that the weak force and the GUT force lost infinite range when it pervaded our space.

To visualize this, recall from chapter 7 how the Higgs field gives masses to the particles that interact with it, including the exchange particles of the weak and the GUT interaction. The larger the mass of an exchange particle, the smaller the range of the force it transmits. Conversely, infinite range can be realized only by forces that are carried by massless field particles.

THE QUANTUM NUMBERS OF THE UNIVERSE

We know there is a conservation law for electric charge; this simply means that the sum of these charges in any process remains constant. If a positive charge pops out of the vacuum, a negative one must accompany it, to keep the overall charge zero, as the vacuum demands. To the best of our knowledge, the sum of all charges in the universe is zero—and so it must have been from its origin. But why? How was the sum of all charges fixed in the first place? And how about

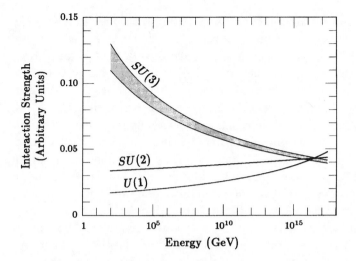

Figure 89 The strengths of the different forces depend on energy caused by vacuum polarization (discussed in chapter 7). Figure 89 illustrates how the strengths of the weak, electromagnetic, and strong forces depend on energy. It is only at very low energies that the weak and electromagnetic forces can be characterized as such. As energy increases toward 100 GeV (where the curves of the figure start), we see that these forces are manifestations of what physicists call U(1) and SU(2) forces. SU(3) stands for the strong color force. What is important about the energy dependence of the strengths of the forces is that all three become equal at a certain energy—at 10^{16} GeV on the GUT energy scale. If they didn't become equal, there would be no unification. It is a further prerequisite of unification that above the meeting point on the GUT energy scale, the three curves continue as a single curve until at least 10^{19} GeV. There, at the Planck scale, the curve describing gravity comes in from below, unifying all forces every after. It should be obvious that this scenario is pure speculation. That the three curves of the figure meet at precisely one point (rather than only within error margins of 10%) is a consequence of the theoretical model—called the *minimal supersymmetric extension of the standard model*—used to construct them. (From Alan H. Guth, *The Inflationary Universe*, Reading, MA: Addison-Wesley, 1997.)

the sum of all other conserved quantities? What determined the initial values of nonconserved quantities?

One example in the latter category is, or rather may be, the total number of nucleons—of protons and neutrons—in the universe. We believe it to be very large today. We have no experimental indication that it has changed with time. On the other hand, the GUT models imply that it could have changed. If so, the universe might have started with a net number of zero nucleons; in other words, there was an equal amount of particles and antiparticles.

When we write down the balance of nucleons, particles (protons, neutrons) and antiparticles (antiprotons and antineutrons) are entered with opposite signs. Should it turn out that the total number of nucleons doesn't change with time, then at the beginning of the universe it must have had the same value we have today. In the physical vacuum, it must be zero. So, in that case, the universe cannot have developed from the vacuum in accordance with the laws of nature.

Every model of the origin and the history of the universe has to come up with a mechanism that gives the observed quantities their overall value. This can be approached in four fundamentally different ways. The first is this: The world owes its contents to the initial conditions of the universe, and that's that, irrespective of further explanations. Second, there may be laws of nature that fix those quantities once and for all. (That is, to the best of my understanding, Stephen Hawking's approach: Forget about initial conditions, just mind the laws of nature.) Third, the laws of nature may have seen to it that the development of the universe reached its present stage irrespective of initial conditions; it had to reach the state we observe today by satisfying those laws. Fourth, there may be worlds without number, each with its own overall value of the quantities under study, each with its own laws of nature. Out of all these, only a universe with the laws prevalent on Earth has the qualities that permit the development of intelligent life (this is usually called the anthropic principle).

If our world originated from the vacuum in accordance with the eternal laws of nature, then all conserved quantities, such as electric charge, maintain their vacuum value—and have always done so. In Edward Tryon's original model, energy was included among these conserved quantities. In its framework, positive energies, such as mass and motion energy, originated from a fluctuation with overall energy zero; they are compensated by the presence of negative potential (or positional) energies. And this is the way Tryon thinks matters should continue: The total energy of the universe does not change with time.

Along with individual processes that take place inside the universe and respect the laws of energy conservation, the general theory of relativity also permits scenarios wherein the universe increases in size; this implies the creation of space and may violate energy conservation. Space, vacant or filled with matter, has very distinct features according to the general theory of relativity. We can make observations that will tell us whether it shrinks or expands. In the inflationary model of the universe, space originally grew at a very rapid pace. It started out with a tiny initial energy due to one of our vacuum fluctuations. The space that is added as the universe expands has the same properties—curvature, energy density—as the parent space. The inflationary model—or, rather, the dominant version of that model—sees the universe, the space it occupies, and the energy it contains as an essentially self-created entity.

INFLATION VERSUS ROTATION OF THE UNIVERSE

For all we know, the universe as a whole does not rotate. Now, the reader might ask, with respect to what should the universe be rotating at all? With respect to some system that has no centrifugal force, it would appear. To the best of our knowledge, there is no such force in a system that does not perform a rotational motion with respect to what we call the firmament—the visible universe in its entirety. This either may be due to amazing coincidence or it may mean that the universe itself determines what, in the laws of nature, makes up the meaning of rotation. This is the way Mach's principle would have it. But does this make sense? If I sit on a stool that turns, I can use Newton's formulae to calculate my rotational velocity with respect to the universe. If that velocity turns out to be zero, the formula tells me there is no centrifugal force, and in fact I will not notice any. So far so good for Mach's principle.

In the same way, we might ask whether the rotational movement of galaxies with respect to the firmament can explain the centrifugal force that is responsible for their shapes. Could it be that a rotational velocity different from the one we observe for our galaxy, the Milky Way, relative to the firmament, would better explain its shape using Newton's formula for the forces acting on it? And if so, should we infer that whatever system stipulates what rotates and what does not coincides with the system that fixes the rotation of the universe as a whole? In this, and only this case, we would have to infer that the universe as such is rotating.

There are, in fact, theories that use information on the rotational velocity and the spatial configuration of the Milky Way to infer that our fifteen-billion-year-old universe cannot possibly have rotated more than fifty times around its own axis in its entire history. The rotation of the Milky Way itself has caused the galaxy to assume a flat spiraling shape, a bit like a merry-go-round. We do know its rotational velocity with respect to the observable universe. If this universe as a whole doesn't rotate, it is this very rotational velocity of our galaxy that we have to enter into the law of nature that permits our calculation of the flattening of our galaxy from its rotation. Had our universe, on the other hand, revolved about its own axis more than fifty times since its inception, we would find a difference between the calculated and the observed flattening of the Milky Way. Independently, and more precisely, we can determine an upper boundary for the rotational velocity of the universe from the observation that our Earth moves only very slowly with respect to the cosmic background radiation.

We can call on the theory of an inflationary universe to explain the lack of its rotation irrespective of its starting conditions. Think of an ice skater who pulls in arms and legs to rotate more rapidly; or, more relevant to a comparison with the rapidly expanding universe, let the skater fling her arms and one leg outward to slow the rotation. Then have her attach weights to her hands and feet: The rotation becomes even slower. The universe likewise gains in size and, perhaps, mass during inflationary expansion by huge factors. And that must mean that any initial rotation has been slowed down to imperceptibly small values when inflation is over.

Figure 90 The traveler in the desert knows that the oasis is surrounded by extremely coarse-grained sand. This allows her to determine the number of sand grains per square centimeter; and she will walk in the direction in which this number decreases most strongly.

If there were a uniform size and density of those grains all over the desert, she would not be able to glean any useful information on her location and on the direction in which she moves: On the average, the desert would be translationally and rotationally symmetric.

ONCE AGAIN: CHAOS, ORDER, SYMMETRY

The cosmological principle is nothing but an assumption about a symmetry of the universe as a whole. If we average over sufficiently large regions, no location takes precedence over, or is distinguishable from, another one. An observer ignorant of local variances to the average values has no way of finding out where he is located. The same is true for the direction in which he looks or moves—none is preferable to any other. Mathematically, we say that the universe as a whole is both translationally and rotationally symmetric.

This statement must not be misread to mean that the galaxies are arranged in some orderly or symmetric fashion, like the stars in Kepler's model (see fig. 45). Quite the contrary: The symmetries are due to a chaotic distribution of the galaxies throughout space. Regardless of the cosmological principle and any specific feature of the location of galaxies, we should retain this important fact: As long as we limit ourselves to average values, there is no way of telling complete chaos and perfect symmetry apart.

Think of a hiker in the desert who tries to find her way (see fig. 90). Looking at individual grains of sand, she has no way to find useful information. She might do better counting the number of grains per square inch, maybe in the knowledge that the sand in the region of the oasis she is headed for is characterized by coarser granularity. That knowledge will persuade her to move in the direction of a diminishing number of grains per square inch.

In the absence of any variation of this number density of grains, the hiker has no way to find her direction, as the model desert of figure 91a makes clear. The numbers in the figure indicate the contents of individual boxes at the heights indicated; they are about the same for the three cases picked. This would be true for all boxes of the given size irrespective of their location or their direction, meaning that the hiker has no way to tell one location, one direction, apart from

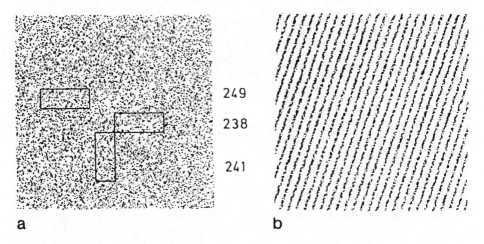

Figure 91 For an interpretation of these figures, see the text.

any other. In this sense, the distribution of points in figure 91a has perfect translational and rotational symmetry. This statement, of course, also assumes that we can ignore the outer boundaries of the figure, as though it covered the entire observable desert plain. As the hiker enters a configuration like that of figure 91b, she can find out certain information about the location and the direction of her motion from an observation of local average densities of sand grains. She might, for instance, observe the direction in which sand dunes are configured by prevailing winds. We see that figure 91b no longer has the high degree of symmetry with respect to any and all translational or rotational movement. The symmetry has been considerably reduced.

In the same fashion, the chaotic distribution of molecules in a gas is completely symmetric with respect to either translation or rotation. The same holds for the molecules in a fluid: All the relevant properties of a gas or a fluid are defined in terms of averaged properties of their atoms and molecules. These move about in random ways—more rapidly as the temperature increases—making only averages observable. But as soon as a crystal appears in the fluid, as ice will do in freezing water, the perfect symmetry of chaotic motion will be reduced. The ordered configuration of molecules in the crystal structure limits the symmetries of the configuration to translations by certain lattice constants, rotations by certain angles. Our perception of crystals starts from their beautiful symmetries, but in reality crystals are a great deal less symmetrical than chaotic systems.

This is quite generally applicable. We cannot possibly achieve the high degree of symmetry of chaos, or of empty space, by ordering individual parts of a system. Any ordering may show us some observable symmetry, but in reality it means the breaking of higher symmetry. When the fading image of a television screen at the end of scheduled programming deteriorates into "white noise," the screen displays complete symmetry. Were it to order its granular chaos into one of the beautifully symmetric images so ingeniously designed by M. C. Escher, the symmetries of the blurred screen would be brutally reduced, as are those of water when ice starts forming.

SPONTANEOUS SYMMETRY BREAKING

It is quite possible that the laws of nature define several states of the lowest energy. But only one of these can be realized in each individual case. The choice nature makes is a random one.

The observer is usually acquainted with various excited states of his particular vacuum, but he usually does not know the ground state. If he wants to find out whether a given transformation is a symmetry transformation, he will apply it to the excited states, but not to the ground state. When the symmetry of the laws of nature is spontaneously broken with respect to the transformation in question, this usually means that the transformation would change a permitted sequence of excitations into a forbidden one. But had the observer tried success-fully to transform the vacuum state together with its excitations, he would have changed one possible sequence of excitations into another possible one. The trouble is that frequently it is not possible to subject the vacuum state to a given transformation. Which observer, we might ask, would venture to transform the ground state on which the entire universe is built? When nature, as it must, out of a variety of equivalent ground states related by symmetry picks a particular one and realizes it, the symmetry of that variety of ground state is, as we say, *spontaneously broken.* Even though the true laws of nature retain their symmetries under such a spontaneous breaking—they are, after all, not affected by the mere selection of a state—the realizable sequences of excitations may obey "effective" laws that lack the *spontaneously broken* symmetry of the true laws. It is obvious from the example of the whole universe that only events in the neighborhood of the chosen ground state will be realizable.

THE FERROMAGNET AS A PARADIGM

We have mentioned the example of water molecules oscillating about their rest position in an ice crystal. We can pick a more telling example when we look at the excitations of a *ferromagnet* (which is just a fancy word for a magnet made out of iron). When we heat a magnet above a particular temperature, its so-called Curie temperature of 768 degrees Celsius, it loses its magnetism. As soon as we cool it below that temperature, it regains its magnetic properties.

Imagine an infinitely large block of iron that fills all space in ideal crystallized form. If it is magnetic, it will have a north pole and a south pole, like every other

Figure 92 A magnetic chunk of iron has a built-in preferred direction in which its magnetization points. This direction is fixed in a spontaneous symmetry breaking process. In a nonmagnetic chunk of hot iron, the spins of the electron and of the iron atoms point in all directions with equal probability. If the alignment of elementary magnets due to the laws of nature that are valid for them were not opposed by thermal motion, all spins would be aligned in one direction, no matter which one it happened to be. As the temperature of the iron block cools below the critical temperature of 768 degrees Celsius, the elementary magnets will align themselves in one direction chosen at random. This is the preferred direction of magnetization.

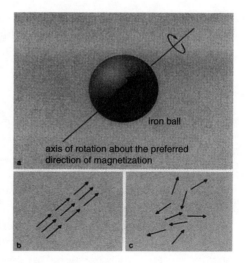

iron ball

axis of rotation about the preferred direction of magnetization

magnet. By implication, magnetized iron picks one out of all possible directions in space—the one that connects its two poles. Let us symbolize that by an arrow in figure 92a—it stands for the direction of the magnetization of this iron volume, drawn here in the shape of a gigantic ball. The arrow can, in principle, point in any direction. We know that the nonmagnetic iron above the Curie temperature is not magnetically ordered, knows no direction from another; as it cools down, the incipient magnetization *spontaneously* picks some direction for the net magnetization. This means that it is only below the Curie temperature that we can draw the arrow into our figure. If the cooldown of the initially warm magnet happens in an otherwise empty space devoid of all magnetic fields, there is no telling which way the magnetization will point prior to the *spontaneous* choice of the direction of magnetization.

In this process of choosing a direction, the magnetization breaks the initial rotational symmetry of the iron volume, as indicated in the figure: The iron sphere will undergo no noticeable changes when we rotate it arbitrarily at temperatures above the Curie temperature, as long as these rotations happen about an axis through the center of the sphere. After cooling below the Curie temperature, the symmetry remains exclusively for rotations about the axis that runs parallel to the magnetization, from the north pole to the south pole.

In this phenomenological description of magnetization seen from the outside, the only rotational symmetry we are concerned with is that of physical objects. If we want to understand what the appearance of a net magnetization, a preferred axis, implies for the rotational symmetry for the laws of nature, we will have to look at the elementary process involved, and at the laws of nature themselves that are operative here. The iron ball consists of a vast number of elementary magnets that are attached to individual iron atoms; leaving the other elementary magnets out of our consideration for a minute, we can rotate any individual one in arbitrary directions. We don't even have to discuss the physical identity of the elementary magnets (we would have to discuss the atomic electrons and their spin). For our purposes, we can just mark them down as little arrows. The iron will have a net macroscopic magnetization if, and only if, the totality of the elementary arrows show a preferential direction in space. Maximal magnetization is synonymous with the (hypothetical) case in which all elementary magnets are arranged in parallel, as illustrated in figure 92b. Conversely, the equally hypothetical case in which the elementary arrows are evenly distributed among all possible directions simply means that the magnetization is precisely zero (see fig. 92c).

The only thing that counts, as far as the net macroscopic magnetization of the iron is concerned, is whether the countless elementary magnets in its interior show any preferential directional arrangement—sufficiently so that this direction can be noticed by an external measurement. The relative potential energy of elementary magnets is minimal when they all point in the same direction, whatever direction that is. Consequently, this is the state toward which they will try to converge. Thermal motion will keep this from actually happening: Collisions of electrons with the oscillating lattice structure will see to it that the distribution of elementary magnets is as chaotic as prevailing energies will permit. That is why it depends on the temperature whether all arrows point in random directions, for a net lack of magnetism—the state prevailing above the Curie temperature. As the iron cools, the potential energy preference for an alignment of elementary magnets wins out: As we pass through the level of the Curie temperature, magnetization takes over.

But in which direction will the net magnetization arrow point? This cannot be predicted. We might imagine that there is a general dispute among all elementary magnets concerned as we pass through the Curie temperature from above. If we look in more detail, the dispute has already been settled in small individual regions, which are like islands of order in a sea of an overall chaotic distribution of arrows. This preordering of small regions may happen in any randomly picked direction. As the temperature decreases, individual islands of order grow in size; elementary magnets or smaller islands in their vicinity adopt the arbitrarily chosen preferred direction of the larger pilot region. It might be that two large islands collide and "go to battle" to impose their chosen directions of magnetization.

Ultimately, it is the speed of the ordering process that decides the outcome. If there is enough time for any particular island to impose its direction on the smaller surrounding ones, the larger magnet will have one net preferred direction of magnetization. If the process happens rapidly, the cooldown may have *frozen* smaller regions of different magnetization directions into the overall volume.

DEFECTS IN CRYSTALS

Spontaneous symmetry breaking applies to one such domain of aligned magnetization arrows; we might take it to be the entire crystal. But it is also important to study the boundary planes between neighboring regions, where magnets with different preferred directions meet.

These discontinuities are called *defects;* they exist in every crystal. They come in various shapes—points, lines, planes—and we don't need to discuss them individually. The most interesting defects are what the physicists call *topologically* stable: They may migrate across the crystal, but they cannot vanish—a bit like a kink in a very long garden hose.

Every defect implies a concentration of energy; the natural law that prefers the lowest possible potential energy, as realized when all elementary magnets are arranged in parallel, opposes such defects with unaligned magnets. They cannot vanish, as we said. That means these topologically stable defects cannot set their energy free; rather, it is frozen at least up to the point where the wandering defect might chance upon a fitting companion—just as a left-handed kink in the garden hose might meet a right-handed one.

A related phenomenon is the delayed boiling of regions in a fluid; it can appear in other phase transitions as well as in that from liquid to gas. When we heat water, it will start boiling at 100 degrees Celsius (assuming normal ambient air pressure). More precisely, I should have said that it *may* start boiling at that temperature; it is possible to heat water beyond its boiling temperature without starting the phase transition implied by the boiling process. For this to happen, water must be very clean, and the heating process has to be realized with great care, avoiding all stirring and all discontinuities such as rough walls of the kettle that contains it.

This delayed boiling process implies that water remains in its liquid phase at temperatures that are usually associated with the steam phase. The boiling process—more precisely, the phase transition from liquid to gas—will start at some location inside the overheated fluid, and will then spread almost explosively. If its volume is sufficiently large, the liquid will burst into violent boiling in other locations almost simultaneously. Steam bubbles will appear here and there, and will join together into larger gas volumes.

Although less familiar than this phenomenon, there is a similar effect of delayed freezing: It is equally possible to cool water to a temperature well below zero without the formation of ice. Just as in the delayed boiling process, the phase transition to ice crystals in the cooled-down water volume can start more or less simultaneously in several locations. The ice crystals, just like the smaller regions of magnetization inside an iron volume, pick preferred directions for their axes. An ice crystal growing from one particular point of phase transition will have one and the same set of preferential directions. But when there are several points of initial crystallization, the preferred directions of crystalline regions growing together are usually not the same. That means that, unlike the case of the individual steam bubbles that join to form a larger uniform gas volume, the resulting larger ice volume will not be homogeneous: There will be points, lines, and planes of discontinuity. The final ice block will consist of distinct regions of homogeneous crystalline consistency; the discontinuities, the *defects*, will be frozen into it.

THE EFFECTIVE LAWS OF NATURE

But let's return to the phase transitions of the vacuum. As the temperature decreased after the Big Bang, the ground state of the universe went through a sequence of phase transitions that can be described in somewhat similar terms— even though, of course, the physical vacuum at that stage didn't contain material units to take the part of the elementary magnets. In all cases, the basic laws of nature remained unchanged through the process of symmetry breaking: What changed is the ground state, not the laws of nature. But inhabitants of a world whose ground state does not display the symmetries of the laws of nature may not be able to tell the difference. They will describe the phenomena that govern their world in terms of what appears to outsiders as modified effective laws of nature, adapted to the asymmetries of the ground state they experience and the outsiders observe.

As far as our approach to the fundamental laws of nature is concerned, we are in the position of hypothetical inhabitants inside a ferromagnet. Imagine these to live at a time after the symmetry breaking phase transition; the ferromagnet has passed through the Curie temperature as it cooled down and is now magnetic. Thus there is a preferential direction indicated by the compasses of the inhabitants. This fact governs many of the phenomena inside their magnetic domain which are accessible to their observation. If they try to experiment with excitations of the ground state, this can be done only locally. The state in which the magnet they inhabit is rotated so that its magnetic axis is perpendicular to that of their ground state is one possible excitation. But this particular state the hypothetical inhabitants of the magnets cannot make a reality; for this they would have to rotate the entire magnetic world they inhabit—a task that is obviously well beyond their capabilities.

On the other hand, they have no trouble experimenting with local excitations, all of which are influenced by the preferred direction of magnetization in their world. They have to add energy to move an elementary magnet away from the direction in which it originally pointed. After doing so, they might notice that as soon as they set it free, it will oscillate about the magnetic axis of their world; it may set neighboring magnets in oscillatory motion about the same axis, thus starting a wave, something we call a *spin-wave.*

As these imaginary inhabitants of the magnetic world formulate the law of nature that describes the wave they initiated, their formulation will include the preferred direction. They have no way of finding out that the truly fundamental laws of nature don't *know* the preferred direction for the motion of their elementary magnets: These laws are symmetric with respect to all rotations—a fact that cannot be observed inside the magnet. As a result, our imagined inhabitants of that world have no way to tell the difference between symmetries of the true laws of nature and those of the ground state of their world. From our own vantage point, we do know about the symmetries governing these laws, and we can determine the influence that the asymmetric ground state of the ferromagnet has on all occurrences in the world it defines. We have therefore, somewhat condescendingly, called the laws that govern the world accessible to those hypothetical creatures inside the magnet *effective laws.*

In the process, we might remember that the ground state, the vacuum state of our world, doesn't display all the symmetries of the laws of nature either. Our relation to the universe is similar to that of the magnet-dwellers to their home. The difference is that the preferred direction of our world is not in the space of our geometric experience; rather, it is an abstract direction in the generalized space made up out of the properties of elementary particles. The fundamental laws of nature in the real world are completely symmetric with respect to rotations in geometrical space.

THE GOLDSTONE THEOREM

It doesn't matter for our discussion that the broken symmetry of our world chooses a particular direction not in actual geometrical space but rather in an abstract space. There is one feature we can carry over from the previous consideration: Spontaneous symmetry breaking implies the existence of waves. The spin-waves in the ferromagnetic world of our model are replaced by excitations in the abstract space of particle properties, where the symmetry breaking actually occurs. In this world, there must be equivalent excitations, waves: The so-called Goldstone theorem says that every symmetry of the laws of nature that is not also a symmetry of the ground state implies the existence of an elementary particle; it even fixes the properties of that particle.

This theorem is predicated on the fact that the broken symmetry—say, a rotational symmetry—is a continuous one. We call a transformation continuous if it can be made up out of arbitrarily many tiny transformations. Rotations in any space, even in an abstract space, are always continuous. Any rotation by, say, 45 degrees about a given axis can always be effected by a sequence of many small rotations of, say, fractions of a degree about the same axis, and in the same direction. On the other hand, mirror transformations are not continuous. We call them discrete transformations: There are no tiny bits of reflections that you might implement one after the other until the full mirror reflection is arrived at.

It is only continuous transformations that we can actually perform; we do so by running through a sequence of arbitrarily small transformations that will, when combined, effect the finite transformation we desire. That is why rotations are accessible to our experimentation, while mirror imaging is not. It is impossible to transform an object into its mirror image: We cannot make a left hand out of a right hand. But I repeat that there is no problem associated with the passage of an arbitrary object into a rotated state of the same object, even though there may be practical obstacles to overcome.

The oscillations—or waves—mandated by the Goldstone theorem originate in the application of symmetry operations to small domains. Recall that the inhabitants of the ferromagnet generated oscillations by rotating one elementary magnet through a small angle from the actual direction of magnetization to a possible one. They were able to do so because rotations are based on continuous transformations. The subsequent identification of waves with particles is inherent in quantum mechanical thinking, and we will not expound on it here.

BURIDAN'S DONKEY

The best-known example of spontaneous symmetry breaking comes to us from the fourteenth-century French Scholastic philosopher Jean Buridan; never mind that his objective at the time was to prove the impossibility of what we call spontaneous symmetry breaking today. He is supposed to have argued that a donkey standing exactly midway between two identical bunches of carrots will starve to death, simply because the animal cannot decide which bunch to eat (see fig. 93). The argument, of course, is fallacious: The donkey cannot remain completely at rest in the central position between the carrot bunches; it will move uncontrollably—say, by itching or by shivering. (We will not take recourse to the uncertainty relation of quantum mechanics!) Once the donkey has moved even the slightest bit closer to one of the two bunches, that one will exert the greater attraction; the donkey will turn toward it, break the symmetry, eat, and survive.

In physics terms, the reason for symmetry breaking is the instability of the symmetric state. As we consider the donkey, it makes no sense to speak of the laws of nature that remain symmetric while the animal makes its choice. The same holds for the group of people arranged symmetrically around the dinner table of figure 94. A different kind of physical example that carries many of the

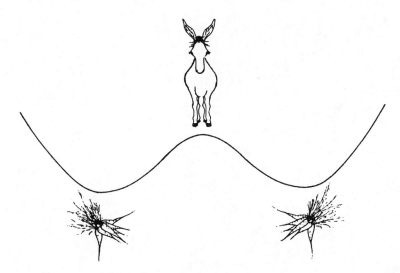

Figure 93 Buridan's donkey sees two identical bunches of carrots at equal distances to either side. We don't know which of these it will choose to feed on—we know only that it will take one of them. The donkey is precisely halfway between the two points, which is only one of an infinity of points it might have chosen; therefore, the probability of having the precisely symmetric initial condition is zero; and even if it were not, inevitable fluctuations would break the symmetry of the unstable state. In any case, whatever happens subsequently completely breaks the symmetry: As soon as the hungry donkey happens to move even slightly closer to one bunch, its attraction for the animal increases and the donkey will eat it. Buridan's invention is arbitrarily unlikely: There is no such thing as two identical bunches of carrots.

The situation changes when we look at the laws of nature, which permit two probabilities that are exactly equal. The theory of spontaneous symmetry breaking then says that the state of the world must be unsymmetric even though the laws of nature don't display any lack of symmetry. The lack of symmetry is due to initial conditions and to fluctuations.

The axiom of *sufficient cause* says the same thing in philosophical terms: Nothing exists, nothing happens, without a cause. Leibniz, in his *Théodicée*, bases many of his proofs on this axiom. In his fourth letter to Samuel Clarke, he argued: "Suppose two mutually exclusive objects are equally good; neither has, either by its own nature or by its relationship to other objects, the slightest advantage over the other. In that case God will not choose to create them." But we find no mention of a separation between the laws of nature and the initial conditions or the fluctuations in Leibniz's writings. To quote his *Théodicée* again: "Equilibrium is never indifferent, never is there complete equality on both sides, so as not to give the slightest tendency for preference to one side." And again: "True, we would have to say the donkey will have to die of hunger if it stands between two meadows offering equal amounts of food. But this problem can certainly never occur in reality, unless God makes an effort to realize it on purpose." Last, he says that "it is unthinkable that God would have bothered to create a world unless He had chosen the best possible one."

Figure 94 Is his napkin on the right- or on the left-hand side of the guest? One of the guests will choose for all: Macroscopic spontaneous symmetry breaking can have microscopic causes. But when two guests decide differently independent of one another, their respective neighbors will follow the example. The resulting lack of symmetry spreads around the table. Wherever the disturbances meet, there will be defects similar to those in the crystallization patterns we show in figure 98: One guest on one side of the table will remain without a napkin; on the other side, a napkin is left over. I owe this illustration of a spontaneous symmetry breaking process to the physics Nobel laureate Abdus Salam.

features of spontaneous symmetry breaking in ferromagnets or in elementary particles is a spherical ball in a hilly landscape. In fact, the ball on an appropriately chosen surface, which may even change with time, can well be used to illustrate the symmetry breaking of the vacuum as well as that of the ferromagnet.

THE MATHEMATICS OF SPONTANEOUS SYMMETRY BREAKING

Consider the rotationally symmetric bowl of figure 95a. There is no doubt that the laws governing the behavior of the spherical ball in its center are rotationally symmetric; the same holds for the only stable position of the ball, which it presently holds.

Now imagine that, with decreasing temperature, the bowl becomes broader and flatter; as the cooling-down process continues and reaches the Curie tempera-

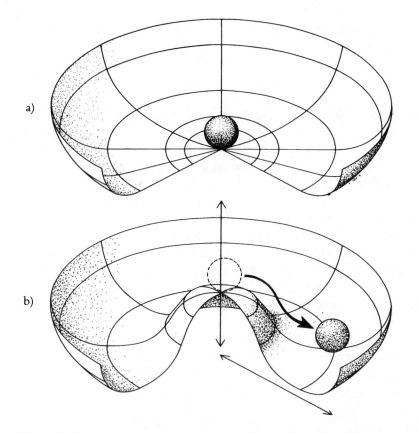

Figure 95 See the text for an explanation of the spherical ball inside the bowls of this figure, and of its importance for the physics of spontaneous symmetry breaking.

ture, a hill will rise in the center of the bowl (corresponding to the chaotic and symmetric state of the individual magnets), surrounded by a circular ditch (which represents the infinitely many ordered states of lower energy)—a bit like the bottom of a champagne bottle. The ball in the center now finds itself on top of a hill (see fig. 95b). It will have to roll down from its symmetric but unstable position. Like Buridan's donkey, it will have to choose a direction in which to roll, which will then determine the position in the ditch where it comes to rest. As it does so, the ball breaks the symmetry still prevailing in the law, as it chooses its new ground state for the system. Note that the laws, which depend exclusively on the shape of the bowl, continue to be rotationally symmetric.

Whenever a state of lowest energy of a given system does not possess all the symmetries of the prevailing laws, several ground states that share the same lowest energy must be accessible to the system. The model in figure 95 shows how such a system can spontaneously change from the initial symmetrical state into any one of a number of equivalent ground states, and in fact must do so.

INFLATION

In the original model of the hot Big Bang, everything in the physical world originated at the moment of the Big Bang out of the vacuum. At time $t = 10^{-33}$ seconds, the matter that originated at time zero had cooled down, had expanded, had taken on different forms; it had changed in consistency and shape, but nothing new had been added. This model, however, cannot expain a number of prominent features of the present universe, with one exception—the disappearance of the cosmological constant, which we have to discuss separately. The model is salvaged by the additional process of inflation.

We recall that the term *inflation* stands for the explosive growth of the universe by a factor of 10^{50} in the time span between $t = 10^{-36}$ and $t = 10^{-33}$ seconds. This is the time sequence suggested by the original Big Bang model but does not necessarily depend on it.

In our present context, it is important to see what triggered the inflationary expansion. The models we mentioned have a vacuum state of our world pass from a symmetric phase into one with reduced symmetries.

At the onset of inflation, some 10^{-36} or 10^{-35} seconds after the Big Bang, the initial era of the universe, when all the forces had the same strength, has long since passed; that ur-state had held only until $t = 10^{-44}$ seconds. Tryon's model includes the possibility that no matter at all existed before the onset of inflation; there was only empty space, but all the laws of nature did exist. In the model of Hartle and Hawking, inflation simply follows what they call the Planck time, the time at which quantum mechanical uncertainty also included space and time. It is at that time that the symmetry of the TOE, the theory of everything, collapsed.

Now back to the start of inflation at $t = 10^{-36}$ seconds: Up to it, and ever since the Planck time, there have been two forces—gravity and the unified forces of the elementary particles. All particles shared mass zero at the onset of inflation; all forces shared range infinity. The universe was, at a temperature of 10^{28} degrees—sufficiently cold to permit the crystallization of a preferential direction in the abstract space of particle properties. This is analogous to the emergence of a direction of magnetization, as we discussed above—with the one difference that we have generalized geometric space to an abstract space.

SYMMETRY BREAKING

The preferential direction fixed by this spontaneous symmetry breaking will now determine which among the elementary particles will share in the strong

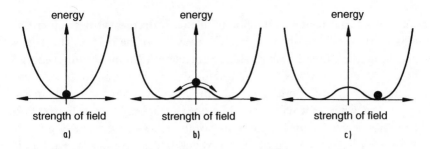

Figure 96 This illustration reduces the two-dimensional surface of figure 95 to a one-dimensional line.

interaction and which will not. The gist is that the spontaneous symmetry breaking sees to it that, in addition to gravity, the one unified force that knew no preferential direction in the abstract space of particle properties is now split into two distinguishable forces—the strong force and the electroweak force.

The symmetry that broke down in this phase transition is what we call GUT symmetry. Formally speaking, GUT symmetry breaking, which permits us to tell the difference between the strong force and the electroweak force, is equivalent to symmetry breaking in the ferromagnet. As one direction in space is spontaneously chosen as a preferential one, a field emerges and, simply by differing from zero, points in some given direction, breaking the previous symmetry. In the ferromagnet, this field is the magnetization; in our cooling universe, it is the Higgs field.

THE HIGGS MODEL OF SYMMETRY BREAKING

Neither example of symmetry breaking, in fact, affects the laws of nature; both apply to the ground state alone.

In figure 96, we redraw the energy landscape of figure 95 for an illustration of the Higgs field. The strength of this field, both negative and positive, is plotted horizontally, while the vertical axis gives the energy of the space it inhabits. The directions the field vector might take are directions in the abstract space of particle properties. Only two of these are shown, affixed to the symmetric point that has field strength zero.

To start with, the temperature of the system is high, and the spherical ball of figures 95 and 96 is in the stable minimum position of figure 96a: The field vanishes, and the energy of the universe has the smallest allowable value. As the system—be it a ferromagnet or our universe—cools down, it will pass through a critical temperature, below which the energy as a function of the field will look like figure 96b. The field still has zero value at the location of the ball, like the universe without a Higgs field; but the potential energy decreases toward increasing strengths of the field (in absolute value), away from the origin. The Higgs field has to assume the strength that the figure correlates with the minimal energy of the universe. As a result, it has to choose some specific direction in the abstract space of particle properties; in figure 95, this might be one out of infinitely many possible directions, whereas it will have to be one out of only two in figures 96b and 96c. The ball—which stands for the system or universe we are discussing—will roll down from its hillock and settle in the valley that surrounds it. And as it does so, the system passes from the symmetrical state of field strength zero into an asymmetric one where the field does not vanish.

The symmetries of the laws of nature remain intact but become hidden, since the corresponding symmetry transformations now also change the vacuum and can therefore not be affected. To implement them would amount to changing the state of the total world—all of the ferromagnet, in the model. Were it possible to subject not only the elementary particles that participate in a collision but also the vacuum itself to a symmetry transformation of the laws of nature, it would turn out that the transformed sequence of events within the transformed vacuum would agree with the laws of nature, just as the original sequence of events within the original vacuum agreed with them.

The Higgs field remains the same everywhere in space, even after the breaking of the symmetry. It cannot be otherwise, since the breaking of GUT symmetry doesn't touch the spatial symmetry of the universe: Its state of lowest energy is both rotationally and translationally invariant. One further symmetry that remains intact during the symmetry breaking process is the symmetry under changes of velocity. Therefore, observers will not be able to determine their velocity with respect to the Higgs field. The vacuum that contains Higgs fields may be a complicated medium, but there is no way we can observe motion with respect to this field. This equals the situation of blackbody radiation at zero temperature: The Higgs field does not oppose any resistance to motion, just like blackbody radiation at zero temperature, whereas blackbody radiation at finite temperatures will.

THE VACUUM ENERGY

The curves drawn in figure 96 show the energy contribution of the Higgs field to the total energy of the universe as a function of the field strength, at two different temperatures. Above the temperature of symmetry breaking, the situation shown in figure 96a will prevail. There is an energy minimum at field strength zero—the physical vacuum (that is, the state of lowest energy) is, at the same time, empty space—at least as long as we consider only the energy supplied by the Higgs field fixed at that value. There will, however, always be quantum corrections, because the Higgs field will always fluctuate about the values shown in our figure. The space is therefore not entirely empty in the absence of the Higgs field; it will still be swarming with virtual Higgs particles. This is just another contribution to the vacuum fluctuation we are familiar with by now.

Space is as empty as it can be when it contains neither real particles nor fields nor energy. Always expecting quantum corrections, we can eliminate particles without untoward consequences. If, however, the energy of the universe depends on the field strength in the way shown in figure 96b, the two remaining properties that define empty space are contradictory: The energy becomes smaller as the field strength increases in absolute value from zero. The Higgs field, therefore, propagates in space, just as the field "frozen crystal" expands in water. The actual vacuum of physics will, in the process, remain the state of lowest energy assumed by the universe, as in figure 96c.

A STATE OF NO ENERGY?

We cannot start discussing the universe as a whole without taking recourse to the general theory of relativity. Like Newton's mechanics, it takes into account both positive and negative energy values. But it differs from Newton's mechanics by singling out one particular energy as having the value zero. Gravity will determine which value that is: Objects with energy zero do not exert any gravitational pull. This definition has permitted us to enter a horizontal line in figure 96 and call it *energy zero*. What we don't know at this point is at what height in the plot to draw this line; we are dealing with fields here, and the only field known to the general theory of relativity is its own field, which defines time and space.

We have said that in this theory masses curve space. It therefore makes sense to say that in the absence of masses there will be no curvature. The straight line of energy zero then means that there are no masses. But things might turn out to be more complicated: The general theory of relativity also permits curved spaces from which all masses have been removed.

We might ask where to stash away the masses that we decided to take out of the universe. This universe we are presently discussing is only a mathematical model of the real universe and depends on parameters we can play with. One of these parameters is the mass of galaxies: As we change them, the curvature of space will be affected.

Let's turn to a curvature of space that persists after all masses have been removed from it. This is the curvature of empty space. We will explore it by means of probes whose masses are so small that they essentially don't act on the space's curvature. If the probes move apart, space expands; as they approach each other, it collapses. This makes the curvature of empty space an observable phenomenon. We can associate an energy with it—the energy of empty space, of the vacuum of physics.

This energy is inherent in empty space; it acts differently from the energy associated with bodies we might introduce into empty space. The gravitational force emanating from all of its usual sources is always attractive, and always decreases with distance. The gravitational force that we associate with empty space, on the other hand, may be either attractive or repulsive—and it *increases* with distance. It is attractive when the energy of empty space is negative and repulsive when that energy is positive.

How could that come about? The only honest answer to this question is that the relevant equations say so. A lot of scientific and popular writers' ink to the contrary, there is no better explanation. The equations that say so are as compelling as those that make two plus two equal four. The term that contains the energy of empty space in the equations of the general theory of relativity is the *cosmological constant*. Einstein introduced it in 1917, and we will discuss it below.

While the question of a reason for the universal repulsion issuing from the positive energy of empty space doesn't lead us anywhere, it makes sense to ask how this repulsion will manifest itself. If energy is contained in empty space, the diminishing gravitational pull of masses moving apart would be accompanied by an increasing repulsion. But we see no manifestation of this effect in the universe we observe; to the best of our observational powers, Newton's gravity law is correct. Neither is there any sign of a curvature of space that might be due to the energy content of the vacuum.

We can always theorize that countervailing effects cancel each other, and I will discuss that later. There is no sign of any kind of hidden energy in the empty space of today's universe. Since the latter is in the state corresponding to figure 96c, its energy should be zero. That, however, implies that it would have been positive before the breaking of the symmetry, as figure 96b illustrates; the universe then was subject to repulsive forces.

These forces, sometimes called antigravitation, are at the basis of the inflationary model of the universe. Tiny massive probes that we insert into an empty space without masses that could compensate a general repulsion, but with a positive energy density, will move away from each other. The first person to suggest a model characterized by an empty, expanding universe and by motion without matter was Willem de Sitter, in the year 1917. One difficulty lies in the fact that the well-known Hubble expansion of the universe is being slowed down by gravity, but that there is nothing to brake the expansion of an empty universe with positive energy. Quite the contrary: Such a universe will expand more rapidly the larger it is. Its repulsive forces increase with distance. And its energy density does not decrease, as in a customary Hubble expansion; newly added empty space has *ab initio* the same positive energy density as its parent space. This would imply that expansion will only accelerate and continue to infinity—unless the universe went through the phase transition that changes figure 96b into figure 96c, breaking the symmetry.

THE MECHANISM OF INFLATION

There cannot be inflation unless the universe spends at least some measure of time in a state that is characterized by a positive energy value for empty space. This state, which we sketched in figure 96b, is also called a *false vacuum*. It is not really a state of lowest energy; rather, it is empty space with positive energy. To make sure that the universe will spend at least some given time—the time of inflation—in this state, before making the transition from the false vacuum to the real one of figure 96c by means of spontaneous symmetry breaking, we can apply small modifications to the energy plots of these two figures.

THE END OF INFLATION

At the end of inflation, the universe has grown by a factor of 10^{50}; it is almost entirely empty, and quite cold. The matter and the radiation it contained at the start of the inflationary process has been enormously diluted and cooled down. This is so since every expansion, including that of the inflationary process, is opposed by normal gravity. Expansion moves the masses of the universe to greater relative distances, and this costs energy. As a result, there is barely any normal matter or radiation left in the universe toward the end of inflation; instead, the inflationary universe has its store of vacuum energy by dint of its mere existence— or, more precisely, by dint of its origin in the false vacuum state of positive energy. This energy is freed when the Higgs field appears: Losing its high original symmetry in a phase transition, the universe inherits the formerly hidden vacuum energy as normal energy as it passes from the false to the real vacuum state. Out of infinitely many states with the same lowest energy, the real vacuum has become "the" ground state of the world by what appears to be a free choice of the Higgs field.

There are different models for the inflationary process, marked by varying concepts of the duration and the final stages of inflation. In the model discussed here, inflation ends the way water cooled below the freezing point turns into ice: From individual nuclei of crystallization inside the inflated universe, there is a growth of domains in which the false vacuum has passed into the true vacuum state (see fig. 97). The mean value of the Higgs field is distinct from zero in the true vacuum, whereas it is not in the false one: After the phase transition, *something*—the Higgs field—exists, whereas *nothing* did before.

In each of the individual domains, the Higgs crystallization leads to the choice of some preferred direction in the abstract phase of particle properties. At the interfaces between individual domains, there will be discontinuities and other vacuum defects, just as in water at temperatures below freezing, where individual domains with different crystallization directions grow together. The illustration in figure 98a may help to explain that. As a result, the ideal case of

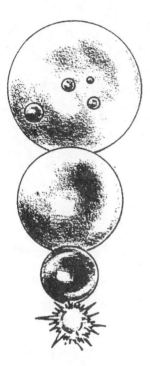

Figure 97 This figure illustrates the fate of the false vacuum. After its rapid growth by inflation, tiny bubbles of *real* vacuum will appear and grow through crystallization.

a spatially symmetric universe we advocated before has not materialized in reality: Since energy is concentrated in the "topological defects" of figure 98, the universe has not arrived at its state of lowest energy. Along with real fields and particles (as opposed to virtual ones), the defects must be counted as "something" in the universe.

FROM INFLATION TO ANOTHER HOT BIG BANG

Clearly, inflation is a very complex phenomenon. It depends not only on quantum mechanics but also on two rather exotic aspects of the general theory of relativity. One of these is the fact that empty space may display curvature—that is, may contain either positive or negative energies even in the absence of all masses. The other is the way in which this energy exerts its influence on space. We mentioned that the general theory of relativity differs from Newton's mechanics, among other features, in that it elevates the energy level zero to a state of signal importance. But that doesn't tell us where exactly that zero energy level is to be found in the real universe.

The model of the inflationary universe dictates that the energy of the false vacuum before spontaneous symmetry breaking is not the same as this fundamen-

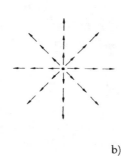
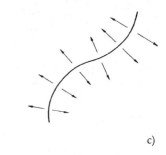

a)   b)   c)

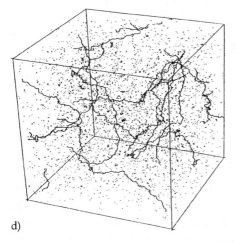

d)

Figure 98 Individual domains, in which spontaneous symmetry breaking led to the definition of preferential directions, grow until they meet other domains. At their interfaces, discontinuities appear as defects—much like the one we described in the example of the dinner table of figure 94. (a) In two dimensions we have linear interfaces; in three dimensions, the interfaces are planar. (b) A point defect. (c) A linear defect. (d) The result of a computer simulation: The linear defects stand for cosmic strings that might exist in the universe, representing extreme mass concentrations.

tal energy zero. Rather, it puts the energy of the real vacuum, after symmetry breaking, in this position. In the false vacuum state, space is empty by virtue of not containing the Higgs field: The positive energy of the unbroken symmetric state is vacuum energy. It differs from normal energy, which attracts and is attracted by normal energy somewhere else, in that it repels and is repelled by its equals—just as electric charges of the same sign repel each other. They are driven apart *within* space, whereas the forces by which vacuum energy repels vacuum energy *act on space itself*—forcing it to blow up in the inflationary process we already discussed.

The energy that was hidden in the false vacuum is liberated in the course of this phase transition of the universe into the real vacuum; it now becomes normal energy, just like the thermal energy liberated in the process of crystallization of ice in water. The Higgs field takes the place of the ice crystals as the physical vacuum develops in the inflationary model. The energy generated in the inflationary process, which originates in the false vacuum ground state, manifests itself as motion energy in the Higgs field. We might say that it makes the Higgs field shiver. It then produces radiation and massive particles that move about randomly. In this thermalization process, it increases the temperature of the universe; it spreads evenly throughout all space. Inflation ends some 10^{-33} seconds after its own Big (or Small!) Bang in the same state the universe would have reached in the hot Big Bang model without inflation. This coincidence is at the basis of our model's success. We believe we can understand all further developments in terms of various theories, some of which are quite speculative, but none of which clash with the inflationary model. The success of the inflationary scenario when compared with that of the hot Big Bang is its implied explanation of the properties of the universe after $t = 10^{-33}$ seconds; these properties have to be entered into the hot Big Bang model as unexplained initial conditions.

At the end of inflation, just as after the hot Big Bang, the universe is hot. It is, we might say, self-created by dint of its explosive growth: The inflationary process generates space, and the energy that space contains, from essentially nothing. To repeat Alan Guth's dictum, "The universe may be the ultimate free lunch."

All serious models of the origins of the world agree that there was a period of GUT symmetry breaking and that this period included inflation, which in turn resulted in a vastly extended size and heating up of the universe. How that came about is, however, vigorously contested among different models. In the conventional Big Bang model with inflation, the universe started with infinite temperature. It then cooled off gradually until the breaking of GUT symmetry began at $t = 10^{-36}$ to 10^{-34} seconds. The development in the Hartle-Hawking universe can be described similarly but lacks the initial singularity. Tryon's model includes the possibility that the universe originated from nothing at all at the moment of GUT symmetry breaking by means of a vacuum fluctuation; this might have produced a little universe in a false vacuum state, which subsequently grew through the inflationary process, acquiring energy as we described above.

Let me stress that all these models of the development of the universe from nothing at all—or, according to Hartle and Hawking, from some point on the space-time surface—have to be seen for what they are: *models,* devoid of compelling experimental verification. The scenarios we develop from them are possible, and they illustrate various features we can follow up on, but none is ultimately persuasive. It is hard even to draw demarcation lines between theories and scenarios.

ANOTHER PHASE TRANSITION

Can we rest assured that our universe has already arrived at the lowest vacuum state? Not really. If it has not, there might be another phase transition in our future. This would be the ultimate catastrophe. It might happen spontaneously anywhere—it might even have already happened. If so, we would learn about it by the arrival with the velocity of light of a wall that destroys literally everything—the large negative vacuum energy would see to it that everything shrinks to almost infinite density. Such a transition could also be triggered by a very high energy concentration occurring when elementary particles collide. Shouldn't we therefore stop building ever more powerful particle accelerators? The British astrophysicist Martin Rees has pondered this question together with colleagues and gives an all-clear signal for the present: Particles of cosmic rays collide in space frequently at energies that are at least one hundred times larger than those of the most powerful accelerator to be built in the foreseeable future—the large hadron collider (LHC) at CERN in Geneva. But far above this energy Martin Rees sees potential danger: "The possibility is not absurd—in our present state of ignorance about unified theories, we would be imprudent to disregard it," he writes. Indeed, caution should surely be urged (if not enforced) on experiments that create energy concentrations that may never have occurred naturally. We can only hope that extraterrestrials with greater technical resources, should they exist, are equally cautious!"

WHY A FLAT UNIVERSE?

Of the two or three connected conundrums that curved space presents to us, inflation can solve one and only one. On the average, the curvature of space is so small that it becomes noticeable only on a cosmic scale, if at all. If space were strongly curved, we would be unable to look straight ahead; rather, we would essentially look around the corner, since light rays will follow the curvature. But then it is also possible that space is not curved at all, that space is *flat*. Our experimental information on the matter is not precise enough to tell.

To be flat, or reasonably flat, today, space must have chosen precisely the correct curvature right after the Big Bang. It is clearly possible that space actually had this curvature as one of its initial conditions. The parameter describing the curvature of the universe is called *omega*. To make space essentially flat, this omega parameter must have had a numerical value extremely close to 1 right after the Big Bang. If it is exactly 1, the value of omega is much easier to understand than a value of 1 plus or minus 10^{-55} would be. Omega $= 1$ is a possible initial condition of the universe; and in the hot Big Bang model there is no other explanation for the fact that the universe is nearly flat today.

Rather than just stating that it as an initial condition chosen out of nowhere, it would be nice to have a theory that explains the origin of the numerical value of omega; just saying that the value is what it is because it was what it was cannot satisfy us. Inflation provides us with a reason for the present-day flatness of the universe. The explanation is given in figure 99. If we inflate a balloon by some

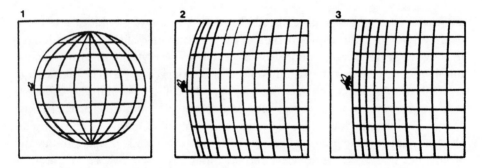

Figure 99 As we inflate the balloon, its surface becomes flatter and flatter.

enormous factor, its small, heavily curved surface areas will grow into large, essentially flat regions.

The inflationary model sees to it that its gigantic magnification factor of 10^{50} makes our universe flat in the sense of figure 99c, to arbitrary precision; for all practical purposes, the curvature sinks to zero.

DARK MATTER

We know that the curvature of space is logically intertwined with the energy it contains. The general theory of relativity tells us that the gravitational force, which has one energy act on another, is nothing but a different name for the curvature of space. Space altogether can be flat only if it contains just the right amount of energy—or if there is a conspiratory cancellation of different effects, a possibility not discussed here. But this condition is not met as long as we consider only the energy of all matter that, to our knowledge, is present in the universe: The *visible* matter—that is, that matter that is observable through its light emission or reflection—contains only about 1 percent of the "critical" mass that is needed for an overall flat space. "Light" in this connotation stands for electromagnetic radiation of any wavelength that reaches Earth—from optical to gamma rays.

It is not only the theory of inflation that tells us we don't know of all of the masses in the universe. In addition, there are observed facts that tell us the same. One of these is the chaotic motion of galaxies inside galaxy clusters. Galaxies don't appear singly, but rather inside such larger groupings. They are kept from escaping by the gravitational pull of the mass of the cluster. Individual galaxies move inside a cluster with velocities that are measurable by means of the Doppler effect (see figures 57 and 100). That's how we know these velocities to be so large that the gravity of the visible mass of the cluster does not suffice to hold the galaxies it contains together: Were there only the cluster's visible mass, its

Figure 100 This figure shows galaxies as they move randomly in a galactic cluster.

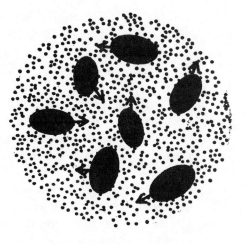

Figure 101 Our galaxy, the Milky Way, has a form roughly like the galaxy shown here called NGC 628 M74. If spiral galaxies contained only the luminous matter that we can observe, gravity would not be able to contain their constituent stars. Centrifugal forces would have long since driven them apart.

galaxies would have had to fly apart long ago. To keep all of them within the cluster, there must be about one hundred times more mass present than what is noticeable to us as visible matter.

Even the most remote stars close to the outer rim of a galaxy like ours, the Milky Way, orbit the center of the galaxy at a velocity that would make them take off into intergalactic space if that galaxy contained only the matter that is visible to the astronomer (see fig. 101). Just to hold the stars within a galaxy, we need the gravitational pull of at least ten times more mass than is visible.

There are arguments in favor of the assumption that the invisible matter that holds the galaxies together consists of the same atoms as visible matter. We call it *baryonic matter*. Baryons (after the Greek word for *heavy*) are the massive constituents of nuclear matter, mostly the protons and neutrons. (The reader may recall that the rest of atomic matter resides in the electrons that orbit the nuclei, and that the electron mass is about two thousand times smaller than that of a proton.) The planet Jupiter could serve as an example for a fairly large concentration of dark baryonic matter that becomes noticeable mainly through its gravitational pull. Burnt-out stars and nonluminous intergalactic clouds of dust or gas may also contribute to dark matter. There are a number of good reasons for astrophysicists to assume that all luminous galaxies are embedded in large spherical regions filled with a ten-times-larger mass of such composition.

The process has to be repeated on the next larger scale: To keep galactic clusters from flying apart, it takes more dark matter. The larger the regions we take under consideration, the more dark matter is needed for a flat universe. All in all, we come to the conclusion that no more than 1 percent of all matter in the universe falls in the visible category.

WHAT DOMINATES THE UNIVERSE?

When we combine the information gleaned from observational astronomy with the model of the inflationary universe, the notion that 99 percent of its mass is *dark matter* becomes compelling. But there is no intimation of the dark matter's nature. If the universe is flat, about 10 percent of all the matter it contains must be baryonic; otherwise there would be a discrepancy with one of the signal successes of the standard model of the early universe—its determination of the abundance of the light elements in our world.

But what makes up the remaining 90 percent? Nobody can tell at this time. Candidate substances have to live up to a number of criteria. For one thing, they must not interact too much with normal matter—otherwise they would long ago have been detected. Next, their very existence has to be compatible with the genesis of structures in the early universe. Among well-established particles, neutrinos meet both of these demands and are therefore frequently mentioned as dark matter candidates. Just like the cosmic background radiation, there is a background population of neutrinos that pervades the universe. But the standard model of particle physics, in agreement with the best experimental values presently obtainable, says that neutrinos have no mass. If they are massless, they cannot constitute dark matter. This point has added interest to the search for neutrino masses, presently very active. Until the question is settled, neutrinos must be seen as dark matter candidates.

There are many more particles, which so far exist only in speculative theories of particle physics, that can, and have, also been advanced as potentially dominant dark matter candidates. We must be content with some name-dropping: In addition to WIMPs (for Weakly Interacting Massive Particles), such as axions, supersymmetric particles, and monopoles, there are cosmic defects—in particular, strings—and of course black holes of any size and makeup.

THE COSMOLOGICAL CONSTANT

The energy of the vacuum in the universe presents us with a dilemma: We know, on the one hand, that this energy must be minuscule, since, as we have discussed, we would otherwise be unable to look straight ahead. All possible contributions to the energy of the vacuum must (almost) cancel out. The dilemma is that we know of many sources of vacuum energy that, to the best of our knowledge, are *independent* of each other; it would be hard to understand why their contributions should cancel. While the contributions of different sources to the total energy of the vacuum have positive as well as negative signs, such that they might cancel, at the same time each of them is huge compared with the experimental limit on their sum. This is the problem of the cosmological constant.

Recall that the cosmological constant owes its name to the contribution of gravity to the energy of the vacuum, in Einstein's original definition. Einstein himself noticed in 1917 that he was able to enter it into the equations of the general theory of relativity without violating any of their tenets. In those days, he believed the universe to be in a steady state. But according to general relativity, a universe in which gravity is the only large-scale force between masses cannot be static. It has a final nonstatic fate that is determined by its total energy content. If there is an additional, repulsive force that grows with increasing distance between celestial bodies, a universe that otherwise might collapse under attractive forces can now become static. The equilibrium required for this process must, however, be very carefully arranged. Should the masses move closer, attractive gravity increases while the repulsive force decreases—and the universe will collapse. On the other hand, if an opposite fluctuation makes masses move apart by only a bit, repulsion will prevail, sending masses off into infinity. The fact that ordinary (attractive) gravity decreases with the distance between massive objects, while the repulsive vacuum energy increases under the same relative motion, makes the universe unstable under fluctuations about the equilibrium point: All the forces acting outside the equilibrium collude to remove the universe even more from that point.

Einstein introduced the cosmological constant in order to obtain a static universe as a solution to his equations. When Edwin P. Hubble discovered in 1928 that the galaxies move apart, he demonstrated that the universe is, in fact, not static; rather, it expands. Smitten by this discovery, Einstein is quoted as berating himself for having committed the greatest stupidity of his life when he introduced the cosmological constant: If he had trusted his original equation in the absence of the cosmological constant, he would have predicted the expansion of the universe. He had, in fact, never liked the constant that he had needed for the static solution. In a 1923 letter to Hermann Weyl, the great mathematical physicist, he wrote: "If there is no quasi-static world, then away with the cosmological term."

But in physics, opinions (should) mean nothing; only new arguments can make a difference. You cannot simply change your opinion; and so there was no way to remove the cosmological constant from Einstein's equations of general relativity simply because its originator had lost faith in it. Einstein's original discovery—that the cosmological constant doesn't violate any of the tenets of his general theory of relativity—is not vitiated by the fact that he came to dislike it. Note that neither he nor anybody else ever found a reason why it should *not* be there.

The experimentally observed curvature of space is the result of many contributions, of which Einstein's cosmological constant is only one. Another one is due to the masses that actually exist. To them, we have to add the energies due to the fluctuations of the elementary particles in the vacuum. We have no way to know how big these contributions really are, since the best approach we presently have for computing them simply sends them to infinity—but we can be sure that each contribution is way above the experimental limit on their sum. This does not necessarily lead to contradictions, since there are positive as well as negative contributions to the cosmological constant, so that they might ultimately cancel. The trouble with this is that the cancellation must produce a tiny sum of very large numbers, although the individual contributions, to the best of our present knowledge, are completely independent of each other.

To list some of the terms that contribute to the energy of the vacuum we have, for starters, that of the Higgs field, which we have to add in the absence of any knowledge of the zero-level energy of the general theory of relativity. Then, there is the zero-point electromagnetic radiation of photons, which manifests itself in the Casimir effect. Add to this the electromagnetic term that describes the vacuum fluctuations into electron-positron pairs and other particle-antiparticle pairs, which we know to be responsible for the experimentally observable effects of vacuum polarization. Analogously, there are energy terms due to zero-point radiation of the gluons of the strong color interaction, and to the vacuum fluctuations of quark-antiquark pairs.

All these terms must add up to zero. If we attempt a calculation of all the terms, the sum we come up with differs from the experimentally admissible upper limit by the grotesquely large factor 10^{100}. This implies that the mutual cancellation of contributions to the ultimately effective cosmological constant, the sum of all terms, must proceed to the precision of 1 part in 10^{100}—obviously, an unimaginably tall order!

The disappearance of an effective cosmological constant that, according to our present knowledge, can happen only through this kind of cancellation provides, I believe, the strongest argument against the mutual independence of all the terms that make up the energy contained in the universe. It argues that there must be a comprehensive symmetry principle that constrains all contributing terms to add up precisely to zero.

CHAPTER 9

EPILOGUE

•

PHYSICS AND METAPHYSICS

•

READERS, WHEN STARTING THIS BOOK, MAY HAVE THOUGHT OF EMPTY space as nothing but the stage on which our cosmic performance takes place; and yet they may have had a suspicion that things could not be quite that simple. I surmise they would start from their notions of matter when trying to understand our world; matter could be seen as the actor on the cosmic stage that tells us about *something* and *nothing*. They probably had no doubt that matter does exist; but less certainty that there really is empty space. Having accompanied me through the eight chapters of this book, they will agree that the line that separates *something* from *nothing*, matter from empty space is blurred—just as we are accustomed in our daily lives to an indistinct separation of subject from object. The universe is but one immense unit; it cannot be separated into spatial domains that are totally empty and others that are completely filled with matter. Matter and space can be distinguished from each other, but where we draw the blurred line between them is largely a matter of taste.

DOMAINS

Our notions of *something* and *nothing*, of matter and empty space, cannot be separately discussed; throughout history, there have been corrections in how they were perceived. Physicists, in their attempt to understand the world, can be successful only if *not* everything depends on everything else—if there are domains that can be studied individually, without recourse to the rest of the world. They tend to see matter as one such domain, empty space as another. The latter is, in modern scientific terms, the vacuum of physics.

The hiker whose fascination upon detecting a satellite lit by the sun in the dark sky causes him to stub his toe on a rock in his path will not agree that the entire world is one holistic unit. He will make a strict difference between matter—that of the rock and his toe—and the space in which the satellite moves.

When Democritus said that there is nothing but atoms and the otherwise empty space in which they move, he did away with the notion that matter extends continuously across the universe. We have come to realize that this notion of continuity is wrong: If we quantify matter by its mass, we know that almost all the matter of the atom, to within a few parts per thousand, is concentrated in its nucleus, and that the nucleus fills only a minuscule fraction of the atomic volume. Translating from the atomic size to that of our solar system, the nucleus is no larger than our Sun on that scale.

Given the knowledge we have today of the physical properties of our world, Democritus's concept gives a fair answer to some of our questions, but breaks down completely when challenged by others. When Democritus asks how matter can dissolve in water, how knives can cut through substances, his answer—that the empty space between the atoms renders both possible—is still valid. In terms of the molecules and atoms of present-day chemistry and physics, these constituents of matter remain the same throughout the mixing or cutting processes. They don't change; they are merely redistributed in space. Democritus was certainly ignorant of chemical reactions; but even those can be approximately described in terms that don't contradict his notions of atoms in empty space. Whereas the outer electronic shells of real atoms are getting inextricably interleaved and intertwined when molecules form, the atoms of Democritus are today's atomic nuclei: It is they that are being rearranged in matter-free space while remaining individually intact.

MODIFICATIONS OF THE VACUUM

When we now ask whether we can further divide up the atomic nucleus into a core plus empty space, we hit bottom with this procedure. The building blocks of nuclei—the protons and neutrons—are configured in an almost continuous way; they essentially touch. But we do get back to the notions of Democritus when we inquire about the makeup of protons and neutrons out of quarks. Quarks are a peculiar bunch altogether. If we investigate the distribution of their electrical charges experimentally, quarks appear to be suspended in a mostly empty space: Their electrical charges are concentrated in regions so tiny that we cannot tell them apart from points. But this is only the outside view of the quarks. As far as their relations to each other within a given proton or neutron are concerned, their electrical charges are irrelevant. It is only by their color charges that they notice each other's presence. Their interactions are mediated by the color field, not the electrical field that surrounds them. This field modifies the vacuum surrounding them much more drastically than electrical fields already do—so much so that there is no distinction possible between the role of the quarks and that of the space surrounding them, as we model the quark-quark interactions: The stage and the actors are of equal importance, and cannot reasonably be separated.

Let's think back to Newton's mass points: According to mechanistic philosophy, they would not be able to influence each other. It is beyond all thinkable probability that one point will actually collide with another point. Their mutual influence is exclusively due to the gravitational fields that surround them. The same holds for pointlike electric charges: Electrons do not repel each other as pointlike charge hits pointlike charge. Their fields actuate their mutual repulsion.

The modification of space between the quarks is such that quarks cannot exist singly. We have experimental knowledge of their behavior down to distances of 10^{-16} centimeters. If we try to penetrate to even greater precision, the various forces blend together to a unified ur-force; on this scale, the quarks and the space that surrounds them are no longer separable.

Seen in this way, the emptiest space known to physics—that is, space that holds as little energy as the laws of nature will permit—is not really empty. Albert Einstein tried to finesse this point when he said that the ability of Newton's space to "confer inertia" shows that this space is *something*, not *nothing*—that it is an "ether of mechanics" that acts like a substance although it is not a substance. It is not a substance because it cannot possess a state of motion. We may recall that the Eleatic school of early antiquity objected to the notion of empty space simply because it cannot be a substance, an objection carried through the ages by no less than Aristotle and Descartes. It is found to hold true in our modern notions, albeit for reasons they could not have dreamed of: Space is, in fact, *something;* but something of a kind that cannot be assigned a velocity. The space of the general theory of relativity has properties that can be expressed in terms of fields, and that have no way of distinguishing the flat—and therefore empty—space from another space that is curved and therefore not empty. As far as general relativity is concerned, all spaces with different curvature are equivalent. Thus space itself can assume properties, and, as a consequence, "empty space" does not automatically qualify as "nothing." Theories that include quantum mechanics don't admit any empty space at all, simply because the uncertainty relation and the special theory of relativity force onto it vacuum fluctuations that will fill space. In its state of lowest energy, the real space of physics at the low temperatures prevailing long after the Big Bang contains Higgs fields that no longer have symmetries of the fundamental laws of nature. Not that every *something* breaks the symmetries of those laws; but certainly there is *something* beyond the *nothing* when the world is less symmetric than the fundamental laws of nature.

We are therefore vindicating Aristotle in the final count: His most important dictum, that there cannot be empty space, turns out to be true. It may not have had any experimentally accessible content in centuries past, but it turns out to be of prime importance for today's physics. Democritus's ideas were good enough to explain what they were invented to describe, plus some. The fact that his empty space is not truly empty does not affect the atoms; they can propagate across the space surrounding them whether or not it is truly empty. This means that, in terms of classical physics, which describes our everyday world (though without explaining it), the objections against Democritus's notions of atoms surrounded by empty space are irrelevant.

If we took these objections seriously on this level, we would deny the existence of a number of features we in fact observe, among them the mixing and separation of substances. The true laws of nature that we consider fundamental in our day are approximated by Democritus's tenets. But to this day we cannot precisely define the conditions that have to be met so that the laws of classical physics that contain a separation of empty space and of massive atoms as valid approximations follow from the basic laws of quantum mechanics. Was it merely one in many possibilities, or was it the only admissible development of the minuscule, hot, quantum mechanical world right after the Big Bang that led to our immense, cold, quasi-classical universe?

NECESSITY AND CONTINGENCY

Something that is logically necessary is not subject to change. If the world with all its present properties corresponds to this definition, it has to remain the same throughout the ages. Properties we call *contingent* are those that might also be different. Contingencies are subject to change and cannot be uniquely captured by thought. Without having any empirical basis for this themselves, the pre-Socratic philosophers believed that the world can be understood because deep down it is logically necessary. For them this meant that there are laws behind every apparent change. Some went as far as to deny that there can be any change at all. And, indeed, a logically necessary world cannot be subject to change. There must be a level beyond which it cannot look one way here and a different way somewhere else. It will have to be the same at all times, in all places, in all its aspects—it must be one continuous entirety. In this context, one set of basic experiences should fix the framework fully. This basic experience, to the earliest Greek philosophers, was the existence of *something*. They elevated the matter of the experience to that which *is* and denied that there might be something, even in thought, which *is not*. Expressing the same trend of thought in terms of a time sequence, this meant that *something* cannot come from *nothing*. Spatially, it means that there cannot be any such thing as empty space. The temporal, spatial, and logical unity of the world was the dominant feature of pre-Socratic thinking. It took the atomists to define the possibility of change of that which *is*, the atoms, by their regrouping inside empty space—which therefore also has a real existence.

Taking these trains of thought beyond a clashing of words, where one man's nothing is someone else's something, there remains the question about what in our world is necessary, what is merely contingent. The natural sciences of modern times have been able to base their successes on a strict separation of the laws of nature and of the initial conditions. This separation has held up against many challenges; but there are fundamental objections to our defining the laws of nature as necessary, the initial conditions as contingent.

True, treating the laws of nature and the initial conditions in these terms has led to a number of successes. This framework permits change in the world. The world's state at any given time may be contingent—that is, could just as well be different—but how one state follows from another is strictly subject to the immutable laws of nature. The world is then subject to *conditional necessity*. Its logically arbitrary state at one time fixes its actual state for all times.

There are many substantive objections to a model that divides the world into realms of coincidences on one side and basic laws on the other, where those laws would make its fate a logical necessity once the initial conditions have been fixed. These objections have transcended the purely philosophical level with the advent of the analytical methods of modern physics. Newton, in his day, never thought of the laws of mechanics that he discovered in terms of rules that govern the development of the real world around us; he regarded the world as unstable in the same way as a bunch of needles standing on their tips would be. To keep this world from collapsing, the otherwise automatic sequence of events needs remedial interference by God from time to time. In a letter Leibniz wrote to Clarke, this divine intervention is likened to that of a somewhat incompetent watchmaker, who has to repair his watches from time to time. It took another generation and the advent of the Marquis de Laplace to elevate Newton's laws, with their fairly far-reaching ability to compute important features of our world, to the level of a philosophical principle. In so doing, we might say that Laplace assigned the position of *retired engineer* to God for all time after creation.

Objections to Laplace's views assign an important part in the development of our world to chance. Leaving aside distinctions between classical and quantum mechanically random developments, we can state that as a system changes, somewhere in its development, from one coincidentally realized alternative to another by means of what mathematicians call a *bifurcation* (an unpredictable choice among two logically equivalent possibilities) chaos will ensue. If the system can develop according to the relevant laws of nature after bifurcations, structures will emerge. This is how galaxies made their appearance from the homogeneous soup after inflation; this is how life issued from another sort of soup on our Earth; how the molecules of life acquired a preferred sense of rotation; how a particular species of bird wound up with its beak bent in a given way by some mutation that was only strengthened by further development.

Shapes emerge from chance decisions and from cooperative action of various components. We might say that they appear from nowhere, rather than as images of predetermined ideal forms, such as Plato would have it. Their existence is permitted by the basic laws of nature, but not predetermined by them. Shapes as ultimately realized are contingent, not necessary. Their further development is governed by laws that originate along with them. Whether these *effective* laws follow by necessity from the fundamental laws once the new shapes have been entered as a contingent initial condition, we need not decide. Maybe so, maybe not: Effective laws cannot be uniquely deduced from the basic ones, as the relatively simple example of atomic physics amply demonstrates.

The shapes that originate from chance choices generate the laws that govern their behavior: As a dog wags its tail, the cat expects it to attack. Creation myths that add the development of shapes to the creation itself are correct when they classify this formation as a further act of creation. The physics of chaos and of the development of structures toe the same line. Taking this cue, we can say that creation did not stop with the Big Bang. In its immediate wake, the universe was structureless and homogeneous. Creation then continued. The formation of structures, the appearance of objects and of the laws that govern their specific behavior—none of this was predestined. Along with the structures, the laws might have turned out differently. Our universe and its laws are what we have called contingent. The same might be said of the laws of Ptolemaean astronomy, which took Earth for the center of the universe. We tend to distinguish some of the laws of nature from others by calling them fundamental, by assuming that they have to be the way they are in order to describe fundamental truths that are independent of our location in the universe. What right do we have to make this distinction?

James Hartle, in a remarkable article, uses the term *ballast* for conditions that apparently must be fulfilled for the laws of nature to hold—conditions that are, however, of restricted validity only. Some of the most notable achievements of science are based on recognizing this kind of ballast. In the words of Murray Gell-Mann, winner of the 1969 Nobel Prize in physics: "In my field an important new idea . . . almost always includes a negative statement, that some previously accepted principle is unnecessary and can be dispensed with. Some earlier correct idea . . . was accompanied by unnecessary intellectual baggage and it is now necessary to jettison that baggage." The transition from the worldview of Ptolemy to that of Copernicus may serve as an example. The dependency of the laws of thermal dynamics on the direction of time flow is another. Once we take into account the rather simple initial conditions that must be assumed to derive these laws, the fundamental laws of nature at their basis must no longer refer to a "direction of time." Or is the presence of an arrow of time in the truly fundamental laws of nature a necessity because the universe *had* to start out in some simple initial state that has been deteriorating ever since? As with gases in an otherwise empty space, the arrow of time shows up only because we are not able to follow the motions of individual atoms.

The apparent law of the *horror vacui* is another, slightly different example. The real law is nothing but the statement that under the conditions prevailing on the surface of Earth it is difficult, but not impossible, to produce a region without air. When Torricelli succeeded in doing so, he destroyed an ancient prejudice and replaced it with a law that is valid on Earth's surface.

What then is truly fundamental about the apparently fundamental laws of nature? Is it the fact that the constants they contain have particular numerical values? That the mass of the electron is roughly one two-thousandth of the mass of the proton? If the constants occurring in the laws of nature differed from their actual values by only small amounts, the universe could not have developed in remotely similar ways. Its specific form of creativity would be forbidden by its very laws. The universe might, for instance, be forced to contain nothing but radiation. It might have to change so rapidly that life could not have developed. And so it would go for many other contingent phenomena. It is obvious to us—as we ponder the numerical values of the constants marking the world around us—that we perforce live on a planet with the properties of Earth. On another planet—one that orbits an extremely hot star, say, in a very eccentric orbital path—intelligent life could not have developed. If we were unaware of our special situation, we might consider the laws of *our* nature, which include the specifics of this situation, as fundamental. In the same fashion, it is not a coincidence that the constants in the laws of nature have their specific values: Were these values different, we would not exist. Do we consider these laws with these numerical values fundamental merely because we don't know where they came from, what more fundamental rules they are based on? Do we think so simply because we are incapable of looking beyond the rim of the plate that is our universe?

Why should it be that only the numerical values of the constants of nature are ballast—why not the form of the laws of nature themselves? Why not the fact that there are three dimensions for space, one for time? Or the difference between space and time? We do believe that the fundamental laws of nature are those of quantum mechanics; does that mean the laws of classical physics, which describe the orbits of planets so beautifully, are nothing but ballast? Shouldn't we grant that the world is basically a quantum mechanical system—without classical features? The fact that this is apparently not so might be merely *contingent*, a consequence of the initial conditions of our specific universe.

Each of the key words in this argument is loaded with connotations—the laws of nature, the initial conditions of the universe, the universe itself, *our* specific universe. When we continue to use the word *universe* to describe all physical reality including arbitrarily complicated structures of space and time, we have to coin another expression—*our specific universe*—to denote the relatively flat part of the total universe that we inhabit. This, *our*, universe we can imagine as just a bubble inside another, all-encompassing universe; it may be connected to other more or less flat bubbles by means of tubes of highly curved space-time regions.

There is another differentiation we might introduce, along with the one between initial conditions and the basic laws of nature: This one is between the physical reality under our control and the broader reality beyond our influence. In our everyday experiments with physical reality, we can fix the initial conditions more or less as we please. The laws of nature are, by contrast, simply the way they *are*, well beyond our control. We are tempted to include this difference in the very definition of initial conditions and natural laws: In this framework, the universe as a whole is subject to its own laws, but not to any specific initial conditions. This would suggest that since we cannot initiate any experiments dealing with the universe, it makes no sense to attempt to distinguish the initial conditions from the laws that govern the universe—both are unique and inevitable. In other words, the initial conditions of the universe are part and parcel of the laws—but if this is so, how can we even ultimately apprehend these laws?

WHENCE THE QUASI-CLASSICAL WORLD?

In trying to define the notions we just discussed, in approaching questions we can barely ask but are fundamentally unlikely to solve, the words we use lose their familiar meaning. We meddle with the meaning of words in order to describe new concepts; without them we would not know how to categorize ideas on the progress of physics—ideas that transcend the language created for familiar notions. In this sense, the epilogue to this book may serve as a prologue for the physics of the future. Progress in physics is needed to meet the answers we are posing here; specifically, how will quantum mechanics and the general theory of relativity be unified into one convincing framework? Outside such a framework, there will be no understanding of the genesis of the universe. The quantity that has to be defined and explained is the *wave function of the universe*. At its inception, the universe was so small that it was subject to quantum mechanical notions—but not outside the laws of the general theory of relativity. This includes, now, that energy fluctuations linked up space and time, changing, creating, annihilating quanta of both. There is no telling whether space and time as we know them are a precondition of quantum cosmology, or whether it is but one of several possibilities at that stage. Maybe quantum mechanical fluctuations initiated not only the *stuff* our world was made of prior to inflation but also space-time itself. Maybe the true *vacuum*, the true *nothing*, of philosophy and religion should be seen as a state wholly innocent of laws, space, and time. This state can be thought of as nothing but a collection of possibilities of what might be. It might develop into arbitrarily many realizations—space with eleven dimensions, four dimensions, two dimensions. . . . But once we determine that quantum mechanics applied to the early universe, the rules were defined, and these rules established numbers for the probabilities that in its further development the universe would take specific forms in specific sequences.

In their model, Hartle and Hawking started with an initial state in which space and time were indistinguishable. This was a purely quantum mechanical initial state. We might ask whether out of this state our essentially classical world *had* to develop as the universe grew large. We might even ask: Did the universe have to become large? Or is all this nothing but ballast—a consequence of nothing but the specific initial conditions of the universe that happens to be observable to us? And again: Is our universe just a bubble inside an all-encompassing superuniverse? If there are other such bubbles, are they all quasi-classical? A purely quantum mechanical universe would clearly be uninhabitable for creatures like humans. Can we ascribe the fact that we live in a quasi-classical universe to the fact that we are here to observe it?

If the cosmological constant differed noticeably from its established range of values, life could not exist in the universe. Among the convoluted arguments that bring us to this conclusion, I will mention only this: A universe with a considerably larger cosmological constant would have developed so rapidly that life would not have had the time to develop.

But quite apart from such anthropic arguments, there remains this question: Can the development of the universe from its quantum mechanical beginnings explain the value of the cosmological constant? Sidney Coleman of Harvard University attempted, in an influential paper published in 1988, to motivate the small value of the cosmological constant in this way. His reasoning is exemplary for the way in which we can deal with the question about the development of our world from its quantum mechanical initial state. Is there a most probable sequence of shapes of our universe that determined the numerical value we find today for the cosmological constant? He answers his own question in the affirmative. To do so, he has to assume that only those shapes of the universe that permit a clear distinction between individual bubbles and the wormholes that connect them contribute to this probability (see fig. 102a). If this is correct, we can disregard the wormholes when computing the probability. Coleman takes only the bubbles into consideration but notes that they have knowledge of each other by dint of the connecting wormholes. When our universe was still small and hot, so that the individual contributions to the cosmological constant were not yet fixed, it communicated with other universes that were large, cold, with cosmological constants close to zero; and the communication was effected through wormholes. The detailed calculation shows that those large and cold universes influenced the development of ours, which was coupled to them. Those large spaces with minuscule curvature have a predominant influence on the develop-

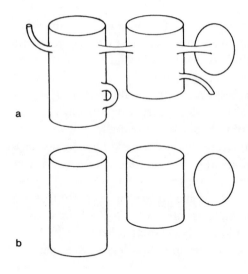

a

b

Figure 102 We suspect that at very small distances the structure of space and time is complex: Both space and time fluctuate; small regions appear, then vanish (compare with figure 84). It is possible that both space and time started out in such a fluctuation. Figure 102a shows a simplified sketch of large bubbles with relatively flat space-time that are connected by tubes: These are *baby universes* and *wormholes.* The wormholes connect the bubbles, so that in all of these we find that the same laws apply, and that the cosmological constant is the same. In the present approximation, this constant is fixed by the bubbles and independent of the wormholes. The latter can be safely removed, as in figure 102b. In Sidney Coleman's original proposal of this model, he states:

> *Even without knowing the details of the argument, we can see how wormhole dynamics has the possibility of resolving one of the questions associated with the cosmological constant, what might be called the question of prearrangement. We believe that ordinary quantum field theory describes physics from shortly after the Planck time to the current epoch. Thus, in principle, all the constants of nature could have been measured very early on, say, during an inflationary phase. How could they have known to adjust themselves so that when everything settled down the cosmological constant would be zero? Wormholes answer this by saying that, on an extremely small scale, our universe is in contact with other universes, otherwise disconnected, but governed by the same physics as ours. Even if our universe is small and hot, these can be large and cool, and see the cosmological constant. Prearrangement is replaced by precognition.*

In a recent book entitled *The Life of the Cosmos* (New York: Oxford University Press, 1997), which attracted much attention, the physicist Lee Smolin argues for a Darwinian history of universes that led to the present small value of the cosmological constant. He takes it for granted that universes are constantly created and that the laws of physics change ever so slightly from generation to generation. To this he adds as a hypothesis of his own that the smaller the value of the cosmological constant is in a particular universe, the more prolific it will be in creating new ones.

What we call Planck time is the extremely brief period after the Big Bang when the universe was so small and so densely packed with matter that its development had to follow the laws both of quantum mechanics and of gravity. It makes no sense to talk about a cosmological constant before the relevant spontaneous symmetry breaking has occurred, fixing the individual contributions that make it up.

ment of the universe, whereas the influence of small, strongly curved spaces is almost negligible. In Coleman's formulation, the flat regions see to it that there is "nothing instead of something."

MODELS

We have digressed into speculation, into new modes of thought. They go well beyond the knowledge that is theoretically compelling, much less experimentally assured. Still, they are essentially formulations of old questions couched in the terms of today's natural philosophy. We recall those questions of pre-Socratic antiquity: How large is the universe? How did it originate? What is the difference between something and nothing? How "empty" can space be? All of these questions are still open for definitive answers.

As the science of physics developed, a fundamentally faulty notion—like that of empty space—often proved to be more fruitful than the correct one would have been at the time. A mistaken notion often opens the way to understanding at a time when a correct statement—such as *there cannot be truly empty space*— could not have been understood. When we change paradigms—as in the passage from the plenum of Aristotle and the Scholastic philosophers to the empty space of classical physics and on to the quantum mechanical vacuum of physics filled with fluctuations—we also change the realms of applicability of these notions. The old ideas remain valid to some approximation. The clashing of concepts of the past with those of our century can be traced back to discussions among pre-Socratic schools—the Eleatics against the atomists—in antiquity.

It remains to be admitted that to this day we don't have a convincing notion of why there is *something*, why not merely *nothing:* What "spiritus rector breathes fire into the equations and makes a universe for them to describe?"—raising the question, What do the laws of nature permit beyond what actually exists?

NOTES

•

ix . . . *saw the world as a rather complicated mechanism* . . . Schrödinger, 1996, p. 57.

ix . . . *a book of this nature has to evoke* . . . Eddington, 1935, p. 265.

1 . . . *Physics and Metaphysics.* . . . Goethe's *Faust,* in Trunz, 1952, p. 88 (part 1).

6 . . . *identity of the indiscernibles* . . . a formulation by Ray, 1991, p. 107.

9 . . . *state of motion* . . . *state of acceleration* . . . *physical properties of space.* Einstein, 1924, pp. 87–89.

9 *The circular motion of two bodies* . . . *here is a paradox that resolved.* Kant, 1921, p. 662.

10 *The vacuum is* . . . Maxwell quoted in Hiley, "Vacuum or Holomovement," in Saunders, 1991.

10 . . . *special stuff* . . . Goethe's *Faust,* in Trunz, 1952, p. 58 (part 1).

14 . . . *a thing is symmetrical* . . . Feynman, 1963, p. 52–1.

16 *The void as such* . . . *is unable to differentiate.* Aristotle quoted in Sambursky, 1965, p. 132.

16 . . . *state of rest* . . . Newton quoted in Barbour 1989, p. 575.

27 *Aristotle's matter is certainly not* . . . *and thus to enter into physical reality.* Heisenberg, 1959, p. 120.

33 *Nothing, Nobody, Nowhere, Never.* Schmidt, 1960, p. 9.

33 . . . *chaos, from which all else issued* . . . *black-winged night is delivered of a still birth* . . . *Eros couple with the winged nocturnal chaos.* From Aristophanes' *The Birds,* in Capelle, 1953, p. 35.

33 . . . *violent* . . . *grotesque* . . . *baudily fantastic.* Capelle, 1953, p. 26.

33 . . . *deeply vortexed Okeanos.* Hesiod's formulation in Capelle, 1953, p. 27.

34 *If there is a beginning.* . . Chuang-tze, 1972, p. 46.

34 *There was chaos* . . . quote from Kojiki in Leach, 1956, p. 227.

34 *Heaven and Earth emerge* . . . *didst create them.* Augustine, 1960, p. 611.

34 . . . *one cannot even say* . . . Chuang-tzu, 1972, p. 46.

35 *In the beginning was the Word* . . . *and it dwelt among us.* Initial description of the Creation in the Gospel according to John.

35 . . . *Sense, Force, Deed* . . . Goethe's *Faust,* in Trunz, 1952, p. 44 (part 1).

35 *In the beginning, there was darkness* . . . Maori creation myth quoted in Leach, 1956, p. 172.

35 *Not a breath—not a sound . . . with light.* Tzakol the Creator and Bitol the Former, in Cordan, 1962, p. 29.

35 *. . . created Heaven and Earth . . .* Description of the creation in Genesis.

35 *I would rather judge . . . a shapeless almost-nothing.* Augustine, 1960, p. 681.

35 *But Thou, O Lord, didst create both . . . matter took shape with no time loss in between.* Augustine, 1960, p. 837.

36 *. . . the evening of October 23, 4004* B.C. Ferris, 1989, p. 185.

36 *What did God . . . had been idle and inactive before the creation?* Augustine, 1960, p. 621.

38 *. . . mighty Fiat of the Creator Spirit . . .* Pope Pius XII, 1952, p. 169.

40 *One axiom says that there are at least four . . . will be located between two others.* After Mangoldt, 1956, p. 212.

44 *. . . a space that is not taken up . . . such objects.* Aristotle quoted in Guericke, 1968, p. 61.

45 Poincaré, 1952.

47 *. . . Einstein juxtaposed these two concepts . . .* Jammer, 1992, p. xiii, in prologue by Albert Einstein.

49 *The world . . . sciences possible.* Wigner, 1949, p.3.

51 Plato, *Timaios*, in *Sämtliche Werke*, vol. 5, 1989b.

51 *. . . is not quite the same . . .* formulation by Heuser, 1992, p. 77.

51 *. . . has to atone for . . .* Anaximander quoted in Capelle, 1953, p. 82.

51 *. . . like putty . . .* Plato quoted in Dijksterhuis, 1965, p. 92.

52 *. . . the universe remains immutable . . . always remaining at rest.* Xenophanes quoted in Capelle, 1953, p. 123.

52 *You cannot step into the same river twice . . . its currents moving apart and reuniting perpetually.* Heraclitus quoted in Capelle, 1953, pp. 132–133

53 *Change alone makes not a law.* Heraclitus quoted in Cappelle, 1953, p. 133.

53 *One of these has true existence . . .* quoted in Capelle, 1953, pp. 169.

54 *Only the being . . . reality.* Parmenides paraphrased in Cappelle, 1953, p. 165.

54 *How will you invent the origin . . .* Parmenides quoted in Capelle, 1953, pp. 166–168.

55 *. . . love and discord. . .* Empedocles quoted in Capelle, 1953, pp. 197–200 (for example).

55 *The universe . . . existing elements.* Empedocles quoted in Cappelle, 1953, pp. 196, 197, 222.

56 *. . . intermingling and separation . . .* Capelle, 1953, pp. 196–197.

56 *The originators . . . denied the existence of empty space . . .* Capelle, 1953, p. 223.

56 *The ether . . . as a true vacuum . . .* formulated by Sambursky, 1965, p. 38.

56 *The physicists who see pores . . . such as air.* Aristotle quoted in Capelle, 1953, p. 223.

57 *. . . air wildly replaces . . .* Empedocles quoted in Capelle, 1953, p. 277.

57 *It seemed to me . . .* Plato quote from Phaidos, in Sambursky, 1987, pp. 82, 83.

58 *. . . we have to realize that when we take all things together . . . how could hair grow out of matter that is not hair? Flesh out of nonflesh?* Anaxagoras quoted in Capelle, 1953, pp. 261–262

58 *Among things small . . . there is always something larger.* Anaxagoras quoted in Capelle, 1953, p. 267.

60 *There is no such thing as empty space.* Anaxagoras quoted in Capelle, 1953, p. 268.

61 *Nothing can spring into existence . . . nothing can vanish into the void . . .* Sambursky, 1965, p. 147.

61 *The atoms have mass; they are not part of the void.* Sambursky, 1965, p. 155.

61 *. . . that all atoms are made up of the same basic substance and differ . . .* Sambursky, 1965, p. 150.

61 *. . . Sphere of the Being.* Formulated by Dijksterhuis, 1956, p. 9

61 *. . . certain shapes in preference to others . . . they assumed that all shapes must be possible.* Epicurus quoted in Sambursky, 1965, p. 152.

61 *The former takes them . . . macroscopic size.* Capelle, 1953, p. 401.

62 *There is no more existence . . . in its existence as matter.* Capelle, 1953, p. 293.

62 *Only seemingly does a thing have color . . . only with atoms and empty space.* Democritus quoted in Capelle, 1953, p. 399.

65 *Democritus took . . . could possibly have created it.* Capelle, 1953, p. 415.

65 *Democritus was one of the philosophers who took the cause of the heavens . . . they brought order to matter . . .* Capelle, 1953, p. 413.

65 *Some worlds are still growing . . . they disappear.* Democritus quoted in Capelle, 1953, p. 416.

66 *The atomist identification of real . . . in paradox and enigma.* Grant, 1981, p. 3.

67 *The Pythagoreans . . .* Aristotle quoted in Robinson, 1968, p. 75.

69 *. . . is artfully shaped such that it acts and suffers everything by its own devices.* Plato, 1989b, p. 157.

69 *. . . intended . . . the most beautiful of all ideas . . . thus he created this visible, living world.* Plato, 1989b, p. 156.

69 *. . . movable image of immortality.* Plato, 1989b, p. 156.

73 *Every body . . . upon it.* Newton, *Principia,* quoted in Cohen, 1985, p. 152.

73 *Nobody can give a reason . . . unless it is hindered from doing so.* Aristotle quoted in Sambursky, 1975, p. 103.

73 *. . . they do so in relation . . .* Aristotle quoted in Sambursky, 1975, p. 104.

73 *At the same time, it is clear that beyond . . . is no body there, but there could be one.* Aristotle, 1987, p. 86.

75 *. . . a place that contains no body, but that could contain one.* Aristotle quoted in Guericke, 1968, p. 61.

75 *There is no getting around the alternative . . . no empty space.* Aristotle, 1989, p. 101.

76 *. . . circular motion is the only motion . . . .* Aristotle paraphrased in Sambursky, 1975, p. 111.

76 *A fixed location is defined as the first immobile limit . . .* Descartes quoted in Dijksterhuis, 1965, p. 41.

76 *. . . shape nor form . . . containing body.* Aristotle quoted in Grant, 1981, p. 5.

78 *. . . the inner limit of a larger body surrounding it.* Dijksterhuis, 1965, p. 41.

78 *And thus we can speak of the Earth . . . is not further surrounded.* Aristotle, 1989, p. 94.

80 *If I am at the extremity . . .* Archytas quoted in Grant, 1981, p. 106.

80 *Let's assume, for now . . .* Lucretius quoted in Harrison, 1984, p. 166.

83 *Furthermore, the universe consists of . . .* Epicurus quoted in Sambursky, 1975, pp. 117–119.

83 *On the Nature of Things . . . De rerum natura,* Lucretius cited in Lukrez, 1973.

84 *First, let me make sure . . .* Lucretius quoted in Jürss, 1977, p. 381.

84 *We cannot say that bodily objects fill space all over . . .* Lucretius quoted in Jürss, 1977, p. 388.

84 *In the beginning of all rational* . . . Lucretius quoted in Jürss, 1977, p. 382.

84 *Let me add, that we cannot find* . . . Lucretius quoted in Jürss, 1977, p. 390.

84 *. . . but space . . . could manifest itself to our senses or to our minds.* Lucretius quoted in Jürss, 1977, p. 391.

85 *Maybe space itself . . . seeks the location and the correlations for which it is destined.* Theophrastus quoted in Sambursky, 1965, p. 380.

86 *. . . there cannot be . . . that is indeed so.* Strato quoted in Simonyi, 1990, p. 85.

86 *We have thus proven* . . . Strato quoted in Simonyi, 1990, p. 86.

87 *Its essence can be described* . . . Plotinus quoted in Dijksterhuis, 1965, p. 53.

87 *The very idea of matter implies absence of form.* Plotinus, 1991, p. 93.

87 *Absolute matter must take its magnitude* . . . Plotinus, 1991, p. 101.

87 *Matter . . . is understood to be a certain base* . . . Plotinus, 1991, p. 92.

87 *There is, therefore, a matter* . . . Plotinus, 1991, p. 94.

87 *. . . ready to become anything* . . . Plotinus, 1991, p. 98.

87 *. . . becomes a thing of multiplicity* . . . Plotinus, 1991, p. 101.

87 *The distinctive character* . . . Plotinus, 1991, p. 104.

87 *Matter is therefore* . . . Plotinus, 1991, p. 10.

88 *. . . ending with the grin* . . . Carroll, 1962, p. 9.

88 *Plotinus's idea exceeded . . . those of Plato's school.* Dijksterhuis, 1965, p. 54.

89 *. . . non-being . . . to attain.* Amaldi, 1991, p. 20.

89 *. . . perennial longing to enter into reality.* Plotinus quoted in Dijksterhuis, 1965, p. 53.

89 *. . . space as container* . . . Jammer, 1960, p. xiv, Albert Einstein's foreword.

90 *Space has a threefold significance . . . comprised by nothing.* Philo quoted in Sambursky, 1965, p. 382.

92 *. . . not filled by bodies . . . Either the sides of the sky . . . vacuum.* Albert of Saxony quoted in Grant, 1974, pp. 324–325.

93 *Arguments demonstrating* . . . Grant, 1981, p. 357.

93 *As for the physical extent* . . . Newton, 1988, p. 35.

94 *. . . if space . . . infinite.* Grosseteste quoted in Crombie, 1971, p. 99.

94 *. . . not a flow* . . . Bacon quoted in Crombie, 1971, p. 146.

94 *There exists space* . . . Nicholas of Autrecourt quoted in Crombie, 1959, p. 275.

94 *. . . "dense" and "rare"* . . . Nicholas of Autrecourt quoted in Grant, 1981, p. 74.

97 *Problems with Nothingness.* Goethe's *Faust* in Trunz, 1952, p. 48 (part 1).

100 *. . . without permitting the air to escape. . . taken up by water. . . in the balloon.* Galileo, 1985, p. 72.

100 *. . . we want to study the motion . . . they will vanish in an actual vacuum.* Galileo, 1985, p. 65.

101 (Figure 29 caption) *Mr. Desaguliers shew'd the experiment* . . . Guerlac, 1977, p. 126.

104 *Not even twenty horses* . . . Buridan quoted in Grant, 1981, p. 82.

106 *We take it for granted that. . . imply that these insights have actually been held.* Dijksterhuis, 1956, p. 82.

108 *If a light vessel . . . filled.* Hero quoted in Cohen, 1948, p. 250.

108 *The particles of air . . . between the particles of air.* Heron quoted in Cohen, 1948, p. 250.

108 *Bodies . . . in contact.* Hero quoted in Cohen, 1948, p. 250.

108 *When wine . . . the water.* Cohen, 1948, p. 253.

109 *We must give up the view* . . . Gassendi quoted in Sambursky, 1975, p. 331.

110 *. . . space cannot act . . . whether the mind thinks of them or not.* Gassendi quoted in Grant, 1981, pp. 209–210.

112 *Many people have said* . . . . Torricelli quoted in Sambursky, 1975, p. 337.

115 *Above the mercury* . . . Pascal quoted in Gerlach, 1967, p. 100.

116 *. . . empty space is as absurd as happiness without a sentient being who is happy.* Descartes quoted in Russel, 1965, p. 87.

120 *. . . useless creations of God* . . . Noel paraphrased in in Mehlig, 1971, p. 49.

120 *This experiment will probably show* . . . Pascal quoted in Mehlig, 1971, p. 89.

121 *Does nature abhor the vacuum* . . . Pascal quoted in Sambursky, 1975, p. 345.

121 *Everything in existence* . . . Guericke, 1968, pp. 69–70.

121 *. . . more precious than gold* . . . Guericke, 1968, p. 70.

122 *Scholars whose conclusions are* . . . Guericke, 1968, pp. xxii–xxii.

124 *. . . barely be moved by two powerful helpers.* Guericke, 1968, p. 83.

125 *. . . pikes spit out smaller fish.* Guericke, 1968, p. 103.

127 *It is the weight of the air that causes the horror vacui . . . is governed by an external condition.* Guericke, 1968, p. 109.

127 *. . . an experimental result that I did not at all expect.* Guericke, 1968, p. 109.

127 *We see that air pressure* . . . Guericke, 1968, p. 95.

131 *Most assuredly, Descartes was a great mathematician* . . . Hund, 1978a, p. 106.

132 *It is obvious that there cannot be* . . . Descartes quoted in Heidelberger, 1981, p. 175.

134 *Galileo should first have determined* . . . Descartes quoted in Damerow, 1992, p. 43.

136 *. . . called a mechanical or mechanistic approach.* Dijksterhuis, 1956, p. 1.

137 *Even though the world is not infinite* . . . Nicholas of Cusa quoted in Koyré, 1980, p. 21.

138 *. . . palace of bliss* . . . Digges quoted in Koyré, 1980, p. 44.

138 *Our world is an infinite sphere* . . . Pascal quoted in Reeves, 1989, p. 124.

138 *There are many who speak of a finite world* . . . Bruno, 1968, p. 53.

139 *Bruno's space is empty* . . . Koyré, 1980, p. 47 (footnote).

139 *In this case, if something can be, it must be.* Koyré, 1980, p. 54.

139 *The medieval fear* . . . Grant, 1981, p. 191.

139 *. . . those immense and varied lights . . . maybe even by the void . . . the Moon is below the horizon.* After Gilbert as quoted in Koyré, 1980, pp. 60–61.

140 *. . . bears some hidden terror* . . .. Kepler quoted in Koyré, 1980, p. 65.

140 *There is no firm knowledge we have . . . them.* Kepler quoted in Koyré, 1980, p. 79.

141 *If anyone asks what would occur if God* . . . Descartes, 1991, p. 48.

145 *Matter in the Void.* Leibniz's third letter to Clarke. Quoted in Schüller, 1991, p. 40.

146 *. . . a tough imposition.* Einstein, 1985, p. 107.

147 *Shut yourself up* . . . Galileo as quoted by Barbour 1989, p. 395.

149 *The causes by which* . . . Newton quoted in Barbour, 1989, p. 626.

153 *. . . distancing himself from that of an ether.* Gernot Bohme in Newton, 1988, p. 9.

153 *Water resists the motion of a projectile* . . . Newton, 1988, p. 73.

154 *If motion is nothing* . . . Leibniz quoted in Jammer, 1960, p. 130.

154 *. . . an ordering of things that exist next to one another.* Leibniz quoted in Schüller, 1991, p. 27 and 37.

154 *. . . more matter affords God more opportunity to exert his wisdom and his might.* Leibniz quoted in Schüller, 1991, p. 27.

154 *No matter how small* . . . Clarke quoted in Schüller, 1991, p. 32.

154 . . . *every substance that has been created* . . . Leibniz quoted in Schüller, 1991, p. 40.

154 *All the reasoning* . . . *hair-splitting.* Leibniz quoted in Schüller, 1991, p. 61.

154 . . . *a difference* . . . *with respect to other objects* . . . Leibniz quoted in Schüller, 1991, p. 98.

154 *I'm not saying that matter* . . . *as such has no absolute reality.* Clarke quoted in Schüller, 1991, p. 102.

154 . . . *death kept Mr. Leibniz from replying to my most recent letter.* Clarke quoted in Schüller, 1991, p. 163.

158 *I frame no hypotheses.* Newton quoted in Barbour 1989, p. 622.

159 *The Mechanization of the World Picture.* Dijksterhuis, 1956.

159 *To me* . . . *carrier.* Newton quoted in Koyré, 1980, p. 163.

159 . . . *all material objects* . . . Bentley quoted in Koyré, 1980, p. 167.

159 . . . *some enthusiasm* . . . *consists largely of empty space.* Koyré, 1980, p. 165.

159 . . . *issues from an immaterial and divine source.* Bentley quoted in Koyré, 1908, p. 167.

159 *But if matter is uniformly distributed* . . . Newton quoted in Koyré, 1980, p. 168.

163 *And against filling the Heavens* . . . Newton, 1979, p. 368.

163 *Is not the Heat* . . .*?* Newton, 1979, p. 349.

164 . . . *much after the manner* . . *force of gravity.* Whittaker, 1989, p. 19.

166 . . . *so close to the observed velocity of light* . . . *the laws of electromagnetism.* Maxwell quoted in Sambursky, 1975, p. 567.

170 . . . *it acts, but it cannot be acted upon. This runs counter to scientific reason.* Einstein quoted in Berry, 1989, p. 38.

175 . . . *the spatial character* . . . *without a field.* Albert Einstein's foreword to Jammer, 1960, p. xv.

179 *Movement All Around.* Buchner, 1958, p. 67.

186 . . . *Lamoreaux* . . . *performed* . . . Lamoreaux, S. K. "Demonstration of the Casimir Force in the 0.6 to 6mμ Range," *Phys. Rev. Lett.* 78, no. 5 (1996).

193 *The first actual transformation* . . . Phys. Rev. Lett. *79* 1626 (1977).

204 *Seeing things your way* . . . Einstein, 1991, p. 270.

206 (Figure 61 caption) *Phys. Rev. Lett. 79,* 1626 (1997).

209 *Spontaneous Creation.* Goethe's *Faust,* in Trunz, 1952, p. 245 (part II).

217 *If no particular state* . . . *alongside of space.* Einstein, 1934a, p. 68.

227 *Let Nature Be as She May.* Goethe's *Faust,* in Trunz, 1952, p. 306 (part II).

228 *The scientist has an obligation* . . . Waldgrave quoted in after D.D. 1993.

232 *Those near the door hear of it first* . . . David Miller quoted in, *Physics World, 1993.*

242 (Figure 76 caption) *A recent experiment* . . . Levine, 1997.

257 *Nothing Is Real.* From the Beatles song "Strawberry Fields Forever." Aldridge, 1969, p. 39.

285 *Suppose two mutually exclusive objects* . . . Schüller, 1991, p. 55. *Equilibrium is never indifferent* . . . Schüller, 1991, p. 169. . . . . *that God would have bothered* . . . Schüller, 1991, p. 173.

297 *The possibility is not absurd* . . . Rees, 1997, p. 217.

302 *If there is no quasi-static world* . . . Einstein quoted in A. Pais, *Subtle Is the Lord* (Clarendon Press, 1982), p. 288.

307 . . . *space to "confer inertia."* Einstein, 1924, p. 88.
307 . . . *ether of mechanics* . . . Einstein, 1924, p. 87.
309 . . . *retired engineer* . . . Dijksterhuis, 1956, p. 549.
314 *Even without knowing* . . . Coleman, 1988, p. 646.
315 . . . *nothing instead of something.* Coleman, 1988, p. 633.
315 . . . *breathes fire* . . . Hawking, 1996, p. 232.

FIGURE SOURCES

•

Figures 1, 2, 5, 9, 10, 15, 21, 23a (after Simonyi 1990), 48, 76a, 79, 92: drawings by Jutta Winter, Oberammergau. First printed in Genz 1984a. Courtesy of Verlags R. Piper, Munich.
Figures 3, 11, 13, 14, 37, 41: Sexl 1980, vol. 1.
Figures 4, 17, 18b, 20a–c, 22a–c, 23b–c, 38, 39, 40a–b, 49, 51a–d, 54, 55a–c, 56a–b, 58a–b, 60, 61a–d, 62a–b, 64a–e, 73a–c, 76b–c, 77a–b, 78a–d, 80a–b, 82, 84, 85a–b, 86a–b, 87, 90, 95, 96: drawings by Fritz E. Urich, Munich.
Figure 6: Science Photo Library.
Figures 7, 8, 81, 89, 98a: Genz 1987.
Figures 12, 38: Lederman 1989.
Figures 16: MacLagan 1977.
Figure 18a: Drawing after Stückelberger 1979 by Fritz E. Urich, Munich.
Figure 19: computer drawing by D. Dunham, University of Minnesota, Duluth.
Figure 24a: Neuser 1990.
Figure 24b, 55: Harrison 1981.
Figure 24c, 95: Fang Li Zhi 1989.
Figure 25: drawing by John Tenniel. Carroll 1962.
Figure 27, 42b: Burke 1978.
Figure 28: Galilei 1985.
Figure 29, 67, 78a, 96: Ballif 1969.
Figure 30: Stückelberger 1979.
Figure 31: Grant 1981.
Figure 32a–b, 35: Fraunberger 1984.
Figure 33: Hund 1987a. Figure 33a drawn by Fritz E. Urich, Munich.
Figure 34: Wußing 1987.
Figure 36: Heidelberger 1981.
Figure 42a: Krafft 1978.
Figure 42b–h: Mulder 1986.
Figure 43: Harré 1981.
Figure 44a–d: Descartes 1991.
Figure 45: Koyré 1957.
Figure 50: Sambursky 1974.
Figure 52: Rindler 1977.

Figure 53: Wheeler 1990.

Figure 54c–d: M. Malvetti, Karlsruhe University.

Figure 63: Carreras 1983.

Figure 66: Genz 1982.

Figure 67: Sexl 1980, vol. 3.

Figure 68a–b: Pohl 1957.

Figure 68c–e: Segrè 1984.

Figure 69: Scientific Information, CERN, Genf.

Figure 70a, 71, 72: DELPHI-Sekretariat CERN, Genf.

Figure 70b: after Schopper 1989.

Figure 74: Experiment NA35 at the CERN SPS.

Figure 75: Nigel Hawtin, *New Scientist,* 17 August 1991.

Figure 83: after Hawking 1996.

Figure 91: Genz 1991.

Figure 93, 94: Mulvey 1981.

Figure 97, 99: Parker 1988.

Figure 98b–c: Trefil 1983.

Figure 98d: Riordan 1991.

Figure 100: Gribbin 1989.

Figure 101: Hale Observatories.

Figure 102: Coleman 1988.

BIBLIOGRAPHY

•

Abbott, Larry. The mystery of the cosmological constant, *Scientific American* 258 (May 1988):82.

Aitchison, Ian. 1985. Nothing's plenty—The vacuum in modern quantum field theory. *Contemporary Physics* 26: 333–391.

———. 1969. The unbearable heaviness of being, *Physics World*, 2 (July 1989): 29.

Aldridge, Alan, ed. 1969. *The Beatles illustrated lyrics.* New York.

Amaldi, Ugo. 1991. The physical vacuum and the metaphysical nothing. Proceedings of the Royal Society, January 1991; and CERN-Preprint DELPHI 91-07 PHYS 83, 12 February 1991.

Amaldi, U. et al. Phys. Lett. B260, 447 (1991).

Anonymous. Where does the zero-point energy come from? *New Scientist* 2 (December 1989): 36.

Aristotle. 1977. *Die Hauptwerke.* Stuttgart.

———. 1989. *Physikvorlesung.* Berlin.

———. 1987. *Vom Himmel.* Munich: Olof Gigon.

Arnol'd, V.I. 1990. Huygens & Barrow, Newton & Hooke. Basel.

Arp, Halton C. 1993. Der kontinuierliche Kosmos. In Ernst Peter Fischer, ed. *Neue Horizonte 92/93.* Munich.

Augustine. 1960. *Confessiones—Bekenntnisse.* Munich.

Baeyer, Hans Christian von. 1992. *Taming the atom.* New York: Random House.

Ballif, Jae R., and William E. Dribble. 1969. *Conceptual physics.* New York: John Wiley & Sons.

Barbour, Julian B. 1989. *Absolute or relative motion.* Cambridge Univ. Press.

Barrow, John D., and Frank J. Tipler. 1986. *The anthropic cosmological principle.* Oxford.

Barrow, John D. 1988. *The world within the world.* Oxford.

———. 1991. *Theories of everything.* Oxford: Clarendon Press.

———. 1992. *Pi in the sky.* Oxford.

Berry, M. 1989. *Principles of cosmology and gravitation.* Bristol: Adam Hilger.

Börner, G. Das Echo des Urknalls, *Physikalische Bltter* 48, no. 6 (1992): 464.

Boyer, Timothy H. The classical vacuum, *Scientific American* 253 (August 1985): 56.

Brecht, Bertolt. 1961. *Flüchtlingsgespärche.* Berlin.

Breuer, Reinhard. 1984. *Das anthropische Prinzip.* Frankfurt.

———, ed. 1993. *Immer Ärger mit dem Urknall.* Reinbek.

Bruno, Giordano. 1968. *Zwiegespräch vom unendlichen All und den Welten.* Darmstadt.

Brush, Stephen G. How cosmology became a science, *Scientific American* 267 (August 1992): 34.

Büchner, Georg. 1958. Dantons Tod. In *Werke und Briefe*. Frankfurt.

Burke, D. L. et al. *Phys. Rev. Lett.* 79, 1626 (1997).

Burke, James. 1978. *Connections*. Boston.

———. 1985. *The day the universe changed*. Boston.

Bynum, W. F., E. J. Browne, and Roy Porter, eds. 1983. *Macmillan dictionary of the history of science*. London.

Cantor, G. N., and M. J. S. Hodge, eds. 1981. *Conceptions of ether*. Cambridge.

Capelle, Wilhelm. 1953. *Die Vorsokratiker*. Stuttgart.

Carreras, Rafel, and Guy Hentsch. 1983. *How energy becomes matter*. CERN Genf.

Carroll, Lewis. 1962. *Alice's adventures in wonderland and through the looking glass*. Penguin Books.

Cohen, I. Bernhard. 1985. *The birth of a new physics*. New York and London: W.W. Norton.

Cohen, Morris R., and I. E. Drabkin. 1948. *A source book in Greek science*. New York.

Coleman, Sidney. Why there is nothing rather than something: A theory of the cosmological constant. *Nucl. Phys.* B310, 643 (1988).

Cordan, Wolfgang, ed. and trans. 1962. *Das Buch des Rates Popol Vuh*. Düsseldorf-Köln.

Cotterell, Arthur. 1989. *The illustrated encyclopedia of myths & legends*. Marshall Editions Limited.

Crabb, Charlene. Casimir and Polder were right, *Discover* (January 1994): 102.

Crombie, A. C. 1959. *Augustine to Galileo*. London and Cambridge, Mass.

———. 1971. *Robert Grosseteste and the origins of experimental science*. Oxford.

Crosland, Maurice. 1992. *The science of matter*. Chemin de la Sallaz.

D. D. Simplify science? I'll drink to that, *Nature* 362, 781 (1993).

Damerow, Peter, Gideon Freudenthal, Peter McLaughlin, and Jürgen Renn. 1992. *Exploring the limits of preclassical mechanics*. New York.

Davies, P. C. W. 1982. *The accidental universe*. Cambridge.

Davies, P. C. W., and J. R. Brown, eds. 1986. *The ghost in the atom*. Cambridge.

Davies, Paul. 1991. What are the laws of physics? In John Brockman, ed. *Doing science*. New York.

Davies, Paul, ed. 1989. *The new physics*. Cambridge.

Davies, Paul. 1992a. *The mind of God*. New York.

Davies, Paul, and John Gribbin. 1992b. *The matter myth*. New York.

Descartes, René. 1991. *Principles of philosophy*. Dordrecht.

Dewdney, A. K. 1984. *The planiverse*. London.

Dickson, David. Boson competition lifts prospects for LHC, *Nature* 365, 96 (1993).

Diels, Hermann. 1951. *Die Fragmente der Vorsokratiker*, vol. 1. Berlin.

Dijksterhuis, E. J. 1956. *Die Mechanisierung des Weltbildes*. Berlin.

Dirac, P. A. M. 1978. In H. Hora and J. R. Shepanski. *Directions in physics*. New York.

Dirlmeier, Franz, ed. 1949. *Platon Phaidon*. Munich.

Dschuang Dsi. 1972. *Das wahre Buch vom südlichen Blumenland*. Düsseldorf-Köln.

Earman, John. 1989. *World enough and space-time*. Cambridge.

Eckstein, Franz. 1955. *Abriß der griechischen Philosophie*. Frankfurt.

Eddington, A. S. 1935a. *New pathways in science*. Cambridge: Cambridge Univ. Press.

Einstein, A. Die Grundlage der allgemeinen Relativitätstheorie, *Ann. D. Phys.* 49, 81 (1916).

———. Über den Äther, *Schweizerische naturforschende Gesellschaft Verhandlungen* 105, 85 (1924).

———. 1934a. *Essays in science*. New York: Philosophical Library.

———. 1985. *Über die spezielle und die allgemeine Relativitätstheorie*. Braunschweig.

Einstein, Albert, and Max Born. 1991. *Briefwechsel 1916–1955*. Munich.

Englert, Berthold-Georg, and Herbert Walther. Komplementarität in der Quantenmechanik, *Physik in unserer Zeit* n. 5 (1992): 213.

Fahr, Hans Jörg. Der Begriff Vakuum und seine kosmologischen Konsequenzen, *Naturwissenschaften* 76, 318 (1989a)

———. 1989b. The modern concept of vacuum and its relevance for the cosmologicals models of the universe. In Paul Weingartner and Gerhard Schulz, ed. *Philosophie der Naturwissenschaften*. Vienna.

———. 1992. *Der Urknall kommt zu Fall*. Stuttgart.

Fang Li Zhi, and Li Shu Xian. 1989. *Creation of the universe*. Singapore.

Farmer, Penelope, ed. 1979. *Beginnings—Creation myths of the world*. New York.

Fauvel, J. et al., ed. 1988. *Let Newton be!* Oxford: Oxford University Press.

Ferris, Timothy. *Coming of age in the milky way*. 1989. New York: Anchor.

Feynman, Richard P. 1965. *The character of physical law*. Cambridge: The MIT Press.

Feynman, Richard P. et al. 1963. *The Feynman lectures on physics*. Reading: Addison-Wesley.

Fierz, Markus. Über den Ursprung und die Bedeutung der Lehre Isaac Newtons vom absoluten Raum. *Gesnerus* 11, 62 (1954).

———. 1988. *Naturwissenschaft und Geschichte*. Basel.

Frauenberger, Fritz, and Jürgen Teichmann. 1984. *Das Experiment in der Physik*. Braunschweig.

Fritzsch, Harald. 1983. *Quarks*. Harmondsworth: Penguin.

———. 1983. *Vom Urknall zum Zerfall*. Munich.

Galilei, Galileo. 1982. *Dialog über die beiden hauptsächlichsten Weltsysteme*. Stuttgart.

———. 1985. *Unterredungen und mathematische Demonstrationen über zwei neue Wissenszweige die Mechanik und die Fallgesetze betreffend*. Darmstadt.

Genz, H., F. Kaiser, and H. M. Staudenmaier. 1982. *Klassiche Teilchen und Wellenpakete 2: Resonanz und Tunneleffekt*. Film, series 7, no. 34. IWF Göttingen.

———. 1983. Quantenmechanische Interferenzen. Film, series 8, no. 19. IWF Göttingen.

Genz, Henning. 1984. Natur, Naturgesetze—und Bilder. In Sigmar Holsten, *Kosmische Bilder in der Kunst des 20. Jahrhunderts*. Kunsthalle Baden-Baden.

———. 1992. Symmetrie—Bauplan der Natur. Serie Piper.

———. Symmetrie und Symmetriebrechung in den Naturwissenschaften, insbesondere der Physik, *Naturwissenschaften* 75, 432 (1988).

———. 1989. Symmetries and their breaking. In Paul Weingartner and Gerhard Schulz, eds. *Philosophie der Naturwissenschaften*. Vienna.

———. Über Wägen, Massen, Gewichte und fünfte Kraft, *Die Weltwoche* 26 (July 1990).

Genz, Henning, and Roger Decker. 1991. *Symmetrie und Symmetriebrechung in der Physik*. Braunschweig.

Genz, Henning. 1994a. Etwas und Nichts—Die Symmetrien des Vakuums und der Welt. In Ernst Peter Fischer, ed. *Neue Horizonte 93/94*. Munich.

————. 1994b. Gestalt und Bewegung in der Quantenmechanik. In C. L. Hart-Nibbrig, ed. *Was heißt darstellen?* Frankfurt.

Gerlach, Walter. Das Vakuum in Geistesgeschichte. Naturwissenschaft und Technik, *Phys. Blätter* 22, sect. 2 (1967): 97.

Godfrey, Laurie R., ed. 1983. *Scientists confront creationism.* New York.

Goss Levi, Barbara. COBE measures anisotropy in cosmic microwave background radiation, *Physics Today* (June 1992): 17.

————. New evidence confirms old predictions of retarded forces, *Physics Today* (April 1993): 18.

Grant, Edward, ed. 1974. *A source book in medieval science.* Cambridge, Mass.

————. 1977. *Physical science in the middle ages.* Cambridge: Cambridge University Press.

————. 1980. *Das physikalische Weltbild des Mittlelalters.* Zürich/München.

————. 1981. *Much ado about nothing.* Cambridge.

Greenberg, Jack S., and Walter Greiner. Search for the sparking of the vacuum, *Physics Today* (August 1982): 24.

Greiner, Walter, and Joseph Hamilton. Is the vacuum really empty? *American Scientist* 68, p. 134

Greiner, Walter, and Heinrich Peitz. Ist das Vakuum wirlklich leer?, *Physik in unserer Zeit* n. 6 (1978): 165.

Greiner, Walter. 1980. *Is das Vakuum wirklich leer?* Weisbaden.

Greiner, Walter et al. 1986. *Elementare Materie, Vakuum und Felder.* Heidelberg.

Gribbin, John. Recreating the birth of the Universe, *New Scientist* 17 (August 1991): 31.

Gribbin, John, and Martin Rees. 1989. *Cosmic coincidences.* New York: Bantam Books.

Guericke, Otto von. 1968. *Neue Magdeburger Versuche über den Leeren Raum.* Düsseldorf.

Guerlac, Henry. 1977. *Essays and papers in the history of modern science.* Baltimore.

Gutbrod, H. H., and R. Stock. Suche nach Quark-Materie: Elementarteilchenphysik mit Schweren Ionen, *Phys. Blatter* 43, n. 3 (1987): 136.

Halliwell, Jonathan, J. Quantum cosmology and the creation of the universe, *Scientific American* 265 (December 1991): 28.

Harrison, Edward Robert. 1981. *Cosmology, the science of the universe.* Cambridge: Cambridge University Press.

Haroche, Serge, and Daniel Kleppner. Cavity quantumelectrodynamics, *Physics Today* (January 1989): 23.

Haroche, Serge, and Jean-Michael Raimond. Cavity quantum electrodynamics, *Scientific American* 268 (April 1993): 26.

Harré, Rom. 1981. *Great scientific experiments.* Oxford.

Harrison, Edward Robert. 1981. *Cosmology, the science of the universe.* Cambridge University Press.

————. Newton and the infinite universe, *Physics Today* (February 1986).

Hartle, James B. Classical physics and Hamiltonian quantum mechanics as relics of the big bang. Lecture at the Nobel-Symposium on "The birth and early evolution of the universe, " Gräftavalem, Sweden, June 15, 1990. Preprint UCSBTH-90-53 of the Department of Physics, University of California, Santa Barbara.

————. 1991. Excess baggage. In John H. Schwarz, ed. *Elementary particles and the universe—Essays in honor of Murray Gell-Mann.* Cambridge.

Hawking, Stephen. On the rotation of the universe, *Mon. Not. R. Astr. Soc.* 142, 129 (1969).

Hawking, S. W. 1984. Quantum cosmology. In Bryce S. Dewitt and Raymond Stora, eds. *Les Houches XL 1983, Relativity, groups and topology*. Amsterdam.

———. The boundary conditions of the universe, *Pontificiae Academiae Scientarium Scripta varia* 48, 563 (1982).

———. The quantum mechanics of the universe. In G. Setti and L. Van Hove, eds. *First ESO-CERN Symposium Proceedings*, CERN, Genf (November 1983).

———. 1989. *The edge of spacetime*. In Davies.

———. 1996. *The illustrated a brief history of time*. London: Bantam Press.

Heidelberger, Michael, and Sigrun Thiessen. 1981. *Natur and Erfahrung*. Reinbek.

Heisenberg, Werner. 1959. *Physik und Philosophie*. Frankfurt.

———. 1984. *Schritte über Grenzen*. Munich.

Hermann, Armin. 1978. *Lexikon der Geschichte der Physik*. Köln.

Herrmenn, Joachim. 1979. *dtv-Atlas zur Astronomie*. Munich.

Heuser, Harro. 1992. *Als die Götter lachen lernten*. Munich.

Hiley, B. J. 1991. Vacuum or Holomovement. In Saunders, p. 221.

Hofstadter, Douglas R. 1980. *Gödel, Escher, Bach: An eternal golden braid*. New York: Vintage Books.

Honigswald, Richard. 1957. *Vom erkenntnistheoretischen Gehalt alter Schöpfungserzählungen*. Stuttgart.

Hörz, Herbert et al., eds. 1991. *Philosophie und Naturwissenschaften*. Berlin.

Hügli, Anton, and Poul Lübcke, eds. 1991. *Philosophielexikon*. Reinbek.

Hund, Friedrich. 1978a. *Geschichte der physikalischen Begriffe*, Teil 1. Mannheim.

Hund, Friedrich. 1978b. *Geschichte der physikalischen Begriffe*, Teil 2. Mannheim.

Hutter, K., ed. 1989. *Die Anfänge der Mechanik*. Berlin.

Isham, C. J. 1988. Creation of the universe as a quantum process. In Robert John Russell et al., eds. *Philosophy physics, and theology*. Vatican Observatory, Vatican City State.

———. 1991. Conceptual and geometrical problems in quantum gravity. In H. Mitter and H. Gausterer, eds. Recent *aspects of quantum fields*. Berlin.

Jammer, Max. 1960. *Das Problem des Raumes*. Darmstadt.

Jammer, Max. 1969. *Concepts of space*. Cambridge: Harvard University Press.

Jürβ, Fritz, Reimar Müller, and Ernst-Günther Schmidt, eds. 1977. *Griechische Atomisten*. Leipzig.

Kant, Immanuel. 1960. Von dem ersten Grunde des Unterschiedes der Gegenden im Raume. In Immanuel Kant. *Werke in sechs Bänden*, vol. 1. Wiesbaden.

———. 1921. Metaphysische Anfangsgründe der Naturwissenschaft. In Immanuel Kants *sämtliche Werke in sech Bäden*, vol. 4. Leipzig 1921.

Khoury, Adel Theodor, and Georg Girschek. 1985. *So machte Gott die Welt*. Freiburg.

Kleppner, Daniel. With apologies to Casimir, *Physics Today* (October 1990): 9.

Koyré, Alexandre. 1957. *From the closed world to the infinite universe*. Baltimore: John Hopkins Press.

———. 1968. *Galileo*. London: Chapman & Hall.

Krafft, Fritz. 1978. *Otto von Guericke*. Darmstadt.

Krauss, Lawrence M. 1989. *The fifth essence—The search for dark matter in the universe*. New York.

Kunz, Jutta. 1990. *Dunkle Materie im Universum*. Oldenburger Universitätsredem Nr. 40. Odenburg.

Leach, Maria. 1956. *The beginning—Creation myths around the world*. New York.

Lederman, Leon M., and David N. Schramm. 1989. *From quarks to the cosmos*. New York.

Lederman, Leon, and Dick Teresi. 1993. *The God particle*. New York: Houghton Mifflin Company.

Leibniz, Gottfried Wilhelm. 1985. *Philosophische Schriften*, vol. II, 2nd half, Herbert Herring, ed. and trans. Darmstadt.

Leslie, J. 1988. How to draw conclusions from a fine-tuned universe. In Robert John Russell et. al. *Philosophy, physics, and theology*. Vatican Obervatory, Vatican City State.

Levine, I. et al. *Phys. Rev. Lett.* 78, 424 (1997).

Linde, Andrei. 1990. *Particle physics and inflationary cosmology*. Harwood Chur.

Luminet, Jean-Pierre. 1992. *Black holes*. Cambridge.

Lukrez. 1973. *De rerum Natura—Welt aus Atomen*. Stuttgart.

Mach, Ernst. 1988. *Die Mechanik in ihrer Entwicklung*. Darmstadt.

Mac Lagan, David. 1977. *Creation myths—Man's introduction to the world*. London: Thames and Hudson.

Maddox, John. The case for the Higgs boson, *Nature* 362, 785 (1993).

Mangoldt, H. V., and Konrad Knopp. 1956. *Einführung in die höhere Mathematik*, vol. 1. Stuttgart.

Maxwell, J. C. 1954. *A treatise on electricity and magnetism*. New York: Dover.

May, Gerhard. 1978. *Schöpfung aus dem Nichts*. Berlin.

Mehlig, Rudolf. 1971. Die Überwindung der mittelalterlichen Tradition in der Geschichte des Vakuums im Werk Blaise Pascals. Dissertation. Frankfurt.

Meyers grosses Universallexikon. 1981. Mannheim.

Milonni, Peter W., and Mei-Li Shih. Casimir forces, *Contemporary Physics* 33, 313 (1992).

———. 1994. *The quantum vacuum*. Boston.

Mulder, Theo. 1986. *Otto von Guericke—Leben und Werk*. Köln.

Mulvey, J. H., ed. 1981. *The nature of matter*. Oxford.

Mutschler, Hans-Dieter. Zwei Kulturen sind besser als keine—Wider die physikalischen Ganzheitslehren, *Merkur* 536, 998 (1993).

Neuser, Wolfgang. 1990. *Newtons Universum*. Heidelberg: Spektrum.

Newton, Isaac. 1963. *Mathematische Prinzipien der Naturlehre*. Darmstadt.

———. 1979. *Opticks*. New York.

———. 1988. Über die Gravitation . . . Frankfurt.

Nicolai, H., and M. Niedermaier. Quantengravitation—vom Schwarzen Loch zum Wurmloch, *Physikalische Blätter* 45, n. 12 (1989): 459.

Nigel, Hawtin. *New Scientist* (August 17, 1991).

Pagels, Heinz R. 1983. *The cosmic code*. Berlin.

———. 1985. *Perfect symmetry*. New York: Simon and Schuster.

———. 1988. *The dreams of reason*. New York.

Pais, A. 1982. *Subtle is the lord*. Oxford: Clarendon Press.

Parker, Barry. 1988. *Creation*. New York.

Parmenides. 1981. *Über das Sein.* Stuttgart.

Pedersen, Olaf. 1993. *Early physics and astronomy.* Cambridge.

Penrose, Roger. 1989. *The emperor's new mind.* Oxford: Oxford University Press.

Pius XII, Pope. Die Gottesbeweise im Lichte der modernen Naturwissenschaft, *Herderkorrespondenz* 6, 165 (Heft 4, Januar 1952).

Plato. 1989a. *Sämtliche Werke,* vol. 4 (*Parmenides und Sophistes*). Reinbek.

Plato. 1989b. *Sämtliche Werke,* vol. 5 (*Politikos, Philebos, Timaios, Kritias*). Reinbek.

Plotinus. 1991. *The Enneads.* London.

Pohl, R.W. 1957. *Elektrizitätslehre.* Berlin.

Poincaré, H. 1952. *Science and hypothesis.* New York.

Polkinghorne, John. 1989. *Science and creation.* Boston.

Pöppel, Ernst. 1987. Die Rekonstruktion des Raumes. In Venant Schubert, ed. *Der Raum.* St. Ottilien.

Pretzl, K. P. Bringing dark matter in from the dark, *Europhys. News* 24, 167 (1993).

Puthoff, Harold. Everything for nothing, *New Scientist* 28 July 1990, p. 52.

Rafelski, Johann, and Berndt Müller. 1985. *Struktur des Vakuums—Ein Dialog über das Nichts.* Frankfurt.

Ramsauer, Carl. 1953. *Grundversuche der Physik in historischer Darstellung.* Berlin.

Ray, Christopher. 1991. *Time, space and philosophy.* London.

Rees, Martin. 1997. *Before the beginning.* Reading: Helix Books.

Reeves, Hubert. 1989. *Die kosmische Uhr.* Düsseldorf.

Rehmke, Johannes. 1983. *Geschichte der Philosophie.* Wiesbaden.

Reidemeister, Kurt. 1957. *Raum und Zahl.* Berlin.

Rempe, Gerhard. Atoms in an optical cavity: quantum electrodynamics in confined space, *Contemporary Physics* 34, 119 (1993).

Renteln, Paul. Quantum gravity, *American Scientist* 79, 508 (1991).

Rindler, W. 1977. *Essential relativity.* New York.

Riordan, Michael, and David N. Schramm. 1991. *The shadows of creation.* New York.

Robinson, John Mansley. 1968. *An introduction to early Greek philosophy.* Boston.

Russell, Bertrand. 1965. *History of Western philosophy.* London.

Sambursky, Shmuel. 1974. *Physical thought from the presocratics to quantum physics.* Hutchinson & Co.

———. 1982. *The concept of place in late neoplatonism.* The Israel Academy of Sciences and Humanities, Jerusalem.

———. 1987. *The physical world of the Greeks.* Princeton: Princeton University Press.

Satz, H. The quark plasma, *Nature* 324, 116 (1986).

Saunders, Simon, and Harvey R. Brown. 1991. *The philosophy of vacuum.* Oxford.

Schischkoff, Georgi. 1991. *Philosophisches Wörterbuch.* Stuttgart.

Schmidt, Arno. 1960. *Kaff auch Mare Crisium.* Frankfurt.

Schmitt, Charles B. Experimental evidence for and against a void: The sixteenth-century arguments, *Isis* 58, 352 (1967).

Schopper, Herwig. 1989. *Materie und Antimaterie.* Munich.

———. *Physikalische Blätter* 47, 907 (1991).

Schrödinger, Erwin. 1979. *Was ist ein Naturgesetz?* Munich.

———. 1996. *Nature and the Greeks and science and humanism.* Cambridge: Cambridge University Press.

Schüller, Volkmar. 1991. *Der Leibniz—Clarke Briefwechsel.* Berlin.

Sciama, D. W. 1973. The universe as a whole. In J. Mehra, ed. *The physicist's conception of nature.* Dorderecht.

Segrè, Emilio. 1984. *From falling bodies to radio waves—Classical physicists and their discoveries.* New York: W. H. Freeman & Co.

Sexl, Roman, and Hannelore Sexl. 1979. *Weiße Zwerge—Schwarze Löcher.* Braunschweig.

Sexl, Roman, Ivo Raab, and Ernst Steeruwitz. 1980. *Einführung in die Physik* (3 volumes: *Das mechanische Universum, Der Weg Zur modernen Physik* und *Materie in Raum und Zeit).* Frankfurt.

Sexl, Roman, and Herbert K. Schmidt. 1985. *Raum-Zeit-Relativität.* Braunschweig.

Shapin, Steven, and Simon Schaffer. 1985. *Leviathan and the air-pump: Hobbes, Boyle, and the experimental life.* Princeton.

Silk, Joseph. 1980. *The big bang.* New York: W. H. Freeman and Company.

Simonyi, K. 1990. *Kulturgeschichte der Physik.* Frankfurt.

Smolin, Lee. 1997. *The life of the cosmos.* New York: Oxford.

Smoot, George, and Keay Davidson. 1993. *Wrinkles in time.* London.

Spruch, Larry. Retarded, or Casimir, long-range potentials, *Physics Today* (November 1986): 37.

Stöckler, Manfred, ed. 1990. *Der Riese, das Wasser und die Flucht der Galaxien.* Frankfurt.

Stückelberger, Alfred, ed. 1979. *Antike Atomphysik.* Munich.

Stückelberger, Alfred. 1988. *Einführung in die antiken Naturwissenschaften.* Darmstadt.

Sukenik, C. I. et al. Measurement of the Casimir–Polder force, *Phys. Rev. Letters* 70 (1993): 560.

Trefil, James F. 1983. *The moment of creation.* New York: Charles Scribner's Sons.

———. 1988. *The dark side of the universe.* New York: Charles Scribner's Sons.

———. 1989. *Reading the mind of god.* New York.

Trunz, Erich, ed. 1952. *Goethes Faust.* Hamburg.

Tryon, Edward P. What made the world? *New Scientist* 8 (March 1984).

———. 1987. Cosmic inflation. In Robert A. Meyers, ed. *Encyclopedia of physical science and technology,* 3. Orlando.

Tschierpe, Rudolf. 1949. *Ein Weg in die Philosophie.* Hamburg.

———. The Higgs boson demystified, *Physics World* (Spetember 1993): 26.

Verdet, Jean-Pierre. 1991. *Der Himmel—Ordnung und Chaos der Welt.* Ravensburg.

Waters, Frank. 1982. *Das Buch der Hopi.* Düsseldorf–Köln.

Weigl, Engelhard. 1990. *Instrumente der Neuzeit.* Stuttgart.

Weinberg, Steven. 1977. *The first three minutes: A modern view of the origin of the universe.* New York: Basic Books.

———. The cosmological constant problem *Reviews of Modern Physics* 61, 1 (1989).

———. 1993. *Dreams of a final theory.* New York: Pantheon Books.

Weizsächer, C. F. 1990. *Die Tragweite der Wissenschaft.* Stuttgart.

Westfall, Richard S. 1990. *Never at rest—A biography of Isaac Newton.* Cambridge.

Weyl, Hermann. 1922. *Space—Time—Matter*. London: Methuen.

———. 1949. *Philosophy of mathematics and natural science*. Princeton: Princeton University Press.

Wheeler, John Archibald. 1990. *A journey into gravity and spacetime*. New York.

Whittaker, Sir Edmund. 1989. *A history of the theories of aether & electricity*. New York.

Wightman, A. S., and N. Glance. 1989. Superselection rules in molecules. In Y. S. Kim and W. W. Zachary, eds. *Proceedings of the international symposium on spacetime symmetries* (College Park, Maryland, USA, May 1988). Amsterdam.

Wigner, Eugene P. Invariance in physical theory. *Proceedings of the American Physical Society* 93, n. 7 (1949).

———. 1967. *Symmetries and reflections—Scientific essays of Eugene P. Wigner*. Bloomington.

Will, Clifford M. 1986. *Was Einstein right? Putting general relativity to the test*. New York: Basic Books.

Willis, Bill. Collisions to melt the vacuum, *New Scientist* 8 (October 1983): 9.

Wußing, Hans. 1987. *Geschichte der Naturwissenschaften*. Köln.

Wynn-Williams, Gareth. 1992. *The fullness of space*. Cambridge.

Zahn, Manfred. 1987. Einführung in Kants Theorie des Raumes. In Venant Schubert ed. *Der Raum*. St. Ottilien.

Zichichi, A., ed. 1982. *Gauge interactions*. New York.

Zurek, Wojciech H. Decoherence and the transition from quantum to classical, *Physics Today* (October 1991): 36.

INDEX

•

Numbers followed by the letter f indicate material in figures and figure captions.

Absolute rest, 150, 163
Absolute space, 145, 146–148, 166, 180
Absolute zero, 1, 182–183
    energy at, 202
Acceleration, 8, 146–148
Air, weight of, 99–100, 114
Air pressure
    effects of, 116
    and *horror vacui,* 112–114
    and vacuum, 142–143, 142f
    variations in, 126–128, 130–131,
        131f
Albert of Saxony, 92
Albertus Magnus, 90
Alice's Adventures in Wonderland (Car-
    roll), 88
Amaldi, Ugo, 88–89
Anaxagoras, ix, 57–60, 62, 65
Anaximander, 48, 51
Anaximenes, 48, 52
Anderson, Carl, 192
Angular velocity, 19
Anthropic principle, 259–261
Antiparticles, 192, 194
Aquinas, Thomas, 90, 91, 109
Archimedes' principle, 117f
Archytas of Tarentum, 80–81
Aristotle, vii, 5, 20–21, 26, 32, 44, 56,
        63, 72–80, 81, 83–86, 88, 90, 91,
        92, 93, 94, 100, 102, 108, 109, 136,
        176, 177, 307
Asymptotic freedom, 244
Atmosphere, levels of, 113
Atomists, 32, 61–66, 107
    cosmological view of, 65–66
Atoms
    electrons in, 216–217, 216f
    large, 203
    structure of, 202–204
Augustine, St., 34, 35, 36
Axions, 300

Baby universes, 314f
Bacon, Roger, 94

Baryonic matter, 299–300
Bentley, Richard, 159
Bethe, Hans, 246–247
Bifurcation, defined, 309
Big Bang, viii, 264–265, 271, 288
    forces generated by, 271–272
    immediate aftereffects of, 230, 232
    standard model of, 262
    and time, 36–39
    vacuum and, 30–31
Black holes, 250–256
Blackbody radiation, 1–2
    characteristics of, 181–184
    experiments on, 237–239
    motion with respect to, 187–189
Bohr, Niels, 95, 203
Bohr model of the atom, 216–217,
    245–246
Boiling, of liquid, 281
Boltzmann, Ludwig, 63
Born, Max, 204
Bosons, 218
Boundary conditions, 49
Boyle, Robert, 27, 93, 130–131, 143
Brecht, Bertolt, 195
Bruno, Giordano, 93, 132, 138–139,
    140
Buridan, Jean, 284
Buridan's donkey, 284, 285f, 286

Carroll, Lewis, 88
Casimir, H.B.G., 185
Casimir effect, 181, 183–187, 189, 202
Cavendish, Henry, 156–157
Celestial mechanics, Cartesian view of,
    132, 133f, 134
Centrifugal force, 8, 147, 149
    gravity and, 162
CERN, 28, 222–223, 222f, 228, 239
Change, philosophical investigation of,
    55–56
Chaos, order and, 59, 89
Charge, characteristics of, 217–218
China, creation myths of, 34

Christianity, Aristotelianism and, 90
Chrysippus, 80
Chuang-Tzu, 34
Clarke, Samuel, 85, 90, 154, 285f, 309
Classical mechanics, 209, 211, 291
Closed systems, 16–18
COBE (Cosmic Background Explorer),
    238
Coleman, Sidney, 313, 314f
Color charges, 243–245, 271
Compactification, 258
Conditional necessity, 309
Confessions (Augustine), 35
Contingent properties, 308
Continuum, 58
Coordinate system, defined, 7
Copernicus, Nicolaus, 3, 50, 137–138,
    310
Cosmic background radiation, 169,
    267–268
Cosmological constant, 208, 269, 292,
    301–303, 313
Cosmological principle, 266–269
    and symmetry, 276–278
Cosmology
    Aristotelian view of, 72–76, 79
    atomistic view of, 65–66
Coulomb electrostatic force, 234
Creation myths, 4, 310
    ancient, 33–36
    Genesis, 34–35
Critical temperature, defined, 11
Crystals, symmetry breaking in,
    281–282
Curie temperature, 278

Dark matter, 298–300
    prevalence of, 300–301
De Gravitatione (Newton), 153
De Magnete (Gilbert), 97, 217
de Sitter, Willem, 292
Defects
    cosmic, 301
    defined, 281
    types of, 295f

DELPHI detector, 223–224, 223f
Democritus, 47, 52, 61, 62, 65–66, 109, 202–203, 306, 308
Descartes, René, 6, 9, 76, 93, 131–137, 152–153, 157–158, 217, 307
Desgauliers, J.T., 100f, 236
Dialogue Concerning Two New Sciences (Galileo), 99–100
Digges, Thomas, 74f, 138, 140, 143
Dijksterhuis, E.J., 106–107
Dimensionality, 171–177, 258–259
Dirac, P.A.M., 12, 192, 245
Dirac equation, 245
Dirac sea, 193, 204–205
Discussion among Refugees (Brecht), 195
Divergencies, 247
Domains, 305–306
Doppler effect, 188–189, 188f
Dunham, Douglas, 44f

Eddington, Arthur, ix
Einstein, vii, viii, 6, 9, 21, 22f, 28, 46, 47, 85, 110, 146, 151, 158, 162, 163, 164, 167, 176, 181, 190, 204, 207, 217, 228, 236, 245, 246, 261, 292, 301–303, 307
Elea, school of, 53
Electric charge, conservation of, 272
Electricity, in vacuum, 205–207
Electron, 307
    characteristics of, 211
    probability distributions of, 216–217, 216f
    wave characteristic of, 213f, 214
Electron-positron interactions, 194, 224–226, 225f, 226f
Electron-positron pairs, 236, 261
    spontaneous creation of, 248–250, 249f
Electroweak force, 272
Elementary particles, 209
    field characteristics of, 218, 220
    fields of, 211–212
    interactions of, 210f, 272
    positions of, 209, 211–216
    types of, 218, 220–221
Empedocles, 55–57, 60, 99, 102, 114
Energy
    at absolute zero, 202
    as mass, viii, 190
    forms of, 190–192
    negative, 12–13
    quantization of, 245
    zero level of, 291–292
Entropy, 58
Eötvös, Roland Baron, 157
Epicurus, 62, 64, 83
Ether, vii, 83
Ether of mechanics, 9
Euclid, 40
Euclidean space, 173, 176
Euler, Leonhard, 165

Evacuation, 2
Exosphere, 113f
Expansionary universe, 263–264, 263f

False vacuum, 293
Fermi Lab Tevatron, 228, 232
Ferromagnet, 278
Feynman, Richard, 14, 44f, 45, 206
Feynman diagrams, 206, 206f
Field, 134–135
    defined, viii
    directionality of, 218
    particle as, 218, 220
    representations of, 219f
Field theories, 177–178
Flat universe, 297–298
Fluctuation, 25
Force, 8
    in accelerated systems, 9
    directionality of, 218
    range of, 271
    strength of, 271
Foucault pendulum, 151–152, 151f
Free fall, 160–162
Freezing, delayed, 281–282
From the Closed World to the Infinite Universe (Koyré), 139
Fundamentalism, 35–36

Galaxy, form of, 298–299, 299f
Galileo Galilei, vii, 8, 50, 95, 97–102, 112, 116, 132, 146, 147–148, 153, 156, 183
Gamma rays, 236
Gassendi, Pierre, 84, 93, 94, 109–110, 137
Gauss, Carl Friedrich, 42, 46, 175
Gedankenexperiment, 2f
Gell-Mann, Murray, 310
General relativity, 46, 151, 155, 158, 171, 173–177, 229, 262
Genesis, 34–35
Geometry
    Cartesian view of, 135
    Euclidean, 40–41
    non-Euclidean, 80, 169–177
Gilbert, William, 97, 139, 143. 217
Gluons, 218, 244
God, issues relating to, 89–90
Goethe, J.W. von, 35
Goldstone, Jeffrey, 257
Goldstone theorem, 283–284
Grant, Edward, 66, 93
Gravitational mass, 156
Gravitational waves, 229
Gravitons, 229
Gravity
    centrifugal force and, 162
    cosmological effects of, 159–160
    measurement of, 7
    Newtonian view of, 136–137, 153
    propagation of, 158–159
Great circle, 41–42

Greece
    creation myths of, 4, 33–34
    natural science of, 47–90
Grosseteste, Robert, 90, 93–94
Ground state, 31
    defined, 10
    energy at, 199–200
GSI (German High-Energy Heavy-Ion Research Laboratory), 239, 249
Guericke, Otto von, 27, 43–44, 47, 93, 104, 110, 121–130, 138
GUT symmetry, breaking of, 289
Guth, Alan, 261, 296
GUTs (grand unified theories), 272, 273

Hartle, James, 31–32, 265, 269, 288, 296, 310, 313
Hartle-Hawking universe, 265–266, 266f, 296
Hawking, Stephen, 31–32, 252–253, 254, 265, 269, 288, 296, 313
Hawking radiation, 250–256, 261
Heisenberg, Werner, 27, 205
Heisenberg uncertainty principle, viii, 194–196, 212f
    and perception, 196–199
    and probability, 199
    at short distances, 221
Heraclitus, 34, 52–53, 54, 189
Hero of Alexandria, 85, 91, 94, 97, 105, 106–108
Hero's ball, 106f
Hertz, Heinrich, 217, 218, 219f
Hesiod, 33
Higgs, Peter, 231, 257
Higgs boson, 228–235
Higgs field, 11–12, 24, 229–235, 236, 272, 289–290
Higgs phenomenon, 218
Homer, 33
Horror vacui, vii, 3f, 21, 27, 94–95, 103, 311
    air pressure and, 112–114
    elevation and, 119–121
Hubble, Edwin, 301
Hund, Friedrich, 131
Huygens, Christiaan, 162–165
Hydra experiment, 56–57, 86, 99, 102–103

Ideal gas, 63
Inertia, 171
Inertial mass, 156
Inflation of universe, 262, 288
    consequence of, 294–296
    end of, 293–294
    mechanism of, 293
Inflationary universe, 272, 275
Infrared slavery, 241f, 244
Initial conditions, 49
Inverse beta decay, 249f
Ionosphere, 113f

Japan, creation myths of, 34
Jets, particle, 224
John, St., 35

Kant, Immanuel, 9, 260
Kepler, Johannes, 50, 70f, 74f, 137, 140–141, 143
Kinetic energy, 234
Kojiki, 34
Koyré, Alexander, 139

Lamb, Willis, 221, 246, 247
Laminar flow, 134
Lamoreaux, S.K., 186
Laplace, Pierre-Simon de, 309
Laws of nature, 48–49
    characteristics of, 311
    effective, 282–283, 310
    and space, 10
    symmetries of, 18–19
    and velocity of universe, 168–169
Learned Ignorance (Nicholas of Cusa), 137
Leibniz, Gottfried Wilhelm, 6, 7, 9, 47, 85, 90, 153–155, 158, 177, 285f, 309
LEP (large electron positron) accelerator, 28, 221–223, 222f, 226
Leucippus, 52, 61, 62, 108, 109
LHC (large hadron collider), 228, 231, 297
Life of the Cosmos (Smolin), 314f
Light
    aspects of, 179–180
    as oscillation, 180–181
    propagation of, 162–165, 179
    scattering of, 247–248, 248f
    in vacuum, 205–207
    velocity of, 165–168
Location, 76–79
    Aristotelian definition of, 147
Lorentz, H.A., 217
Lucretius, 64, 80, 81, 82–85, 141, 203

Mach, Ernst, 85, 150, 151, 162, 177, 180
Mach's principle, 22f, 275
Maddox, John, 229
Magdeburg hemisphere, 27, 128–130, 128f
Magnetism, 6, 97, 217–218, 278–281
Manometer, 130, 131f
Maori, creation myths of, 35
Marcus Aurelius, 80
Mass
    as energy, viii, 190
    effect of, 46–47
    point, 157–158, 171
    types of, 156
    weight and, 155–157
Materia prima, 21, 26–27, 77–78, 87–88
Mathematics, Pythagorean, 67–69

Maxwell, James Clerk, 10, 165, 168–169, 218, 219f
Maxwell's equations, 83, 165
Mayans, creation myths of, 35
Mechanistic philosophy, 109–110, 136
Mechanization of the World Picture (Dijksterhuis), 159
Mercury, density of, 114
Metric, defined, 263
Metric field, 173–175
Michelson-Morley experiment, 166–167, 167f
Miller, David, 230, 231, 232
Mind, importance of, 57–58
Minimal supersymmetric extension of the standard model, 273f
Mirror symmetry, 14
Monopoles, 301
More, Henry, 93
Motion
    Aristotelian view of, 75–76
    and empty space, 60–61, 73–76
    matter and, 63–66
    Newtonian view of, 149
Muon-antimuon interactions, 224, 225f
Musical instruments, 190

Neutrinos, 95, 231, 236, 300
Neutrons, 243, 244
New Pathways in Science (Eddington), ix–x
Newcomen, Thomas, 29
Newton, Isaac, vii, 18, 22, 28, 47, 50, 73, 85, 101f, 110, 145–165, 168, 176, 177, 180, 217, 229, 275
Newton's law of gravity, 19, 20
Newton's law of motion, 8–9, 20
Nicholas of Autrecourt, 94
Nicholas of Cusa, 93, 137
Noel (Jesuit), 119
Nothingness
    ancient Greek view of, 5
    philosophical ideas of, 5–6
    and space, 54, 307
Nuclear fission, 191
Nucleons, makeup of, 241f

Omega, defined, 297
On the Nature of Things (Lucretius), 83–84
Optics (Newton), 163
Order, and chaos, 59, 89
Ordered state, 13
Oresme, Nicholas, 93
Oscillation, 180–181, 183–187, 190f, 200

Panta rhei, 52, 189
Parallel lines, Euclidean, 41
Parmenides, 48, 51, 52, 53–54, 61
Particles. *See* Elementary particles; Virtual particles.

Pascal, Blaise, 27, 28f, 93, 110, 115–121, 126, 127
Patrizzi, Francesco, 93
Pauli, Wolfgang, 95
Perception, uncertainty and, 196–199
Périer, Florin, 119, 120f
Phase transition, 11, 13–14, 31, 227–228
    defined, 11
Philo of Alexandria, 85, 89–90, 91, 104–105, 106
Photoelectric effect, 164
Photon(s), 218, 220, 248
    as antiparticle-particle pair, 194
    defined, 207
Pius XII (Pope), 39
Planck, Max, 181
Planck length, 254
Planck time, 288, 314f
Planetary motion, 146
    early investigations of, 50
Plato, 26, 51, 58, 66, 69–72, 86–88, 309
Plotinus, 32, 87–89
Pneuma, 82
Poincaré, Henri, 44f, 45
Point masses, 157–158, 171
Point of rest, 13
Polarized vacuum, 241–243, 273f
Polyhedra, regular, 70–72
Popol Vuh, 35
Poseidonius, 80
Positional energy, 200, 234
Positrons, 192
Potential energy, 200, 234
Principia (Newton), 145, 153, 163
Probe, defined, 6
Proton-antiproton pairs, 236
Protons, 243, 244
Ptolemy, 50, 310
Pumping, 2
Pythagoras, 66–69
Pythagoreans, 67–69

Quantum mechanics, 24, 32, 211, 213–216
Quark-antiquark interactions, 224
Quark-gluon plasma, 236, 241f
    experiments on, 239–241
Quarks, 72, 241f, 243, 306–307
Quasars, 81–82

Radiation
    blackbody, 1–2, 181–184, 187–189, 237–239
    cosmic background, 169, 267–268
    Hawking, 250–256, 261
    synchrotron, 222
    thermal, 235–236
    types of, 181
    zero-degree, 183
Rees, Martin, 297
Regular polyhedra, Platonic view of, 70–72
Relative motion, 149

Relative space, 145, 152, 180
Renormalization, 247
RHIC (Relatively Heavy Ion Collider), 239–240
Ricci, Michelangelo, 112
Rotation, 149
Rotational symmetry, 14

S-matrix theory, 205
Salam, Abdus, 286f
Scattering, defined, 242
Scholasticism, 91–93, 108–109, 121–122
Schrödinger, Erwin, ix, 245
Schwarzschild radius, 251
Seneca, 79
Singularity, 264f
  defined, 30
Smolin, Lee, 314
Solar system, 137–138
Space
  Aristotelian view of, 85–86
  curvature of, 41–46, 208, 269
  defining, 9, 177
  dimensionality of, 171–177, 258–259
  distinguished from nothingness, 307
  effect of mass on, 46–47
  Euclidean, 173, 176
  impossibility of emptiness of, 315
  and laws of nature, 10
  motion and, 60–61, 73–76
  Newtonian, 110, 145–149
  nothingness and, 54
  particles and, 220–221
  properties of, 39–46
  symmetry and, 16, 19–21
  time and, 31–32
  velocity of, 7–10
Sparnaay, M.J., 185–186, 187
Special relativity, 24, 167–168, 176
Specific heat, 202
Spin-wave, defined, 283
Spontaneous materialization, 227
Spontaneous pair creation, 248–250
Spontaneous symmetry breaking, 16, 278, 281–282, 286f
  Higgs model of, 289–290
  mathematics of, 286–290
SPS (super proton synchrotron), 222
Standard model of particle interactions, 272
State function, 212
State of rest, 16
Steinhardt, Paul, 261
Stoics, 79–81, 82–83
Strato, 85, 86, 90, 152
Stratosphere, 113f
String theory, 258, 259
Strong force, 272
Substance, Aristotelian, 77
Sufficient cause, axiom of, 285f
Superstrings, 258, 259
Supersymmetric particles, 301

Symmetry
  breaking of, 15–16, 278, 281–282, 286–290
  cosmological principle and, 276–278
  and empty space, 16
  types of, 14–15
Synchrotron radiation, 222

Taoism, 34
Tempier, Etienne, 90, 91
Thales of Miletus, 5, 10, 33, 47, 48, 51
Théodicée (Leibniz), 285f
Theophrastus, 85, 152, 176
Thermal equilibrium, 182
Thermal radiation, 235–236
Timaeus (Plato), 51, 69, 70
Time
  Big Bang and, 36–39
  Newtonian, 146
  space and, 31–32
TOE (theory of everything), 271
Topologically stable, defined, 281
Torricelli, Evangelista, vii, 10, 93, 94, 95, 102, 112–114, 117, 124, 126, 127, 143, 207, 311
Torricelli experiment, 2–4, 3f, 27, 57f, 110–114, 111f, 143
Trajectories, 171–173, 172f
Transformations, types of, 283–284
Translational symmetry, 14, 18
TRISTAN accelerator, 242f
Troposphere, 113f
True motion, 149
Tryon, Edward, 261, 274, 296
Two-slit experiment, 194

Unified theory, defined, 48
Universe, 311–312
  Aristotelian view of, 141
  baby, 314f
  constants of, 272–274
  effects of gravitation on, 159–160
  energy of, 290–292
  epochs of, 269, 270f, 271
  expansionary, 263–264, 263f
  flat, 297–298
  future of, 296–297
  Hartle-Hawking, 265–266, 266f, 296
  Hubble expansion of, 292
  inflation of, 272, 275
  initial cooldown of, 234
  Kepler's view of, 140–141
  origins of, 261–265
  rotational velocity of, 274–275
  scholasticist view of, 141
  size of, 271
  velocity of, 168–169
  wave function of, 312
Ur-matter, 21, 51–52
  change by way of, 54–56
  divisibility of, ix
  philosophy of, 26–27
  physics of, 24–26

Ussher, James, 36

Vacuum, 4
  acceleration with respect to, 187
  air pressure and, 142–143, 142f
  airlessness of, 124–126
  and Big Bang, 30–31
  charge and momentum of, 221
  color charges in, 243–245
  complexity of, 189–192, 204–205
  defined, vii, 207–208
  early demonstrations of, 102–106
  effects of, 97–99
  energy of, 290–291
  excitation of, 235
  false, 293
  free fall, in, 100, 101
  light and electricity in, 205–207
  modifying, 306–308
  polarization of, 241–243, 273f
  practical definition of, 10
  realizing, 101–102
  scholasticist view of, 93–94
  Strato's experiments on, 86
  structure of, 29–30
  symmetry and, 20
  Torricelli experiment on, 2–4, 3f, 27, 57f, 110–114, 111f, 143
  weight of, 116–118
  within vacuum, 118–119, 118f
Vacuum fluctuations, 25, 25f, 201
Vacuum pumps, 121–124
Vacuum state, defined, 10
Velocity, angular, 19
Virtual particles, 12, 24, 201
  defined, viii
  transformation of, 224, 234, 248–250, 249f
Viviani, Vincenzo, 98
Void
  forces in, 21–24
  symmetry in, 16, 19–21
  See also Nothingness; Space; Vacuum.

Waldgrave, William, 228, 229, 232
Water, 13f
  phases of, 11, 13–14, 227–228
  role in creation myths, 5, 33
Waves, 83
Weight, 155–157
Weyl, Hermann, 14, 302
Wigner, Eugene, 12, 49–50, 53
WIMPs (weakly interacting massive particles), 300–301
Wormholes, 313, 314f

Xenophanes, 52

Zeno of Citium, 80
Zero-degree radiation, 183
Zero-point energy, 199–202
Zero-scale, viii
Zwitterbewegung, 246